COMMUNITY, STATE, AND MARKET ON THE
NORTH ATLANTIC RIM

Studies in Comparative Political Economy and Public Policy
Editors: Michael Howlett, David Laycock, Stephen McBride, Simon Fraser University

Studies in Comparative Political Economy and Public Policy is designed to showcase innovative approaches to political economy and public policy from a comparative perspective. While originating in Canada, the series will provide attractive offerings to a wide international audience, featuring studies with local, sub-national, cross-national, and international empirical bases and theoretical frameworks.

Editorial Advisory Board:
Isabel Bakker, Political Science, York University
Colin Bennett, Political Science, University of Victoria
Wallace Clement, Sociology, Carleton University
William Coleman, Political Science, McMaster University
Barry Eichengreen, Economics, University of California (Berkeley)
Wynford Grant, Political Science, University of Warwick
John Holmes, Geography, Queen's University
Jane Jensen, Political Science, Université de Montréal
William Lafferty, Project for an Alternative Future, Oslo
Gordon Laxer, Sociology, University of Alberta
Ronald Manzer, Political Science, University of Toronto
John Ravenhill, Political Science, Australia National University
Peter Taylor-Gooby, Social Work, University of Kent
Margaret Weir, Brookings Institution, Washington, D.C.

Published to date:
1 **The Search for Political Space: Globalization, Social Movements, and the Urban Political Experience**, Warren Magnusson
2 **Oil, the State, and Federalism: The Rise and Demise of Petro-Canada as a Statist Impulse**, John Erik Fossum
3 **Defying Conventional Wisdom: Free Trade and the Rise of Popular Sector Politics in Canada**, Jeffrey M. Ayres
4 **Community, State, and Market on the North Atlantic Rim: Challenges to Modernity in the Fisheries**, Richard Apostle, Gene Barrett, Petter Holm, Svein Jentoft, Leigh Mazany, Bonnie McCay, Knut H. Mikalsen

RICHARD APOSTLE, GENE BARRETT, PETTER
HOLM, SVEIN JENTOFT, LEIGH MAZANY,
BONNIE MCCAY, KNUT H. MIKALSEN

Community, State, and Market on the North Atlantic Rim: Challenges to Modernity in the Fisheries

UNIVERSITY OF TORONTO PRESS
Toronto Buffalo London

© University of Toronto Press Incorporated 1998
Toronto   Buffalo   London
Printed in Canada

ISBN 0-8020-0745-7 (cloth)

Printed on acid-free paper

Canadian Cataloguing in Publication Data

Main entry under title:

Community, state, and market on the North Atlantic rim : challenges to modernity in the fisheries

(Studies in comparative political economy and public policy)
Includes bibliographical references
ISBN 0-8020-0745-7

1. Fisheries – Atlantic Provinces.   2. Fisheries – Norway, Northern.
3. Fishery policy – Atlantic Provinces.   4. Fishery policy – Norway, Northern.
5. Fisheries – Economic aspects – Atlantic Provinces.   6. Fisheries – Economic aspects – Norway, Northern
I. Apostle, Richard.   II. Series.

SH213.2.C65 1998   338.3′727′094843   C97-932817-9

The publication of this book has been funded in part by a grant from the John D. & Catherine T. MacArthur Foundation.

University of Toronto Press acknowledges the financial assistance to its publishing program of the Canada Council for the Arts and the Ontario Arts Council.

Contents

ILLUSTRATIONS vii
FOREWORD ix
PREFACE xiii

Introduction 3

Section 1: Institutional Development
Introduction 23
1 The End of Commercialism 29
2 The Rise of Industrial Capitalism 59
3 The Resource Management Revolution and Market-Based Responses 85

Section 2: Resource Regimes – Co-managing the Commons? The Politics of Fisheries Management in Atlantic Canada and Norway
Introduction 121
4 Managing the Fisheries: Procedures and Politics 131
5 From Procedures to Policies 172
6 Institutional Structures and Management Policies: The Case of Individual Quotas 187
7 Management Reform: The Search for Appropriate Institutions 205

Section 3: Communities and Entrepreneurship – Re-embedding Coastal Communities
Introduction: Parallel Crises 229
8 Community Sustainability, Small Firms, and Embeddedness 235
9 Modernization and Crises 246

10 Traditionalism and Crisis: The Social Bases of Disembeddedness and Re-embeddedness 259
11 Fordism, Neo-Fordism, and Community Disembeddedness 270
12 Post-Fordism and Re-embedding Coastal Communities 284
13 Bugøynes: A Case Study of Community Resistance 297
14 Fogo Island: A Case Study of Cooperativism 307
15 Sambro: A Case Study of Participatory Development 315
16 Re-embedding Coastal Communities: Towards a Localized Globalization? 326

Conclusion: Community, Market, and State: Dilemmas in Fisheries Policies 335

REFERENCES 345

Illustrations

Figures
1 Community, state, and market   8
2 Organizational strategies in Canadian and Norwegian fisheries during the twentieth century   26
3 Number of fishermen in Canada and Norway 1945–1990   61
4 The sector system in Norwegian fisheries during the postwar period: Economic organization   69
5 Sector reforms in Norwegian fisheries   114
6 Sector reforms in Canadian fisheries   115
7 Management procedure in Norway   135
8 Management procedures for groundfish in Atlantic Canada   157
9 Proposed management structure in Atlantic Canada   215
10 Community, state, and market principles of organizations in the fisheries   336

Map
Atlantic Canada and North Norway   xvi

Foreword

This important book documents and explains the historical processes that have generated coastal communities on both sides of the northern North Atlantic, draws a clear picture of the present social organization of fishing, and gives us some very important clues to what the future of the fishing populations might be.

The problematic relationship between *community* and *economic sector* is the dominant theme. Villages and industries may grow and develop in harmony, one serving the other, but unfortunately one cannot take for granted that what is good for the fishing industry is good for Fogo Island or the Varanger Peninsula.

The coastal regions of Maritime Canada and Northern Norway share many characteristics. Not only natural conditions – like cold, stormy winters and rich marine resources – but also historical origins. On both sides of the ocean, fishing communities grew up in coves and inlets and increased their population rather quickly in the long period between the Napoleonic Wars and the Second World War. To the propertyless of southern and central Fenno-Scandia and the British West Country, these regions functioned as a 'frontier' where anybody – with their two empty hands – could eke out an existence, denied them in their rich and landowner-dominated regions of origin. When depression hit the industrial world, the trade-combining, subsistence-based coastal communities were able to retain or welcome back a fair share of their young people.

From the viewpoint of the common man and woman, these peripheral regions competed quite well with the centre in what we might call the *preindustrial* phase of the national societies of which they were part. In the age of landed property, the beauty of Newfoundland and Tromsø was that the land they needed for subsistence *had no alternative uses* of economic interest; most important of all, rich fish resources were freely available to the local population. Many Maritime outports owe their origin to sailors and fishermen who, often forcibly pressed to compulsory service for Bristol shipowners, fled ashore and simply lived from the land. The local economies which they built could hardly have been established in Britain, as the landlords in power rather preferred sheep to tenants, and even claimed to own the fish, as they did in Shetland.

Precisely those advantages that made the coastal communities attractive and successful in the preindustrial phase turned out to be grave disadvantages in *the age of industry* – when governments tried to organize the fishing industry according to 'modern' or – as they are called in this book – *Fordist* principles.

Ecology and technology met in frontal collision: The modern filleting machine demands a predictable volume of raw material every day, whereas to the coastal population, the sea is a garden or a farm – to be harvested cyclically. Nobody expects a farmer to drive a harvesting combine every day throughout the year. The goals of the coastal populations and the food companies moving in were – in the long run – incompatible.

In the period that the governments and the expansive commercial interests had to confront a united coastal population, there was a certain balance of power on both sides of the Atlantic: Massive state support to trawler owners or vertically organized filleting companies had to be *balanced* by state support in the form of individual payments to small-scale fishermen or generous financing of decentralized housing and infrastructure – to maintain political consensus. But underneath this stalemate, dynamic forces were at work: In Norway differentiation processes within fishing communities generated rather small, but capital-intensive and expansive units among a previously egalitarian coastal population – that is, skippers with potential interests in common with larger trawler companies. In Newfoundland most coastal communities are closed out from the sea as a source of income: in the short run because of the almost complete destruction of the cod stocks, and in the longer run by losing the right to harvest the resource that made their forbears settle on these barren coasts.

The introduction of property rights in fish stocks may solve certain overcapacity problems in areas where fishing is but one of a whole range of economic opportunities. To regions like Finnmark or Newfoundland, however, with whole communities and regional populations with no alternatives to the rich marine resources, the opportunities of outside interests to *buy* their resource base are bound to imply serious social problems – not taken into consideration by many consultants to the governments. Transferring prescriptions that may work (at any rate on paper) in Seattle or Vancouver to areas where fisheries are *embedded* in the culture and organization of local communities can lead nowhere but to rural depression and impoverished villages, permanent dependence upon transfer payments, or, in the best cases, to depopulation and ghost towns – if anybody at the centre needs the peripheral population. Certain schools of economic thought are today more of a menace to coastal communities than foreign fleets, parasitic middlemen, and failing export markets ever were.

Our northern coasts could play an important role in our struggle against the most menacing national problems of the *postindustrial* age. Our societies are threatened by chronic unemployment and the growth of a new underclass. Reducing the part of the population engaged in rural occupations can no longer be a goal. In Norway it seems to be far easier to create or maintain labour-intensive jobs in coastal fishing than in advanced technology. But the reinstitutionalization of the fishing industry, which implies closing the marine resources off from the great majority of the coastal population, seems to eliminate the potential of the marginal areas to play their part in the national effort to reduce the problems of the postindustrial age.

This book is a first-rate, objective, and amply documented chronicle of the costly war going on in many coastal regions throughout the past half-century. The team of authors masters a wide range of relevant academic disciplines, which enables them to analyse this field from several points of view. Combining theoretical reasoning, observations, and experiences in the field in an admirable way, they have a very interesting story to tell. The comparative perspective does not only enable the team to throw new light upon the processes which have created the present situation in northern North Atlantic communities. As I read it, this book also contains important practical messages – to the people of the coasts as well as to those who try to govern the fishing industry.

Ottar Brox
Centre for Regional Research
University of Oslo
Norway

Preface

This project grew out of a decade of contact and collaboration among the authors on various topics in maritime social science. We met at academic conferences and during sabbaticals or research trips. In turn, our work was facilitated by pre-existing academic connections between Canada and Norway, particularly those between Memorial University of Newfoundland and Tromsø University in Norway. An earlier generation of social scientists built intellectual bridges based on common interests in Arctic studies and fishing cultures (e.g., Apostle and Mikalsen, 1993).

We began from a general social science concern with the fate of coastal areas in the northern North Atlantic and received practical stimulus from the parallel emergence of resource crises in our two regions as the project unfolded (Jentoft, 1993). In the summer of 1990, on Richard Apostle's initiative, we applied for a seed grant from the John D. and Catherine T. MacArthur Foundation under their program for interdisciplinary studies on environmental change. The foundation was interested in having researchers address 'two fundamental questions: how can we promote economic development that is also environmentally sound and democratically responsible, and how can industrialized and developing countries best cooperate in this effort?' We received support to bring the group together in the summer of 1991, and we met in Tromsø to create a more elaborate proposal for consideration by the MacArthur Foundation, among others. We were very fortunate to be among the small number of applicants who received full grants in the MacArthur competition. Their generous support gave us a three-year period from 1992 to 1994 in which to conduct and complete our investigations.

In our original research statement, we proposed

to investigate the consequences of market oriented policies for sustainable development in regions with a strong dependence on marine resources. Specifically, we wish to examine the implications of common market integration, privatized resource management, and small business development policies for fishery dependent communities in terms of long-term sustainability and participatory democracy. Our approach will be comparative and collaborative. We want to study the effects of such policies for two regions on the North Atlantic rim – North Norway and Atlantic Canada – both of which are now experiencing a severe crisis due to overexploitation of the resource-base.

To carry out our research aims, we delineated three major subprojects dealing with institutional development (Petter Holm and Leigh Mazany), resource regimes (Richard Apostle, Bonnie McCay, and Knut H. Mikalsen), and communities and entrepreneurship (Gene Barrett and Svein Jentoft). While most of our work did indeed occur on a bilateral team basis, there were numerous occasions on which we got together as a partial or complete group. In particular, we had a meeting in the fall of 1993 at Dalhousie University to develop and exchange research plans, and another in the summer of 1994 at the Woods Hole Oceanographic Institute to organize our plans for publication. Subsequent to the Woods Hole seminar, we exchanged basic chapter drafts with other members of the group to increase integration of the manuscript and to improve its quality.

Our project received specific institutional support from the School of Resource and Environmental Studies at Dalhousie University and the Norwegian College of Fisheries Science in Tromsø. We are indebted to Ray Côté and Abraham Hallvenstvedt for providing a broad collection of social scientists with congenial work settings for the duration of the project. Professor Helge O. Larsen, then dean of the Faculty of Social Science at Tromsø University, gave us much personal encouragement in the early stages of preparing the proposal. Richard Apostle was the project coordinator and leader of the research group. We are grateful to Donna Edwards for serving as the project's administrator. Not only did she manage to decipher innumerable claims in two languages and currencies, but she also provided the organizational glue for the project. We thank Lesley Reilly for providing us with a wonderful work area at the Woods Hole Oceanographic Institute for the most important intellectual period in the project's life. Norwood Whynot of the Canada Department of Fisheries and Oceans kindly provided us with the map included in this volume.

We would like to acknowledge a number of individuals for their contributions to specific research initiatives.

Petter Holm and Leigh Mazany would like to thank: Reidar Antonsen, Jane Barnett, Randy Bishop, Ron Bulmer, Nilo Cachero, Bruce Chapman, Gunnar deCapua, Steinar Eliassen, Torben Foss, Pierre Goetschi, Stephen Green, Bryson Guptill, Bjarne Haagensen, Gunvor Holst, Trygve Myrvang, Steve Noseworthy, Svein Nybo, Otto James Olsen, Paul-Gustav Remoy, Richard Stead, Roger Stirling, Karl Sullivan, and Rolf Voldsund.

Richard Apostle, Knut H. Mikalsen, and Bonnie McCay are indebted to the following individuals for providing time from their busy schedules: John Angel, Josten Angell, Chris Annand, Neil A. Bellefontaine, Harald Bollvag, G. Leo Brander, Les Burke, Ghislain Chouinard, John Collins, Bob Cook, Claude d'Entremont, Richard d'Entremont, Eric Dunne, Larry Felt, Ted Gale, Brian Giroux, Rob Goreham, Petter Gullestad, J.E. Haché, Edgar Henriksen, Jan Ingebriktsen, Tore Jacobsen, Jim Jamieson, Christopher Jones, Jim Jones, Glen Jefferson, Halvard Johansen, Jan Jorgensen, Patricia King, Egil Kvammen, Paul Lamoureux, John Loch, Art Longard, Clarrie MacKinnon, Eric Moore, Herb Nash, Rosemary Ommer, Peter Partington, L. Scott Parsons, Roald Paulsen, John Pringle, George Richard, Rickey Rideout, Faith Scattalon, Peter Sinclair, Hans Svendsgard, Quinn Taggart, Nils Torsvik, Gunnar Trulssen, Peter Underwood, and Glen A. Wadman.

Gene Barrett and Svein Jentoft acknowledge the assistance of: Grethe Andreassen, Jennifer Bailey, Jeanette Baker, Earl Chase, Susanne Dybbroe, Sam Ellsworth, Einar Eythorsson, Arne Gruhn, Donnie Hart, Egil Johansen, Hallvard Johansen, Joakim Johansen, Nils Jonny Larsen, Kjell-Olaf Larsen, Laura Loucks, the late Poul Moeller, Kirsten Monrad Hansen, Peter Nielsen, Oddmund Otterstad, Liv Toril Petterson, Lorna Read, Ton Ola Rudi, Peter Sinclair, Thor Skattor, Judy Smith, Hilde Stavdal, Elisabeth Vestergaard, Torben Vestergaard, and Jonathan Webb.

Finally, we would like to thank Virgil Duff at the Press for his guidance in preparing the manuscript, and Kate Baltais and Harold Otto for their meticulous editing.

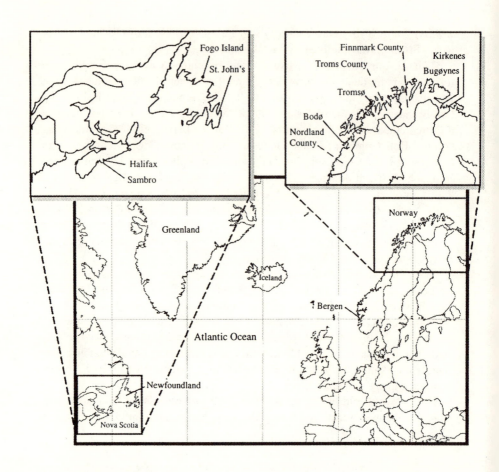

COMMUNITY, STATE, AND MARKET ON THE
NORTH ATLANTIC RIM

Introduction

The fisheries of the North Atlantic are in turmoil. In the spring of 1996, as we complete this book, crisis is widespread in Norway as well as in Atlantic Canada. In both areas codfish is a key resource, and, when it is in trouble, coastal communities and regional economies suffer. In the Atlantic provinces of Canada the collapse of cod stocks has resulted in a moratorium on fishing that has brought many communities to the brink of dependency or abandonment. In Norway, after the fishery recovered from a resource crisis in the early 1990s, a record low price for cod products led to a wave of bankruptcies in the processing sector, causing great repercussions in coastal areas, particularly in the north.

Crisis has been a recurrent phenomenon in the fisheries of Atlantic Canada. This time, however, the situation is more troublesome and the prospects bleaker than before. We are not witnessing marginal shifts or routine variations, but more fundamental changes. There is little prospect that things will settle down or return to normal. Two interrelated but somewhat contradictory processes are at the centre of these changes. First is the development of highly competitive global markets. Second is the diminishing role of state governance in fisheries and the growing legitimacy of market mechanisms. These shifts leave coastal communities more open and vulnerable. As they are increasingly exposed to forces beyond their control, and without the buffering role of government support systems, they are left to fend for themselves.

In this book we will analyse these processes and discuss what consequences they may have for the fishing industries and coastal communities of Atlantic Canada and Norway. In an age of globalization, what does it take to make crisis-ridden industries and communities sustainable? Is the way of life that for ages has characterized coastal communities of the North Atlantic a thing of the past?

Globalization and the retreat of the state are not confined to fisheries; they are features of western capitalism as a whole. Nonetheless, and despite striking similarities between developments in the fisheries and general trends, there are departures. An analysis of the fisheries can both draw on concepts from the study of these more general processes and contribute new insights as to their nature.

Points of Comparison

This is a comparative, interdisciplinary, and international study of crises affecting the fisheries and coastal communities of Atlantic Canada and Norway. The stormy waters of the North Atlantic are the basis of rich and diverse marine ecosystems and cultural adaptations. Historically, fishing has been extremely productive in the coastal margins of the North Atlantic, particularly where warm waters meet cold, nutrient-rich waters, whether off the coast of northern Norway or much farther south on the offshore banks of Atlantic Canada.

Atlantic Canada comprises the provinces of Newfoundland and Labrador, Nova Scotia, Prince Edward Island, New Brunswick, and the oceanic part of Quebec. We refer to Newfoundland and Labrador as Newfoundland, and in our use of Atlantic Canada usually have in mind Nova Scotia and Newfoundland. By Maritimes we mean the provinces of Nova Scotia, New Brunswick, and Prince Edward Island.

Canada and Norway, the foci of this study, are linked by the Gulf Stream, which starts in the Gulf of Mexico. It carries warm water north on the east coast of North America until it mixes with cold water and ice from the Canadian Arctic brought by the Labrador Current. This occurs on the offshore banks of Newfoundland and Nova Scotia, between 45° and 50° North latitude. From there the Gulf Stream veers off to the other side of the Atlantic, carrying its warm waters to the far north of the European continent to mix with the cold waters from the European Arctic, off the shores of Norway, north of 65° latitude.

Although Norway and Atlantic Canada are at very different latitudes and thousands of miles apart, their marine ecosystems are very similar. The marine communities are dominated by cod and other groundfish or demersal species; also abundant are pelagic species, such as capelin and herring, and crustaceans and shellfish. They are also highly productive and at a similar scale. In the late 1980s the Atlantic Canada and Norwegian fisheries had total annual catches worth around Can $1 billion, and groundfish catches accounted for roughly half of the total value throughout the 1970s and 1980s. Cod (*Gadus morhua* L.) has been the mainstay of commercial fisheries and coastal adaptations on both sides of the Atlantic for hundreds of years.

The fisheries crises of Norway and east-coast Canada have been various and complex. The most recent ones were precipitated by the collapse of major cod stocks: the Arcto-Norwegian cod of Norway's Barents Sea, declared near collapse in 1989, and the northern cod of Newfoundland, which came to be protected by a ban on all fishing in 1992.

Understanding the causes and consequences of the fisheries crises requires analysis of the social and economic structures and processes of institutional development and change. For example, important in this comparative analysis is the fact that both fisheries are organized primarily for export. Eighty per cent of the Canadian and 90 per cent of the Norwegian catch is exported. Although Canadian and Norwegian fish companies sell fish in a variety of markets all over the world, both groups are particularly dependent on a single market, although not the same one. In the late 1980s slightly more than half of the Canadian fish export went to the U.S. market, while of the Norwegian fish export almost 60 per cent went to the European Community.

The fisheries of Atlantic Canada and Norway share the problems of economic systems that are dependent on the vagaries of nature on the one hand and unpredictable international markets on the other. On the resource side, the industry is defined by the principle of capture – the resource is not owned by individuals until it is captured. This gives fishing the traits of a lottery, where the chance of success is strongly influenced by weather conditions, fish migration patterns, and stock sizes. On the market side, the industry is not only vulnerable to changes in economic conditions and the wealth of the consumers, but also to political events, protectionist measures, and fluctuating exchange rates. These instabilities on both the input and output sides create fundamental organizational problems.

The vagaries of markets and nature are compounded by the institution of common property, or public claims to natural resources. Fish are 'common pool resources' – mobile, difficult to bound, and with physical properties such that extraction by one person affects what is available to others (Ostrom and Ostrom, 1977). In both Canada and Norway the fisheries are common property, in the sense of being the property of the Crown or the nation which is managed by the state in the public interest. In modern times, most of these fisheries have also had open access: There were no legal restrictions on who could fish, although within territorial waters there might be rules about how, with what, where, and when. The open-access nature of the fisheries was a basis for the dispersed structure of settlement and the traditionally low rate of unemployment found in the two regions (Brox, 1989). The resource itself, the uninhabited land, and the few legal and technological restrictions on access, paved the way for the entrance of large numbers of people into coastal areas. Thus, historically the coastal regions of north Norway and Atlantic Canada were resource-rich 'frontiers' into which people moved, rather than poor regions which people left, as they now have become.

Fish as an open-access common pool resource is vulnerable to overexploitation, as has been clearly demonstrated in both regions. The collapse of the Norwegian Atlantic herring stock in the late 1960s was a dramatic and devastating event that changed the whole perspective on fish: They were no longer a plentiful and resilient resource but a scarce and vulnerable one. Since then the fishing industries of both regions have been subject not only to the problems of natural and market forces but also to the challenges and problems of resource conservation. This enterprise is informed by the paradigm of 'the tragedy of the [open-access] commons' (Hardin, 1968; Ciriacy-Wantrup and Bishop, 1975; McCay and Acheson, 1987), which locates the problems of fisheries in the absence of property rights or strict government intervention.

The more recent collapses of cod stocks in Norway (1989) and Newfoundland (1992) took place in a vastly changed institutional setting, where only the inshore fisheries had any trace of open access left, and all fisheries are regulated. Accordingly, the common property analysis, which has been privileged in government policy documents and economic analyses, is clearly inadequate. Canada's Kirby Report of 1982, which explicitly used the tragedy-of-the-commons metaphor, has been criticized for being insensitive to power relations between large-scale and small-scale players within the industry and for ignoring the fact that not all fishing gear and modes of production are equally damaging to the fish stocks (Barrett and Davis, 1984; Davis and Kasdan, 1984; Davis and Thiessen, 1988).

It is time to evaluate how various government initiatives to manage the fishery have worked and to examine critically the effects of government involvement in the fishery. The resource crises that have been experienced on both sides of the Atlantic came in a situation in which access had been restricted. They appeared after decades of science-based management of the most detailed and comprehensive nature. The tragedy, therefore, as it is now unfolding in North Atlantic fisheries, is not so much a tragedy of the commons as a tragedy of management failure (Marchak, 1988–9) and community failure (McCay and Jentoft, 1996).

The predicaments of the fisheries of the two regions are even more problematic, since they cannot be regarded simply as economies or even economies dependent on common-pool resources. Fisheries are also parts of sociopolitical and cultural systems. To understand the social and cultural importance of the fisheries, their historical roots must be kept in mind. In both Norway and Atlantic Canada the history of the fisheries stretches back centuries. Indeed, the rich fish stocks were the very reason why the coastal regions of Norway and Atlantic Canada were originally settled. On both sides of the North Atlantic coastal communities and coastal cultures were built on fishing, fish processing, and fish marketing. Thus, fishery resources are more than simply an economic basis for occupation and income. They provide a way of life as well as a means to life for thousands of people. The fisheries are cultural 'containers,' carrying and protecting specific technologies, organizational forms, institutions, knowledge, and identities with strong roots in history. Part of diversified coastal adaptations involving locally crafted or simple technology and experience-based skills, as well as traditions of open access, the fisheries also have offered opportunities to 'surplus' labour and provided alternatives to urbanized and industrialized lifestyles.

The fisheries of Norway and Canada are integral parts of modern societies. Pressures to shed the past are enhanced by the need to compete in the marketplace, to compete for a share of the catch on the fishing grounds, and to offer reasonable working conditions and incomes to fishers and fish-plant workers. The fisheries cannot escape such requirements, nor can they escape from the seductions of expensive new technology and the demands of regulations based on scientific information.

The transition to modernity is a precondition for the fisheries' survival as an economic sector in Atlantic Canada and Norway. However, this transition has been filled with problems, dilemmas, and setbacks. The traditional adaptation in the fishery, based on simple technology, occupational pluralism, and few barriers to access, was robust and flexible and thus a sound answer to the fluctuations of resources and markets. Modern technology and institutions are based on large investment and specialization, and they require predictability and control – the lack of which is a defining trait of the fisheries. The fisheries of Norway and Atlantic Canada, thus, are caught not only between uncertain resources and markets, but also between tradition and modernity.

Important and contradictory changes have occurred in the Canadian and Norwegian fisheries industries and their environments. These changes exacerbate the problems we have outlined and shape responses to them. In the international regime of the law of the sea changes have transformed most important fish stocks from a global commons to national properties. The fishing industry itself has rapidly been becoming international. Always oriented towards export trade, fish markets are becoming truly global, as are parts of the fishing fleet and processing industry. Moreover, as the nation state sheds traditional tasks and responsibilities in the fisheries, such as trade regulation, financial support, and industrial development – or relies more on marketlike mechanisms where the retreat of government is difficult – its direct involvement in resource management goes deeper than ever. Traditional ties between firms and communities

Figure 1. Community, state, and market.

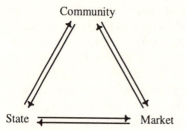

appear to be breaking, as many large firms step into the global arena, although some small firms have been able to combine new opportunities in global markets with close ties to local communities.

The process of modernization in the fisheries is both crucial and precarious. It is crucial because stabilization and rationalization are unavoidable if the resources of the oceans shall be accessible as a source of value within modern society. Given the inherent variability and uncertainty of the fisheries, it is also precarious. Modernization is identified with rationalization and control, and to that extent fishing seems almost by definition to be outside its scope. What are the likely consequences of these inherent contradictions? Will modern techniques of resource management bring fish stocks under control, allowing the fisheries a safe seat on the train of modernity? Or do factors like the globalization of fish markets, the selective withdrawal of the state, and the increasing reliance on flexible specialization point towards a radically new, if not postmodern, social order in the fisheries?

Community, Market, and State: An Analytic Framework

Changes in the fisheries can be analysed as shifts in the relationships among three institutional orders: the community, the market, and the state (Streeck and Schmitter, 1988) (Figure 1). First is the living and lived-in human community, particularly the coastal communities that provide much of the labour, knowledge, skill, and organization for fishing, fish processing, and trade. In sociological terms, many of these communities are characterized by more or less close interpersonal ties, egalitarian social networks that are often multiplex, and shared identities. As such, they contrast with the institutional order of the state. The state, which includes the institutions and structures of policy, law, and governance, is broadly characterized by hierarchical order, bureaucratic structures, and authority relations, and professional and one-dimensional relationships. The third construct for this analysis is that of the market, and it is characterized by competition, economic efficiency, and rationality. In contrast to the community, relationships in markets are fleeting, impersonal, task oriented, and without inherent value. In contrast to the state, the market is characterized by decentralized exchange relations rather than central command and formal authority.

An emphasis on only one of these three dimensions will give simple but unrealistic solutions to the problems of the fisheries. From purely a market perspective, where the fishery is perceived primarily as an economic system, and efficiency is the only relevant criterion, it becomes self-evident that privatization, downsizing, and changes in marketing stand out as solutions to problems in the fisheries. This conclusion, however, disregards the social costs of disembedding the fisheries from coastal communities, not to mention the fiscal and political costs to the state. From purely a community perspective, problems are seen solely in their local and social manifestations, disregarding the interests and claims of other groups in society, and the economic costs and political trade-offs of measures that might be taken to help a coastal community prosper. At the extreme, this perspective leads to an unrealistic vision of coastal communities as premodern and self-sufficient, outside the reach of market forces and state intervention. From purely a state perspective, say that of an administrator or minister responsible for fisheries management, the goal is to maximize order and control, ignoring the unpredictabilities of social life and the complexities of ecological systems as well as the ledgers of the marketplace.

Sociologists and anthropologists are often accused of being romantics in their emphases on communities. However, it should be evident that *each* of these perspectives is romantic in its simplicity and utopianism. Each is also wrong if taken by itself. The community–market–state triad is a vivid reminder that policy-makers must balance conflicting objectives, reconciling economic efficiency with social justice, resource conservation, and other concerns. Fisheries in modern societies are affected and structured by all three institutional orders, although the balance will differ over time and place and will often be the object of political struggles. Thus, as we show in this book, over the past century the balance has shifted from the community towards the state and the market corners of this system. In recent decades the state and market institutional orders have vied for dominance, marginalizing the community perspective. Most recently, the state has begun to withdraw, emphasizing solutions closer to the market pole, creating the flux, opportunities, and uncertainties known as postindustrialism or postmodernism.

The community–market–state triangle facilitates the analysis of pairs of relationships that express the dilemmas faced by the fishing industry, the communities dependent on it, and those empowered to develop and implement appropriate policies. The *community–state* axis invites questions on the impacts on coastal communities of state actions, such as subsidies and regulations. It also directs attention towards the locus of decision-making including calls for greater 'co-management' on the one hand and more reliance on state-controlled science in resource management on the other. The *community–market* axis draws attention to the socioeconomic and cultural effects of the ambiguous ways in which global market forces penetrate local communities (Giddens, 1990). The community–market axis also points to potentials of flexible specialization (Piore and Sabel, 1984) and other strategies available for coastal communities and firms struggling to cope with crises. The *market–state* axis points to the opportunities, problems, and dilemmas of state withdrawal from the fisheries. States have had a more active involvement in fisheries than in most other sectors of the economy. This is largely because of the need for resource management created by the 'common pool' nature of fisheries resources. The withdrawal of the state started in markets and trade, and it has continued in resource management through the introduction of private property rights in the form of individual transferable quotas (ITQs). Because we are modeling an interrelated system, it should be noted that there are critical cross-linkages between each pole in the diagram. For example, the retreat of the state affects the community–state axis both directly, as subsidies and other supports for local enterprises dry up, thus making them more vulnerable to unrestrained market forces, and indirectly through the effects of privatization.

Theoretical Concerns

As already indicated, the changes unfolding in the fishing industry in Norway and Canada are part of a larger and more global pattern. From a postwar period of continuous growth within stable conditions, western society as a whole has entered a period of transition that has affected most industries, including the fishery. Up to the 1970s there was a high degree of compatibility between economic practices, political institutions, and cultural values. In the present situation the relations between the major subsystems are much more problematic. In the following, we briefly explore the nature of the transformations that are currently taking place in the fisheries and their relationships to the broader processes in modern society. The discussion provides theoretical underpinnings for our focus on phenomena such as Fordism, flexible specialization, and globalization, as they are displayed in modern fisheries.

Fordism

In the postwar growth period, the production process of western capitalism was typically organized in large-scale, highly routinized, and specialized enterprises. Henry Ford's assembly-line production of standardized cars represented the prototype, hence 'Fordism.' The Fordist model, applied in industries like cars and consumer electrical goods, became the engine of the postwar economic boom. The rationalization of the fishing industry in Norway and Canada in the 1950s and 1960s was structured along Fordist principles.

The technological superiority of Fordism, however, does not constitute a sufficient explanation for this growth. The Fordist model made mass production possible. But that presupposed the existence of mass markets. The reason for the unprecedented growth during the postwar years was that an economy along Fordist lines not only supplied products, but it also produced mass markets in the form of a relatively affluent working class. However, the Fordist model could not automatically reproduce the conditions for its expansion. To produce sustained growth, the model had to be integrated into a system of regulations at the aggregate level. In the period of postwar growth, these regulations were supplied by interventionist governments.

The technical and economic efficiency of the Fordist model was an important explanation for its success. This trait of the regime is so predominant that it is customary to refer to the postwar economic regime itself as Fordist. However, the term is properly used in a more restricted sense, as a certain way to organize production. Inspired by Frederick W. Taylor's (1967 [1911]) theory of 'scientific management,' a basic postulate of Fordism is that the principles for optimal work organization can be discovered using scientific methods. A fundamental characteristic of Fordism is the high degree of specialization and routinization on the shopfloor. The totality of tasks that goes into making the final product is broken down to simple, prespecified routines, and each worker is responsible for only a few of them. The right to specify routines and design jobs belongs to management. A second crucial feature of Fordism, in contrast to the artisanal organization of production that it replaced, is the location of power over work design at the management level, as opposed to the shopfloor. The Fordist model is a divided one, as it makes a sharp distinction between workers performing routine tasks and managers who design them.

There is a parallel between scientific management as it was perceived by Frederick Taylor for the industrial workplace and current practices of resource management in fisheries. In both instances scientific methods are applied as an integral part of the management approach. They both share the perception that knowledge gained from the scientific analysis of basic processes is sufficient to eliminate uncertainty and facilitate control. They both take for granted that hierarchical control is a necessity to ensure a rational and efficient process of production. In both the human factor is seen as problematic and an influence that managers should reduce.

To see the implications of the Fordist model for the economic system as a whole, and for the fishery as a subsystem, three points should be observed. The first is its rigidity. The high degree of routinization that defines the Fordist mode of production makes it very efficient for the tasks it is designed to perform. It also makes it inflexible and thus vulnerable to uncertainty and change. If the investment in a specialized system of interlocked routines is going to pay, the costs will have to be spread across a large number of units. The Fordist model, then, is a model for mass production of standardized items designed for mass markets. It has to be shielded from uncertainty in the environment (Thompson, 1967), and it will not work well in unstable markets, for one-off products, or when technology is rapidly changing (Stinchcombe, 1990).

The second feature is the conservative relationship of the Fordist model to the structure of society. The two subsystems of the Fordist model, management and workers, tend to be filled with people from different social strata. Typically there are few career paths leading from one subsystem to the other. The Fordist model, hence, reflects and reproduces the structures of a class society.

The third feature is the model's high degree of institutionalization. Fordism is more than a recipe for organizing work. It is an institutionalized social form, taken for granted as the standard way of organizing production, even in contexts where it is not very appropriate, as in the fisheries of the North Atlantic.

Flexible Specialization

The vulnerability of Fordism with regard to environmental fluctuations and uncertainty was not apparent in the period of postwar growth, at least not in the core economic sectors, because of the success of the buffering role of the state. Gradually, however, as a consequence of changes in overall market structures, the weaknesses of Fordism became apparent. These changes included a move from standardized to non-standardized products, an increased number of varieties of products, and shorter intervals between product model changes. The competitive edge no longer rested on price alone but, increasingly, on innovation. Fordist sectors were lagging behind and no longer at the forefront of technical development. The electronics and service industries became the new engines of economic growth. Since the early 1970s there has been a rapid contraction of employment in manufacturing and an equally rapid growth in employment in the service sectors.

These structural changes revealed the inflexibility of the Fordist model. To deal with the situation, two general options were available to Fordist firms. One was to reduce the costs of rigidity through techniques like 'just-in-time' and subcontracting. Frequently, however, the introduction of such techniques met with resistance from the unions. The response of the firm was to relocate to third world countries, where wages are lower and government regulations are often absent. The other alternative was to insert flexibility into the core of the production process, for instance, through (re)skilling of workers. If Fordism deskilled the workers, flexible specialization requires a workforce with multiple skills.

The Fordist model is one of mass production and giant companies. When flexibility and innovation are important criteria, large scale can be a disadvantage. The tendency towards market segmentation and shorter product cycles has opened up opportunities for small- and medium-sized firms. This does not necessarily mean the end of the giant company. In many sectors the premiums on scale in areas like marketing, research, financing, and lobbying are considerable. In some cases, however, a group of small firms forming a network or 'strategic alliance' will be as effective as one large firm. This brings us to the discourse on 'flexible specialization' and the concept of a 'Marshallian industrial district,' as depicted in the seminal work of Piore and Sabel (1984). The Marshallian districts are geographically defined productive systems characterized by a large number of small- and medium-sized firms cooperating on various stages in the production process (Pyke and Sengenberger, 1991). They are all global-market oriented, highly competitive, technically innovative, quality minded, and economically efficient (Becattini, 1990). Firms are also influenced by, and benefit from, the cultural practices and institutional patterns that characterize the districts in which they are located.

Globalization

The problems of Fordism were compounded by changes in national and international economic institutions. The host of techniques and policies loosely summarized under the term 'Keynesianism' lost effectiveness and legitimacy. Increasingly, economic problems could not be contained within the nation state and solved by government intervention. Location has become a matter of choice and as such an important strategic variable for global competitiveness. The enterprises can threaten to move, thus undermining the ability of the state to intervene. This tendency is part of a more general trend referred to as 'globalization' which has also affected marketing. The modern enterprise operates as if the world is one single market. Furthermore, globalization pertains to changes in the world's financial systems. There is now a worldwide market for money and credit supply, as 'banking is rapidly becoming indifferent to the constraints of time, place and currency' (*Financial Times* 8 May 1987; cited in Harvey, 1989). This means, among other things, that the state no longer can regulate credit supplies. The use of trade barriers or subsidies to protect national production is becoming increasingly problematic as they run the risk of being met by sanctions, or threats of sanctions, from other states.

The general trend towards globalization has triggered institutional reforms at supranational levels. Examples include the renegotiation of the General Agreement on Tariffs and Trade (GATT), the strengthening of European cooperation in depth (Maastricht Treaty) and breadth through the establishment of the Extended Economic Area (EEA) and the North American Free Trade Agreement (NAFTA). These reforms restrict the use of measures at national levels that distort free trade, such as tariffs, subsidies, and legal monopolies.

Globalization means that the competitive pressures of the marketplace are felt more directly in traditionally strong export-oriented nations like Norway and Canada. Since trade regulations have played a more important role in Norway than in Canada, and because the EEA agreement is more restrictive than its North American counterpart, the effects are more obvious in Norway. The general trend of globalization is felt in both countries, however, within the fisheries as well as in other industries.

These theoretical considerations allow us to rephrase our research issues at a more general level. For instance, what is 'the modern project' in the fisheries, and will it succeed or fail? Will 'scientific management' of fish stocks finally turn the fisheries into a predictable and stable economic sector? Does Fordism represent the future or the past? How are coastal communities coping with the transformations that are now occurring in western society? What are the policy responses to the conflicting trends? Is the paradigmatic shift from fisheries as a rural way of life to fisheries as a global business a threat to sustainable development or a new opportunity for crisis-ridden coastal communities?

Research Methodology

Our research is multidisciplinary, collaborative, and comparative. We have worked in international teams on the various sections and chapters. Each team comprised one Canadian and one Norwegian; an American also contributed to two of the teams. Sociology, anthropology, political science, and economics are the disciplines represented. Data collection was carried out jointly in a way that allowed us all to work in each other's regions. Thus, we learned not only from our own individual experiences in the field, but also from the other's observations of our own systems. Those things we tend to take as given, that we do not see because they are so familiar to us that they appear as 'natural,' even when they clearly are socially made, were revealed as social constructions by the fresh perspective of the outsider. That, we think, is the great merit of our methodological approach.

Our belief in the comparative method is derived from our own previous research experience (Apostle and Jentoft, 1991) and that of others. The comparative method belongs to the social science toolbox (Holt and Turner, 1970; Ragin; 1987; Øyen, 1990; Glaser and Strauss, 1980) and has a distinguished tradition in research on North Atlantic fisheries. We see ourselves as part of a twenty-five-year history of intellectual and academic exchange between social scientists in Canada and Norway. In the North Atlantic region, there has been a long-standing appreciation of the need to understand the workings of other systems (Cohen, 1985), reflected in annual meetings on what are known as the International Seminar on Marginal Regions. It has also resulted in frequent visits among scholars, international seminars to discuss similar research topics, (for example, the International Seminar on Social Research and Public Policy Formation in the Fisheries held in Tromsø in May 1986 (see MacInnes et al., 1996), and fine studies being done by outsiders, such as Ottar Brox, Mollie O'Neill (1981), and Cato Wadel.

In the early years these connections were primarily mediated by contacts between Memorial University of Newfoundland and the University of Tromsø. The two primary foci for research activity arising from this contact have been Arctic studies and fishing cultures. The early leaders of Memorial University's Institute of Social and Economic Research (Ian Whittaker, Robert Paine, and George Park) spent considerable time in Norway, focusing primarily on the Saami. The Norwegian scholars who spent time in Newfoundland tended to devote their efforts to studies of the fishing industry (Brox, 1972; Wadel, 1969, 1973). Raoul Andersen of Newfoundland and Cato Wadel of Norway developed several transnational collections and syntheses on North Atlantic fishing cultures (Andersen and Wadel, 1972; Andersen, 1974, 1979).

Ottar Brox's work on economic dualism remains significant because of the comparative dimensions of his analysis. Brox emphasized both the importance of extensive variation within capitalist societies and the similarities among Newfoundland and other coastal areas in the northern North Atlantic, including the Faeroes, Iceland, north Norway and Scotland. He also argued that 'comparison is most fruitful when societies that have much in common undergo examination' (1972: 6). Brox's book on the case of Newfoundland is in many ways a follow-up of his previous book on north Norway, which in Scandinavia is seen as a seminal social analysis of rural communities and inshore fisheries (Brox, 1966). The theoretical perspective employed in the Newfoundland study, which Sinclair (1990: 16) has not unfairly characterized as being rooted in modernization theory, postulated a fundamental dualism between traditional and modern sectors in the Newfoundland economy. Following Fredrik Barth's economic anthropology, Brox characterized the relationship between these two sectors as involving 'conversion barriers' that are crucial to understanding the difficulty in making transformations across the sectors. Specifically, Brox recognized the existence of regulatory mechanisms that made it difficult for coastal community ('outport') households to get involved in fish processing and power differentials – which prevented fishermen from controlling the fish pricing system. He maintained that the problem Newfoundland fishermen had with landing prices is attributable both to the limited political power of outport fishermen and the existence of unemployment insurance as a subsidy, which in turn lowered the price of fish (Brox, 1972: 30, 74). Brox further maintained that the Newfoundland resettlement programs were fundamentally misdirected. They did not recognize that urbanization would be unable to provide secure employment or housing for outport families. As an alternative that continues to have relevance, Brox proposed the adoption of intermediate or appropriate technologies that would permit decentralized, bottom-up rejuvenation of Newfoundland's traditional economy (Brox, 1972: 59, 90; McCay, 1976, 1979).

The issues that Brox focused on in the late 1960s and early 1970s are still with us. As we are about to turn into a new century, the problems that coastal communities are having are in many ways more severe and acute than they were then. Now coastal communities in Atlantic Canada are faced with a resource crisis of a magnitude that brings their very existence into question. In the early 1990s north Norway got a glimpse of what a depleted cod stock could mean, but Norway was lucky to escape the situation in which Atlantic Canada now finds itself (Jentoft, 1993).

The ecological, social, economic, and cultural similarities of the situations that prevail in our two fisheries, the identical challenges of globalization, free trade, and problems of resource management, and yet the many differences in the policies and institutions of our two systems, spurred our interest in undertaking this comparative project. As Ottar Brox observed, we believe that there are lessons to be learned from comparing our two systems, particularly because comparison brings up issues that we normally do not consider. Brox contended that the 'comparative perspective has ... made me aware of factors that are not so easily discovered if one examines Newfoundland in a non-comparative fashion' (1972: 6). That is also our experience. Comparisons invite answers not only to the problematic but also to what is familiar. In many ways it is like the experience that tourists have in a foreign country. What tourists see makes them wonder not only about the foreign culture, but also about their own: 'If things work abroad, why not at home?' Ottar Brox has said that he only discovered the significance of the institutions of the Norwegian fisheries by visiting Canada, a country that relies on different institutional principles.

Introduction 17

Book Outline

The book consists of three sections, which are interconnected and address different aspects of the same problem, but which stand by themselves and can be read separately.

In the first section, Petter Holm and Leigh Mazany give a broad overview of the historical and institutional development of the fisheries throughout the twentieth century. The story they tell is one of modernization: how the fisheries in Atlantic Canada and Norway gradually were forced out of their traditional adaptations and into the era of modern industrial capitalism, with all its promises and problems. In their analysis, Holm and Mazany show that the development in the fisheries on the two sides of the North Atlantic has remarkable similarities but also important differences. During this century a series of waves of reform have swept over the fisheries, gradually bringing the sector further in line with the modern social and economic order, shifting the balance away from communities and towards both the state and the market.

Four such waves are identified and discussed. The first occurred in connection with the general economic depression of the interwar period, which occasioned a major crisis in all export-dependent fisheries and attempts at organizational reforms on both sides of the North Atlantic. The second reform wave started during the Second World War, but reached its climax in the first two postwar decades. In this period the ambitions to transform the fisheries from a crisis-ridden and backward sector into a modern and rational industry were tied to trawlers, large-scale mass production, and vertical integration – in short, the Fordist model of industrial organization. The third wave of reform, which rolled in over the fisheries during the 1970s, was set off by the increasingly obvious problem of depletion of resources. Thus, during the next twenty years, the Norwegian and Atlantic Canadian fisheries, like fisheries throughout the world, were preoccupied with the construction and implementation of institutions for the management of fisheries resources. While the problems of resource management are by no means solved, the fisheries by the middle of the 1990s also have been exposed to the forces of globalization. In their last chapter, Holm and Mazany discuss the prospects and problems this represents for the fisheries.

Section 2, by Richard Apostle, Knut H. Mikalsen, and Bonnie McCay, compares the regimes of fisheries management in Atlantic Canada and Norway. Resource management brings the state into new and problematic relationships with both community and market institutions. The focus of this section is on the institutional aspects of management: on the ways in which regulatory decisions are made and enforced, the division of jurisdiction between different levels of government, the role of user groups in management policy-making, and attempts at institutional reform. At the core of the analysis lies a conception of management as a series of trade-offs between bioeconomic imperatives and political considerations; between centralized control, based on scientific knowledge on the one hand, and participation by user groups, justified by democratic doctrines on the other.

Throughout their analysis the authors identify both similarities and differences between the two management regimes as to how these trade-offs have been made. In both countries management has been about developing institutions and policies for dealing with problems of overcapacity and resource depletion. The reliance on science and commitment to fleet rationalization are common themes, as are attempts at balancing the goals of resource conservation and economic efficiency against broader concerns of social equity and regional development. The role of the state, however, is currently under debate, and management policies are increasingly influenced by market models and a renewed emphasis on economic efficiency. This trend, apparent in the controversies over individual transferable quotas (ITQs), is at present more conspicuous in Atlantic Canada than in Norway.

The state–community axis includes struggles over management decision-making power. In both Norway and Canada user participation and consultation are emphasized, although these vary with respect to the scope of executive discretion as well as the forms of involvement of user groups. There is, according to the authors, a paradox here in that the capacity for centralized control seems greater in the federal system of Canada than in the unitary state of Norway. In the former, executive power is anchored in the constitution, underpinned by other legislation, and strengthened by the organizational fragmentation of the industry; in the latter, centralized control and ministerial discretion are circumscribed by well-organized user groups working closely with government officials within the framework of corporatist arrangements.

The third section, by Gene Barrett and Svein Jentoft, with the assistance of Bonnie McCay (who wrote Chapter 14 and contributed to the Introduction to Section 3), brings the analysis to the community level, examining relationships between community and institutional orders of both the state and the market. They argue that if one wants to find out what is rational behaviour under conditions that are chronically unstable and beyond the control of the individual firm and the local community, one should study how they are currently responding to these challenges. In addition, the authors contend that for determining how government could help make communities more sustainable, one should fully understand the day-to-day struggle of individual actors – households, fishers, and fish processors – individually as well as collectively, what their problems are and how they see their future. Barrett and Jentoft start out by exploring conceptual issues in small business management and community organization, with special emphasis on the theory of flexible specialization and the sociocultural embeddedness of economic activity. They go on to describe how the Fordist model of industrial organization was adopted in the fishery in Norway and Atlantic Canada after the Second World War and how Fordist firms have responded to the resource crises in the 1990s. Another organizational model, smaller in scale and more embedded in the local community, has existed alongside Fordism. How such firms managed through the cod resource crises is discussed.

The decline and collapse of cod stocks affected whole communities, and how these communities responded is described in three case studies, two Canadian and one Norwegian. Each captures a re-embedding process, where communities have acted collectively to reassert control of local resources and employment opportunities. These cases stand out in a situation dominated by tendencies for local communities to become less and less relevant for an increasingly disembedded, globalized fishing industry. Barrett and Jentoft argue that sustainability requires a process whereby the fishing industry can resume its anchoring in the local community while at the same time acquiring more global competence.

The concluding chapter, in summing up the analyses and arguments of the book, discusses the future of the fisheries along the rim of the North Atlantic. The fisheries are at a crossroads. While we do not pretend to know all the answers, we hope to be able to point out the main questions. One pertains to the future of modern management of fisheries resources. Does fishery science offer adequate solutions to the fluctuations and uncertainty of the fisheries, finally bringing the fish stocks under rational prediction and control? Or will the complexities of marine ecosystems also defy this attempt at disciplining nature? Where does the process of globalization lead? It is clear that the role of the state is changing, leaving fishermen and fish workers to fend for themselves. But what will fill the void the state leaves behind? Will the retreat of the state mean decentralization and a resurgence of coastal communities? Or will it leave the fisheries open to global firms and expose coastal communities to the brutality of the international economic order? The fisheries are particularly interesting and problematic because they are more than economies. They supply income and employment, of course, but also identities, values, and meaning. The present development trends hold strong pressures for the fisheries to shed all but the economic ethos and to leave the coastal communities and cultures behind, concentrating on the rational utilization of fish stocks. This might exorcise many of the problems and dilemmas that have haunted the fisheries throughout this century. At the same time this would turn the fisheries into an insignificant sector, bringing good incomes and safe employment to a handful of people, but hardly anything else.

We hope that the next century holds a different future for the fisheries; that it will remain a source of identity, meaning, and culture, as well as a source of income. With this book, we hope to show that this not only is possible, but also that it represents the best use we can make of the fisheries resources.

SECTION 1
INSTITUTIONAL DEVELOPMENT

Introduction: Institutional Development

The commercial fisheries of Norway and Atlantic Canada have fundamental traits in common. Both are based on groundfish, particularly cod, which typically display large fluctuations in stock sizes and availability. Both are dependent on export markets, which create uncertainties because of long distances and transportation problems, incomplete information, currency fluctuations, cultural differences, trade barriers, and competition. The instability in the industry's environment has had far-reaching consequences for organizational solutions and economic development in the fisheries. Throughout the twentieth century predicting and controlling such environmental uncertainties have been at the centre of a succession of attempts to modernize the fisheries of the North Atlantic both economically and institutionally.

Uncertainty is the fundamental threat to and focus of modern, rational organization: A successful organization is one that manages uncertainty (Thompson, 1967). In the export-dependent fisheries of Norway and Canada, however, uncertainty has proved more persistent than it has in most other economic sectors. The two factors that constitute the *raison d'être* of the fisheries, fish stocks and export markets, have never been brought thoroughly under control. Until the beginning of the twentieth century the instability problem was handled by way of flexibility in the traditional coastal economy. When the fishing failed or when prices dropped, people could resort to other means of subsistence, particularly agriculture and forestry. The dependence on the fishery, and thereby the vulnerability to fluctuations, was kept low by maintaining alternative occupations. This also meant a low degree of specialization, a broad-based but nonintensive utilization of resources, low efficiency, and, quite often, a fairly low standard of living.

The general modernization process within the western world did not pass by the fisheries of Norway and Canada, however, and the traditional institutions and organizational solutions were gradually undermined: Capital replaced labour; the degree of specialization increased; the utilization of resources became more intensive; fishermen have been transformed from occupational pluralists to professionals; and the household has been replaced by the firm as the dominant economic unit. The fisheries have been disembedded from local communities and other economic activities and turned into a separate and rationalized economic sector. This meant that much of the flexibility of the traditional economy was lost. In step with the modernization process, the vulnerability of the fisheries to fluctuations and uncertainty in fish resources and export markets has increased.

Put generally, the history of the fisheries in the twentieth century can be seen as a series of strategies to overcome the problem of uncertainty. We identify four such strategies that have dominated the sector in four successive periods. This section of the book is organized on the basis of this delineation, and we deal with each period and strategy in separate chapters.

In Chapter 1 we describe the organizational reforms that took place in the fisheries during the interwar period and led to the end of 'commercialism,' the traditional economic adaptation in the fisheries. These reforms were triggered by the market crisis that followed the international economic depression of the time. The nations of the North Atlantic that had export-dependent fisheries converged towards the same solution: the establishment of cooperative marketing organizations, or cartels, particularly within fish exports. The detailed design of these organizations varied. We shall see that there are important differences between Norway and Atlantic Canada as to the extent to which these organizations were entrenched in the fisheries. They all reflected the same general solution to the problem of market uncertainty in the fisheries, however, which was to shield the sector through collective mobilization. In sharp contrast to the traditional way to deal with uncertainty – laying low until the difficulties had passed – the strategy now was to close ranks and build barriers against the problems in the market (Figure 2b).

In Chapter 2 we describe the dominant organizational strategy of the postwar period. Now the market problems of the 1930s were almost gone, and the buffer strategy no longer attracted much attention. Instead, interest was directed towards the organizational form that had been so tremendously successful within industrial manufacturing: the Fordist model of mass production. Instead of building buffers against natural and economic fluctuations, this model relied on bridging strategies, in which the tasks along the production chain were to be coordinated within large, vertically integrated firms. In the fisheries this solution was coupled to huge investments in technological solutions to the uncertainty problem. On the resource side, modern trawlers were to seek out the fish wherever they were located and bring them in regardless of weather conditions and season. On the market side, demand fluctuations were to be solved by forward integration and large-scale marketing efforts right down to the level of the consumer. What remained of variability in the production chain was to be smoothed out by freezing excess product (Figure 2c).

The Fordist model proved less successful in dealing with the twin uncertainties of resource and market fluctuations than its protagonists had hoped. In particular, the huge investments in new, efficient technology were undertaken as if fish resources were unlimited. The gradual realization during the 1960s and 1970s that this was not the case propelled yet another organizational model into focus: management of fisheries resources. This model had much in common with the buffer strategies of the interwar period, but with a different thrust. Instead of focusing on the problems in the export markets, attention was now directed towards the problems of overcapacity in harvesting and scarcity of resources (Figure 2d). In Chapter 2 we describe this strategy, its organizational implications, and how it was implemented in the fisheries of Norway and Atlantic Canada.

The explicit efforts to reduce and control uncertainties throughout this century have been a crucial part of the attempt to turn fishing into a modern economic activity. This has meant explicit mobilization of political and economic actors on the basis of their economic interests in the fisheries, the construction and elaboration of formal organizational structures, the substitution of corporations, for households, and increasing reliance on scientific knowledge and bureaucratic structures. All of this has contributed to a development by which the fisheries have been gradually lifted out of their traditional cultural and social settings.

Towards the 1980s and 1990s these tendencies tied into and were vastly strengthened by the more general trend towards globalization in politics and markets. This brings us to Chapter 3 and the latest strategy to overcome the problem of uncertainty and fluctuation in markets and resources. Globalization is a strategy that allows the firm to step out of its dependency on particular export markets and particular fish stocks. When the prices in one market decline, it can direct its efforts to another. When one fish stock collapses, it can seek out one that is in better shape. In this way globalization is based on flexibility and is similar to the traditional economic adaptation within the fisheries. While the traditional coastal households were flexible in tasks, however, the global firms are flexible in place (Figure 2e). Where the traditional form of flexibility systematically relied on and reinforced the ties between fishing and local communities and local values, flexibility in the global variety is based on the ability to cut such ties.

Section 1: Institutional Development

Figure 2. Organization strategies in Canadian and Norwegian fisheries during the twentieth century.

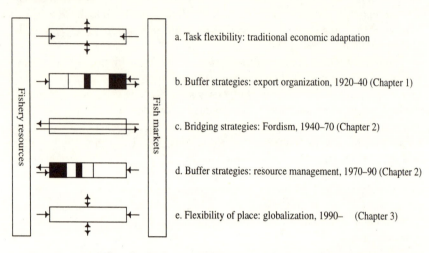

a. Task flexibility: traditional economic adaptation

b. Buffer strategies: export organization, 1920–40 (Chapter 1)

c. Bridging strategies: Fordism, 1940–70 (Chapter 2)

d. Buffer strategies: resource management, 1970–90 (Chapter 2)

e. Flexibility of place: globalization, 1990–   (Chapter 3)

The different strategies for dealing with the problem of uncertainty have been pursued systematically by powerful actors in each period, and they have often inspired large-scale campaigns for reform. They have not, however, always been successful. Implementing new organizational forms will always be subject to political struggle and will therefore be dependent on the pre-existing power structure within a sector. Put generally, the strategies and institutions that exist at any particular time do not conform to any ideal type, but are a loose aggregate of solutions, some left over from previous periods and others newly established to deal with the problems currently perceived to be the most pressing. Underneath the simple scheme on which our analysis is based, we therefore find a jumble of different organizational and institutional forms.

Striking parallels are found in the way the four organizational strategies have made their appearance on the two sides of the North Atlantic. However, the consequences for economic organization and institutional structure have not been the same. An important difference between the fisheries of Norway and Atlantic Canada pertains to the status of the Fordist model. The reason for this difference can be found in the interwar period, when the buffer strategy was pursued with more enthusiasm and success in Norway than in Atlantic Canada. This had repercussions in the postwar period, when the attempts at Fordist reorganization of the fisheries were met with counterforces of different strength in Norway and Canada. Resistance in Canada while, it was fierce in Norway, in particular from small-scale fishermen empowered during the fishery reforms of the 1930s. The fisheries of Canada were hence more influenced by Fordism than were those of Norway, where small-scale fishermen and petty-capitalist structures remained dominant. This difference in organizational structure also affected the direction of the waves of reform triggered by the introduction of resource management during the 1970s and 1980s, and those triggered by the globalization process of the 1990s. On both sides of the North Atlantic, these reform waves were set off by crises in which existing organizational solutions lost legitimacy. Hence, the Fordist model was undermined in the setting of Atlantic Canada, while the petty-capitalist model went into decline in Norwegian fisheries.

This section is an attempt to account for crucial but often ignored or misunderstood differences between the fisheries of Atlantic Canada and Norway. Why did large firms become more dominant in Canada than in Norway? Has this structural difference had consequences for the way uncertainty and fluctuations have been handled in the two settings? Why have Norwegian fishermen managed to unite in one powerful organization, while their Canadian counterparts have remained, in comparison, organizationally divided and politically weak? Why is the Norwegian raw fish market dominated by sales organizations that are protected by law, while the Canadian is not? What are the consequences of fisheries subsidies in Norway primarily having been given in the form of price support, while in Canada they have primarily been given in the form of unemployment insurance? What are the functions of the complex corporatist system of Norwegian fisheries? Why did no such system develop in Canada?

Such questions invite systematic comparison between the fisheries of Norway and Canada at the sector level of analysis. It is not the fisheries as experienced by the individual fishermen, processing company, or state agency that are described in this section. Instead, we try to come to grips with the logic of the fisheries as a sector of society.

The choice of the sector as a basic analytical unit has implications that are worth noting at this point. First, it forces us to take a broad historical perspective, since the shape of the sector today has been defined to a large extent by decisions and processes in the past. Within this historical framework, we are primarily interested in the forces that shaped the fisheries as economic, political, and social systems, and in particular the processes that led the fisheries of Norway and Atlantic Canada in different directions. While historical in outlook, our analysis should not be read as a general history of the development of the fisheries in Norway and Atlantic Canada.

Second, our interest in the development and logic of the fisheries as sector systems invites an unconventional treatment of the management of fisheries resources. As shown in Chapter 3, the institutionalization of resource management as the most important concern for state intervention in the fisheries in the 1980s and 1990s had tremendous effects on organizational structures, decision-making processes, and the distribution of power within the fisheries. Looking at the fisheries as sector systems, we are primarily interested in resource management as an organizational model, containing specific prescriptions for reform and strategies for modernization. As an organizational model, resource management has the same analytical status as the sales organizations protected by law (MSOs) had in the interwar period, or the Fordist model in the postwar period. However, while the political content of the latter two is easily recognized, it is usually ignored or downplayed when management of resources is concerned. This is, we believe, because the resource management reform was legitimized as a rational adaptation to the condition of the fish stocks, as neutrally reported by science. Hence, resource management has been regarded as a pre-eminently technical and scientific issue. If its introduction had any effects on organizational patterns, politics and the distribution of power, this was purely incidental and beside the point.

Our interest in the implications of resource management for the fisheries as a sector system forces us to look beyond the conventional claim that it is nothing but a neutral and rational adaptation to nature. If our account seems unfamiliar, it is because we emphasize the aspects of the resource management reform that are usually ignored. This emphasis should not be interpreted as a denial of the rationalist claims on which the revolution in the management of resources has been legitimized. The point in this section is simply that there is more to it.

The third implication of choosing the sector as the basic unit of analysis concerns the usage of the term 'crisis.' In this section we are not primarily interested in crises as sources of problems and traumas in peoples' lives. Instead, we focus on the organizational and institutional mechanisms developed to handle fluctuations and uncertainties at the sector level. A crisis in the sector system will result if the change in some important variable exceeds what can be handled by established buffering mechanisms. Crises in this sense are important in the analysis here, because they open up for debate the basic organizational and institutional structure of the sector. During crises those features of the sector system that tend to be invisible, because they are taken for granted, are suddenly exposed.

1
The End of Commercialism

The fisheries of the North Atlantic went through a major crisis during the interwar period. This crisis was set off by the combination of an increasing supply of fish in international markets, brought on by, among other things, the mechanization of the fisheries, and falling demand, caused by the international economic recession. The result was severe economic and social distress in coastal communities all over the North Atlantic, as fish prices dropped to historic lows. Characterizing the situation as a 'crisis' is justified by more than these hardships, however.

Concomitantly, a fundamental structural transformation occurred in the fisheries, as the traditional commercialist economic system, dating back to the Middle Ages, was gradually challenged by the organizational principles of modern capitalism, particularly in the emerging fresh and frozen fish industry. Essentially, a crisis of commercialism in the fishery occurred in the 1920s and 1930s. While the final breakthrough of modern capitalism in the fisheries did not happen until after Second World War, the interwar events cleared the ground for it.

The transition from commercialism to capitalism involved conflicts and social struggle, and also in this respect events of the interwar period deserve the 'crisis' label. This was a period of intense political struggle, where different groups strove to influence the definition of the problems of the fisheries, and, consequently, their solutions.

These processes were to a large extent informed by immediate concerns, but they were played out against the backdrop of larger questions regarding the basic organization of the industry. What should be the primary goals and values of the fisheries: profits for the few or livelihood for the many? Who should be its core actors: capitalist firms or coastal people? How should the riches of the sea be utilized: through large-scale efficiency or small-scale flexibility? The transition from commercialism to capitalism raised essentially the same overall questions on both sides of the Atlantic, but the political processes and the answers they generated were different. In Norway the process of reform resulted in a complex corporatist structure, in which the petty-capitalist fishermen had a privileged position and from which they dominated the sector. In Atlantic Canada and Newfoundland the fishermen never managed to assert themselves in a similar fashion. In absence of this, large-scale capitalist enterprises gained a much stronger position in Canada than in Norway. The consequences of this difference in the postwar period will be described in Chapter 2. Here, we will concentrate on the events during the interwar period that led to the end of commercialism in the fishery.

Commercialism versus Capitalism in the Fishery

Salted and dried fish was one of the commodities that were traded within the traditional commercialist economic system, where the merchant was the core actor. Any commodity would do, but the merchants preferred staples like grain, sugar, tobacco, fur, and fish. The merchants were 'engaged in securing differential gains between the prices at which they bought commodities and the prices at which they re-sold them' (Pentland, 1991: 195) The merchant was a capitalist, with trade and speculation at the centre of his interests, not production. In the fishery, the merchant preferred to leave fishing and preservation to the fishermen, while he himself concentrated on trade of the finished product.

Commercialist principles of organization entered the Norwegian fisheries as early as the twelfth century, when the German Hanseatic League gained control over the trade from the rich north Norwegian cod fishery. The Hansa was a mercantile trade network based in northern Germany, with branches extending eastward into the Baltic and westward to Bergen in Norway. Bergen was the most important locus of the fish trade. This was where the north Norwegian fishermen would bring their stockfish, that is, the unsalted, dried cod that had been caught in the winter fishery and dried on wooden racks during the spring. In Bergen they would exchange their stockfish for grain, rope, and other necessities.

Stockfish was relatively easy to produce. It required no costly additives; it was light and robust, and it would keep for a long time. The major drawback was that it could only be produced in locations that were cold and arid enough for the fish to dry instead of rot. North Norway was one such place; Iceland was another. Until other methods of conservation were invented, these two regions dominated the fish trade in Europe.

The introduction of preservation by salting, which was pioneered by Dutch fishermen in the fourteenth century, allowed for an expansion of the fish trade

The End of Commercialism 31

for two reasons. First, it permitted the development of a fish trade in warmer and more humid climates. Second, it allowed for a new and more efficient way of fishing, since the fish in salted form could be stored onboard the ship for a long period. These traits made salt fish extremely important in the development of the new territories across the Atlantic during the sixteenth and seventeenth centuries.

One of the treasures of North America was its extremely rich fishing grounds and the salt fish trade was the tool by which they were exploited:

A three-cornered trade from England to Newfoundland, Spain and the Mediterranean provided a basis for expansion, and gave England an industry with an abundance of shipping, an outlet for manufactured goods and provisions, a supply of semitropical products and specie, substantial profits, and ideal possibilities for the development of a mercantile policy. England was able, in part because of her relatively shorter distance from Newfoundland and in part of the nature of fish as foodstuff, to secure a strong and continuous hold on a product by which she obtained a share of Spanish specie and the products of the Mediterranean. Cod from Newfoundland was the lever by which she wrestled her share of the riches of the New World from Spain. (Innis, 1954: 52)

The salt fish trade formed the economic basis on which the North American colonies were built. It also had an impact on established trade patterns in Northern Europe. When John Cabot returned from North America in 1497, his report of the rich fishing grounds was particularly welcome because it meant that 'this kingdom would have no further need of Iceland, from which place there comes a great quantity of fish called stockfish' (Raimondo di Soncino, 1497, cited in Innis, 1954: 11). Although this observer only mentioned Iceland, he might as well have included Norway, the other large supplier of stockfish. In any case, salt fish from Newfoundland did have a considerable impact during the next four centuries. Thus, a Norwegian fisherman in 1600 had to produce twice as much fish to get the same amount of grain as compared with 1530 (Coldevin, 1938: 60). Partly this was because of higher grain prices, but to a large extent it was also the result of competition from the large supplies of salt fish from overseas (Coldevin, 1938: 68).

The fishery off Newfoundland was conducted, in the early stages, by expeditions starting off from Europe, linking production and trade as two inseparable parts of the enterprise (Innis, 1954). As the new colonies were gradually settled, however, the merchants retreated to trade and left the fishing to self-employed fishermen and their families (Barrett, 1992). Although self-employed, the fishermen were hardly independent. Instead, they were tightly bound to the merchant by the credit, or truck, system. Cash transactions between fishermen and merchants were rare. In 1903, a Labrador fisherman, selling ten fish for 50 cents, remarked that: 'In all my long life, I have never sold a fish for money before' (cited in Innis, 1954: 494). The fisherman bought on credit the foodstuffs and equipment necessary to conduct the season's fishery. In return, he was required to sell his catch to the merchant. Within the truck system, the merchants were able to control the exchange so that the fisherman was never out of debt. This brought a double gain for the merchant. First, he was able to keep the fish prices down and skim off most of the profit from the fishery (Ommer, 1981; Brox, 1984). Second, he was able to keep control over his dependants and thereby secure a steady supply of fish (Ommer, 1981, 1989). Thus, the historian Sverre Steen, commenting on the north Norwegian fishermen's debt to the merchants in Bergen, characterized it as 'the merchants' largest asset' (Steen, 1957: 122).

In Norway a fisherman who was in debt to a merchant was forbidden by law to sell his fish to anyone else. This is but one example of the network of privileges, rules, and regulations in which the commercialist system was embedded (Steen, 1930). Another important example is the decision in 1562 that no one from north Norway had the right to engage in the export trade and that foreigners were not allowed to trade in the region. Two years later, the constraints on trade in north Norway were further tightened when the city of Bergen was granted a monopoly over the trade from the region (Knutsen, 1988: 16). This put Bergen in a key position within the fish trade.

In the eighteenth century, however, Bergen began to lose its dominant position. One important factor in this process was the establishment of new cities along the coast. Molde, Kristiansund, and Ålesund were granted trade privileges during the 1740s, and expanded on the basis of the salt fish, which, at this time, was still a new product in Norway. Around the end of the eighteenth century, north Norway got its own cities. Vardø, Hammerfest, Tromsø, and Bodø were established in 1789, 1789, 1794, and 1816, respectively. The whole trade system was partly changed during this period, by the establishment of a class of local merchants, positioned in between the fishermen and the city merchants. All these developments, justified by the new 'liberalist' ideology so effectively communicated by Adam Smith, weakened Bergen's position in the fish trade and clearly brought to an end the most conspicuous exploitation by south Norwegian merchants of the fishery population in the north. Still, the trade system remained within the commercialist tradition. In the spirit of the Bergen merchants, the local merchants took over the truck system of credit. Their extensive control and unchallenged authority within local communities along the coast is well expressed by their nicknames as 'community owner' (væreier) or 'outport king' (nessekonge).

The salt fish technology originated in the era of commercialism, and the salt fish trade retained strong elements of commercialist institutions and practices far into the twentieth century: It was dominated by merchants, focused on trade as opposed to production, exploited fishermen and their families, and was heavily protected by state privileges.

The fisheries did not remain unaffected by the development of modern capitalism and industrial technology, however. This happened in particular with the development of a new industry based on fresh and frozen fish. In contrast to the commercialist salt fish trade, its focus was not trade and exchange, but production. Instead of the self-employed fishermen held in check by the truck system, the fresh and frozen fish industry was based on formal control and management of production, with backward linkages to fishing and forward linkages to retail. Because of the larger investment requirements that followed the introduction of modern technology in fishing (trawlers) and production (mechanized filleting and freezing), this pointed towards a large-scale and centralized pattern of operation.

The utilization of new technology and new forms of organization within the fisheries were important for the development of the new industry. Developments in other sectors, however, should not be overlooked. Trading fresh and frozen fish required predictable and efficient transportation. The growth of the capitalist fishery was therefore closely connected to the extension of steamship routes and railways. The importance of this factor is reflected in the geographic development of the fresh and frozen fish industry. The main centres were England and New England, both of which commanded large and accessible markets.

In Norway, Newfoundland, and Canada, the transition from commercialism to industrial capitalism, from salt fish to fresh and frozen fish, was more problematic. Unlike England and New England, they were located too far from the markets to make selling fresh and frozen fish products feasible. Further, they had strong commercialist interests not willing to give in without a fight.

The Salt Fish Trade in the Interwar Period

The interwar years were extremely difficult for the export-dependent fishery nations of the North Atlantic. To some extent this was because of the international depression, from which most industries suffered. The fishery was particularly vulnerable, however, in part because of the weakness of its most important component, the salt fish trade. The major salt fish markets were found in relatively poor countries of southern Europe, South America, and the Caribbean. As Gerhardsen (1949: 79) pointed out in his report on the salt fish industry for the Food and Agricultural Organization (FAO), the export-dependent agriculture industries so important for these countries were severely affected by the economic depression. This, again, undercut the demand for salt fish because 'so many salted fish consumers are dependent on agriculture for their income.'

In the face of the economic depression, most nations retreated behind import quotas, tariff barriers, and bilateral barter arrangements. In addition, the usual fish markets were reduced because two of the major fish-importing nations, Portugal and Spain, expanded their cod fisheries in order to increase their self-sufficiency (Gerhardsen, 1949).

This would have been a minor problem if fish producers had been able to turn to other economic activities, or if they had found alternative market outlets. But this did not happen. Instead of lower production, the supply of fish in international markets increased. From 1920 to 1938, for instance, total landings of Atlantic cod increased from approximately 1 million to 1.7 million tons, while the production of salted codfish in the region remained around 250,000 tons (Gerhardsen, 1949: 120, 122). One reason for the increase in landings was that, in addition to increased fish production by traditional importing nations, the period saw two new participants in the fish trade, Iceland and the Faeroe Islands. Another reason was that production capacity in the traditional fish-exporting countries expanded as fishing and processing were mechanized. A third factor had to do with the organization of the fisheries. Capitalist firms were becoming more important, but most fish production still occurred within a traditional economy, in which the household would usually combine cash income from fishing and subsistence farming to get by. Since these households both producted and consumed, they tended to increase rather than decrease their output when prices fell to ensure a constant level of consumption (Brox, 1984). Hence, the overproduction crisis of the interwar period became a vicious circle, with the patterns of response exacerbating rather than resolving the basic problem (Gerhardsen, 1949).

As a result of overproduction, normal trade relations in international markets deteriorated. Thus, a Commission of Enquiry Investigating the Seafishery of Newfoundland and Labrador reported in 1937:

The conditions in foreign markets are such and the selling competition has been so great, that prices in recent years have reached such a low point as to make the selling of salt codfish unremunerative. The low prices are, undoubtedly, largely due to the conditions of the countries to which we export our fish, but much evidence has convinced us that the underselling of one broker by another has been very detrimental. Prices have also been adversely affected by the unregulated arrival of cargoes and by the unsatisfactory quality, in some cases, of the fish. (Newfoundland, 1937: 183).

Gerhardsen (1949) emphasized, in a similar vein, the turbulence that had prevailed in the markets during the 1920s and 1930s:

As soon as a surplus appeared, the lack of export planning was recognized. There was also a noticeable lack of uniformity in prices and other conditions of sale and delivery; even the same freighter could bring to a market commodities which, though equal in quality, were bought under entirely different terms. During the critical years the consignment business was larger than was considered normal. Sales in small quantities and sales direct to the retailer took place frequently, with the result that the large and regular importers were undermined and their interest in the commodity lessened. Export was carried on partly by people who were nothing more than export agents. (Gerhardsen, 1949: 57)

Even though falling prices and cut-throat competition became commonplace in international fish markets, the situation was worse at the producer end of the industry. In general, the prices fell more in firsthand markets than in export markets during the interwar crisis (Giske, 1978). Times were difficult for the exporters. For the fishermen, they were disastrous not least because of the asymmetric relationship between fishermen and merchants, as institutionalized in the truck system. The Amulree Commission, which had been established to investigate Newfoundland's financial situation, reported:

As a result of three successive seasons in which the fishery yielded no return, the winter of 1932-33 found the people living in conditions of great hardship and distress. Privation was general, clothing could not be replenished, credit was restricted, and hardly anywhere did the standard rise above a bare subsistence. Lack of nourishing food was undermining their health and stamina; cases of beri-beri, a disease caused by inferior diet, and of malnutrition were gradually increasing, and were to be found in numerous settlements; the general attitude of the people was one of bewilderment and hopelessness. (Great Britain, 1933: 84)

Although the problem was particularly acute in Newfoundland, similar situations were reported from all the fishery-dependent nations across the North Atlantic. The Income Task Force, established in 1934 to find solutions to the problems of Norwegian fisheries, reported that 'the fishermen's standard of living is now lower than it was prior to World War I and a great deal of the fishing population suffers direct want.' (Government of Norway 1937: 62). Despite the more diverse economic structure of Nova Scotia, the decline of the salt fish trade was also deeply felt there:

This decline meant serious hardship and poverty in many Nova Scotian communities. In the fishing villages it meant decline and deterioration of fishing fleet and curing establishments, a melting away of capital resources of all kinds. Credit, never having been plentiful, became almost impossible to obtain. Had it not been for the lobster and for the growth of small fresh-fish industries on parts of the coast, Nova Scotia's position would have become almost as bad as Newfoundland's. (Watt, 1963: 17)

The cod fishery of Canada, Newfoundland, and Norway at the beginning of this century took on the character of a lottery. On the harvesting side, the unpredictability comprised fluctuations in stock sizes and variability in the migration patterns of the fish. On the market side, fluctuating demands in consumer markets were compounded by dependence on exports which exposed the industry to exchange rate fluctuations and protectionism. The fishermen blamed the merchants and mobilized for organizational and political reform. The strength of the fishermen's attempts to wrestle the control over the fisheries from the merchants was not the same in Norway, Newfoundland, and the Maritimes, however. The pattern of conflict that emerged in each of the three North Atlantic fishery nations depended on the industry's market structure, the organizational power of the interest groups involved, the economic and social importance of fishing, the nature of the political system, and a good dose of pure coincidence.

Struggle and Reorganization in Three North Atlantic Nations

Norway

Like all the export-dependent fishery nations in the North Atlantic, Norway was deeply affected by the interwar economic depression. In contrast to Newfoundland and the Maritimes, however, the reform wave that was triggered by this depression had long-term consequences in Norway. Even today, more than half a century later, organizations and institutions that were created during this period remain core features of the Norwegian fisheries. Many of the struggles in the fishery during the 1920s and 1930s concerned the design and implementation of a legal framework for the sector. As a starting point for our analysis, we describe three pieces of legislation adopted during the period: the 1932 Salt Fish Act, the 1936 Trawler Act, and the 1938 Raw Fish Act. These three acts formed the basis of a sector system which gave the fishermen the dominant position in Norwegian fisheries in the postwar period.

The Salt Fish Act
The Salt Fish Act (Klippfiskloven) was adopted by the Norwegian Parliament in 1932. According to this act, the government could decide that the export of dried salted fish (klippfisk) would be permitted only for members of an officially recognized organization. The Norwegian Association of Dried Salted

Fish Exporters (De Norske Klippfiskeksportørers Landsforening) was granted such approval according to the act in 1933. With this act, the export of salted and dried fish became the exclusive right of the members of a mandated sales organization. To qualify as a member, a firm was required to have a minimum equity of Nkr 10,000, and its owner to have at least five years of experience in fish exporting (Government of Norway, 1932). At first the Salt Fish Act applied only to exports of dried salted codfish. Later in the decade it was extended to stockfish. The act thereby regulated the bulk of the Norwegian fish trade in the groundfish sector.

The Salt Fish Act must be understood in light of the extremely difficult situation that existed at the time in the salt fish market. By way of common strategies and cooperation among the exporters, it was hoped the crisis could be alleviated. However, it was also apparent that the act would not solve all the problems. The oversupply problem was international in character. Some participants in the debate over the organization of the salt fish industry, notably the Director of Fisheries, even argued that such measures would be counterproductive. Simply, there was no internal Norwegian cause for or solution to the crisis. In the absence of international agreements, government intervention in Norwegian fish exports could only magnify the problem. Any restraints on the Norwegian fish exporters would advantage their foreign competitors (Director of Fisheries, 1928).

The counterargument, which won through with the adoption of the Salt Fish Act, was that some form of coordination of Norwegian exports would at least solve the problem of internal competition among the exporters. The export trade was extremely open. There were no formal barriers to entry. The extension of steamship routes and railways had removed the high financial entry costs of earlier times. During the crisis, market access became the critical factor, and, hence, a large number of new and often unskilled participants rushed in, motivated by desperation or easy profits. This ruptured long-term market relations and forced even the professional traders to engage in short-sighted competitive practices (Giske, 1978; Hallenstvedt, 1982). It was argued that the Salt Fish Act would help solve this problem in two ways. First, it reserved the export of salt fish to a relatively small group of 'professional' traders. This excluded the part-timers and the unskilled newcomers, that is, the group that allegedly had caused havoc in otherwise well-ordered export markets. Second, it created an arena for cooperation among the exporters, something that in itself would help forestall unhealthy competitive practices.

The Salt Fish Act and the organization it supported were far from unique. Rather, they were manifestations of an organizational form that appeared in a wide variety of settings during the 1930s. Some form of export organization was established in all the major North Atlantic fish-exporting nations, with the notable exception of Canada. Furthermore, similar organizational structures developed in the major importing countries, Spain and Portugal, where, from 1934, imports of salted codfish were centralized through importer associations (Gerhardsen, 1949). Nor were such trade organizations exclusive to the fisheries. In export-dependent agricultural nations, which faced the same type of problems as the North Atlantic fishery nations during the depression, similar organizational structures were established. Danish agriculture is a good example from the European side (Tracy, 1989). In North America, Canadian wheat exports have been coordinated through the Canadian Wheat Board since 1935 (Alexander, 1977; Lipset, 1971; Sacouman, 1979).

While these efforts were all attempts to solve problems originating in the export markets, they varied as to which interests they were established to protect. Hallenstvedt (1982) has argued that the Norwegian Salt Fish Act essentially encoded in law the existing export practice, allocating state resources to the protection of 'professional' exporters' interests. In contrast, the Danish Export Boards, as well as the Canadian Wheat Board, primarily catered to cooperative organizations formed by self-employed farmers. Hence, they protected exactly the opposite end of the interest spectrum as compared with the Norwegian Salt Fish Act.

Why did the merchants win in the struggle over the organization of Norwegian salt fish trade? Why was the industry not organized on the basis of the fishermen's interests? The policy process leading up to the adoption of the 1932 Salt Fish Act revolved around a struggle between the merchants and the fishermen over the control of the fish trade. Although the Salt Fish Act promoted the interests of merchants, the fishermen were far from absent in the political process that preceded its adoption. Indeed, it was the fishermen who had taken the initiative to have fish exports reorganized. This must be understood against the backdrop of the structure of the primary fish markets, where the merchants held a strong position because of a combination of long distances, small boats, and the fishermen's financial dependence on the merchants (Hallenstvedt, 1982; Brox, 1984). When the international economy slumped, this pattern of inequality and dependence exacerbated the crisis that followed. Because of the merchants' strong position *vis-à-vis* the fishermen, they could, when the international market prices fell, compensate for their losses by reducing prices in the raw fish market (Christensen and Hallenstvedt, 1990; Giske, 1978). The merchants saved themselves at the fishermen's expense (Government of Norway, 1937).

During the economic depression the situation became intolerable, and the fishermen pressed for reform. Their goal was to gain control over the trade through an organization of primary producers. They did not have to look far to find their organizational ideal, however. In 1930, the Norwegian Parliament had adopted the Herring Act, by which an organization of herring fishermen in western Norway had been granted monopoly in the primary herring market. That is, merchants wanting to export herring were forced by law to buy their raw material through a fishermen's organization authorized by the state. The point, of course, was to strengthen the fishermen's bargaining power, which under free-market conditions had been very weak. The herring fishery was seasonal to an extreme; enormous quantities were landed during a few winter weeks. At the beginning of the season, the fishermen usually would get a fair price. When the fishery peaked, however, the bottom would fall out of the market. At this point, the firsthand price no longer stood in a reasonable relation to the price in the export market. The Herring Act changed all this. The sales organization enforced a fixed price, securing the fishermen a reasonable income. At the same time, it created order out of chaos in the herring trade (Hallenstvedt, 1982).

When the fishermen set out to reform the salt cod trade, it was the model of the 1930 Herring Act that they wanted. They did not succeed, however, and the 1932 Salt Fish Act built on the merchants' organizations. One factor behind this outcome was the weak organizational development among the cod fishermen. The complexity of the cod fisheries, covering a large geographic area and a wide range of different technologies and economic adaptations, made it difficult for the fishermen to join forces (Christensen and Hallenstvedt, 1990).

The organizational weakness of the cod fishermen was an important argument in the policy process that resulted in the Salt Fish Act. The government insisted that the policy instrument in question, that is, legal protection of an organization, could only be applied if an appropriate organizational apparatus already existed (Government of Norway, 1932: 3). In the case of the Herring Act, the fishermen's organization had been authorized only after its ability to manage the herring market on a voluntary basis had been demonstrated. In the case of the cod fishery, no organization of fishermen existed. On the other hand, a countrywide organization of salt fish exporters had been established in 1931 (Government of Norway, 1932). The very same principle that had led to the authorization of a sales organization of fishermen in 1930, resulted two years later in the legal support of the merchants.

The Director of Fisheries pointed out that this outcome was highly problematic. While an organization of salt fish exporters was intended to enable them to join forces in the export market, the organizational weapon could also be turned around and used against the fishermen in the raw fish market (Government of Norway, 1932: 5). Rather than improving the situation in the salt fish trade, then, state intervention would exacerbate the basic problem: the unequal bargaining position between fishermen and merchants in the primary fish market. This line of argument did not carry weight in the process that led to the adoption of the Salt Fish Act, however. One reason seems to be that the involved parties, in particular the fishermen and their allies, believed that state intervention in almost any form would be better than nothing. When their primary preference, a sales organization controlled by fishermen, did not gain sufficient support, they, too, went for the majority proposal, the authorization of an exporter organization. The main reason, it seems, was a somewhat optimistic notion as to the amount of control that could be exercised by government authorization (Christensen, 1986).

The 1936 Trawler Act
The process resulting in the adoption of the Salt Fish Act revolved around the conflict between buyers and sellers in the raw fish market. Another important area of conflict closely connected to this involved the introduction of trawler technology. This conflict came to a climax in the debate over the revision of the Trawler Act during the 1930s, but it had roots going back to the beginning of the century. The first Trawler Act, adopted in 1908, banned trawling within Norwegian territorial waters. At that time there were virtually no Norwegian interests tied to the trawler technology, as the Norwegian fishery was overwhelmingly dominated by small-scale, owner-operated vessels, using handlines, longlines, and nets. But the European trawl fishery was expanding, particularly in the United Kingdom, but also in Germany and France (Christensen, 1991). As fixed gear fishing and trawling do not mix well, the trawler question became a hot issue once the European trawler fleets began operating along the Norwegian coast and in the Barents Sea. Hence, the purpose of the 1908 Trawler Act was to protect the fixed gear of the Norwegian coastal fishermen from being destroyed by foreign trawlers.

In 1908 the trawler ban was not controversial in Norway, since the trawler industry was controlled by foreigners. As time went by, however, trawler technology was to become an important dividing line within Norwegian fisheries. This happened as a few fish export companies began to invest in steam trawlers during the first half of the 1930s. Several unsuccessful attempts with trawler technology in Norway had been recorded before this. These had, following the English model, been directed towards the fresh fish sector. However, it was only when the trawl technology was applied to salt fish production in 1933 that it started to pay off. This success was quickly copied by other companies. Only two years later, six trawlers were in operation, and more were on their way. This was taken as the beginning of a major expansion, and created a massive uproar among the fishermen and in the coastal districts. For instance, the fishermen's representatives on the Income Task Force urged the government to 'as quickly as possible take steps to prevent a Norwegian trawl fishery from developing. Such a development will, in our opinion, be the largest misfortune that ever has hit the fisheries population' (Innst.O.xxxiii, 1936: 4, cited in Mikalsen, 1987: 45).

The storm of protests against the trawlers resulted in the 1936 Trawler Act. The ban against trawling in Norwegian territorial waters was maintained. In addition, licences were required for Norwegian trawlers fishing in international waters. This meant that the Norwegian trawl fishery was stopped before it could take off, as no licences were given out except to the eleven vessels already in operation.

The fishermen's success in the case of the Trawler Act can to a large extent be explained by the shift in Norwegian politics that occurred when the Labour Party came to power in 1935. The question as to why the fishermen opposed the trawler technology so strongly remains, however. Several factors contributed to this resistance. One was the fishermen's long experience with foreign trawlers destroying their gear. This formed a background of hostility against which mobilization to ban Norwegian trawlers was easy. But the hostility against this innovation was even more deeply rooted, as it fell into a long-standing tradition of fishermen's resistance to new technology. This stretched back to the seventeenth century, when the hook-and-line fishermen struggled to have nets and longlines banned from the fishery (Steen, 1930). Later, the 'Troll Fjord Battle' of 1891, resulting in a ban against seine fishing in Lofoten, and the struggle against motorization of the fishing fleet at the beginning of this century (Wille-Strand, 1941), are examples of the same type of conflict.

The fishermen's struggle against the introduction of new technology has often been taken as confirmation of their conservative nature. The introduction of new types of gear and other innovations has usually meant social stratification, however. Only the well-off fishermen could afford the new, usually more expensive, equipment. Within the truck system, this also meant increasing dependence on the merchant. In addition, the possibilities of new and more efficient technologies have drawn outsiders to the fisheries. No wonder, then, that technological change has been surrounded by conflict.

The trawler issue represented a constellation of these traditional issues of conflict that united the fishermen in strong opposition. Capital-intensive steam trawlers were outside the financial reach of the ordinary fisherman. They could only be introduced by 'outsiders'; in this case, the merchants. This attempt by the merchants to extend their traditional domain was extremely provocative to the fishermen because it happened in a period with very difficult market conditions. Even a modest trawler fleet would have an adverse effect for the fishermen, since the merchants surely would give priority to their own trawler-caught fish. In this way, the trawler question was intimately tied to the already heavily politicized question of market organization. With the 1932 Salt Fish Act, the merchants had gained a legal monopoly in the salt fish export. But why should fishermen and processors be banned from the export trade, when the merchants were free to go into harvesting? A ban on trawlers would balance the situation, reserving fishing for the fishermen, just as the Salt Fish Act had reserved the export trade for the merchants.

But there was more to the trawler issue than the conflict between buyers and sellers in the fish market. The fishermen conceptualized the fight against trawler expansion not only as one against merchants. It was also a fight against a capitalist reorganization of the fishery. The introduction of trawlers, they argued, would mean that the small-scale, owner-operated model typical of the Norwegian fishery would be replaced by a large-scale capitalist model. The independent fishermen would be reduced to nothing more than wage workers. The nightmare scenario, explicitly based on the industrialized English model, was one where the trawlers were assumed to have completely taken over the Norwegian fishery. It was calculated that a fleet of 200 steam trawlers with a crew of 8,000 men could catch more fish than the existing coastal fleet, which at the time comprised 107,000 independent fishermen (Government of Norway, 1937: 46).

This turned out to be a weighty argument against the trawlers. When the Director of Fisheries supported the tight restrictions on trawling, for instance, his main reason was the problem of 'what should be done with the thousands of fishermen who would be excluded from participation in the cod fisheries as a result of a large increase in trawl fishing.' (cited in Johansen, 1972: 130). This consideration is probably the most important explanation for the trawler ban.

The Raw Fish Act
With the 1932 Salt Fish Act, the merchants had won the battle over the organization of the salt fish trade. But this was only one battle in the more general power struggle within the fisheries, in which the balance during the 1930s gradually tipped in favour of the fishermen. The 1936 Trawler Act was the fishermen's first important victory. Two years later they came out on top in the battle for control over the primary or raw fish markets. This happened with the adoption of the Raw Fish Act (Råfiskloven) in 1938, a key act that shaped the structure of Norwegian fisheries for most of the twentieth century. The act banned the sale of fish outside mandated sales organizations controlled by fishermen. These sales organizations were given substantial prerogatives. First, they were

granted a legal monopoly in firsthand sales of fish, which meant a strong bargaining position *vis-à-vis* the buyers. If necessary, they could unilaterally set the price of raw fish. Second, the sales organizations were empowered to license and control fish buyers. Third, the sales organizations could, if marketing conditions required it, regulate the supply of fish by catch restrictions or by directing landings to specific ports.

The Raw Fish Act was modelled on the Herring Act, giving fishermen-owned organizations monopoly control over the firsthand trade of fish. There was an important difference, though. While the Herring Act was limited to a specific fishery, the Raw Fish Act gave the government a general mandate to authorize fishermen's sales organizations. The first organization to be authorized according to the act was the Norwegian Raw Fish Organization (Norges Råfisklag). This organization, which covered the most important Norwegian fishery, the cod fishery of northern Norway, was established and authorized in 1938. Shortly after the war, similar organizations were established along the whole Norwegian coast, and virtually all marine fish caught in Norway were channelled through organizations controlled by fishermen (Government of Norway, 1948).

The 1938 Raw Fish Act would have important consequences for the fish trade and for the Norwegian fishery sector as a whole. As suggested by the act's popular name, 'the Fishermen's Constitution,' it was the basis on which the fishermen built a strong position within the sector. The reason for the Norwegian fishermen's relative strength will be discussed in a comparative context later in this chapter. As we shall see, Norwegian fishermen were vastly more effective in their struggle for reform in the 1930s than their counterparts in Newfoundland and the Maritimes. An important factor was the Norwegian fishermen's gradually improving access to political resources during this period. After the 1935 election, the Labour Party came into power, and the long period of social democratic hegemony in Norwegian politics began. This had important consequences for the fisheries. The conflict between the merchants and the fishermen corresponded to the main division line in Norwegian politics, between the conservatives and liberals on the one hand and the social democrats on the other. When the Labour party won in the 1935 national election, this shifted power within the fisheries, and gave the fishermen the upper hand (Holm, 1995).

Newfoundland

Like Norway, Newfoundland was deeply affected by the interwar economic depression. In some ways, the result of the depression was more dramatic. Not only did poverty in Newfoundland go much deeper than in any other western nation, the economic depression also created a political crisis that threatened the very existence of Newfoundland as a self-governed political unit. The poverty problem was a direct cause of the political collapse in 1934, when Newfoundland lost dominion status and a British Commission of Government replaced the representative government (Alexander, 1977: 1). Although more severe, the problems facing Newfoundland's fisheries were the same as those of Norway. The all-important cod fishery was dominated by a commercialist salt fish trade, in which the main line of conflict was drawn between St John's merchants and outport fishermen. The latter, who were engaged not only in fishing, but also in fish curing, farming, and forestry, were largely self-employed and had formal independence from the merchants. In practice, they were heavily dependent on the merchants through the truck system of credit. This meant, of course, that the same pattern of exploitation occurred as in Norwegian fisheries, giving the fishermen a small share of the ups and a large share of the downs in the fishery (Great Britain, 1933: 80).

In spite of the similarities between Newfoundland and Norwegian fisheries during the 1930s, the patterns of response were different. The crisis also spawned reform attempts in Newfoundland. Some of them paralleled those we have seen in Norway, in particular the government-assisted coordination of fish exports, which will be described below. The most conspicuous part of the Norwegian fishery reform, however, the fishermen's successful attempt to wrest control over price formation in raw fish markets, had no parallel in Newfoundland.

The Salt Codfish Acts
It was a widely held opinion that the lack of organization among Newfoundland exporters was a main cause of the fishery crisis during the 1930s. One of the strongest proponents of this view was the Amulree Commission, established by the British government to report on Newfoundland's economic problems. The commission blamed the export firms, which, in the face of the crisis, had mainly relied on individualistic strategies of diversification:

An equally obvious and far more effective method would have been for the firms engaged in the business to weld themselves into one organization, which could have instituted arrangements for the packing and grading both in size and quality of the fish bought, for the inspection and regularisation of shipments of fish to foreign markets, and for the sale of the fish in those markets. Few can doubt that, had steps of this kind been taken, the industry to-day would have been in a sufficiently strong position to weather a temporary period of depression without penalising the actual producer, viz., the fisherman, on whom, as ever, the main burden now falls. But to join in any large-scale cooperative effort is precisely what the merchants of St John's have always failed to do. They have insisted on conducting their business on a basis of pure individualism without regard to the interests of the country and without regard to the successes achieved by their foreign competitors. Intent only on outdoing their local rivals in a scramble for immediate profits, they have failed to realize that time does not stand still. (Great Britain, 1933: 108)

In 1933, when the report of the Amulree Commission was published, attempts at coordination along the lines suggested by the commission had already been initiated. The 1931 Act Relating to Salt Codfish had introduced a licensing system for salt fish exports. The act also established a Salt Codfish Exporters Association, open to all firms exporting at least 1000 quintals (50 tons). The association was authorized to adopt rules pertaining to the export trade, for instance, grading, inspection, and sales conditions. These rules were to be implemented and policed by a Salt Codfish Exportation Board, appointed by the government. These Newfoundland regulations closely paralleled those implemented in Norway with the 1932 Salt Fish Act. In Newfoundland, this was only the beginning, however, as government intervention in the salt fish trade would move it closer to complete centralization. The first step on this road was taken in 1933, when amendments to the Act Relating to Salt Codfish sought to increase the Exportation Board's power in relation to the exporter association. Since the board was appointed by the government, this move also strengthened the authorities' position in relation to the merchants (Alexander, 1977: 28).

With the collapse of self-governance in 1934, government control over the salt fish trade was tightened further. The 1935 Act for Better Organization of the Trade in Salt Codfish strengthened the executive board by giving it the right to refuse export licences. In 1936 the move towards centralization was completed with the Act for the Creation of the Newfoundland Fisheries Board. The board, consisting of three members appointed by the government, could regulate all aspects of the production of all fish products, not only salt codfish. The act specifically empowered the board 'to form groups or associations of licensed exporters to various markets and to admit or refuse membership ... in the absolute discretion of the Board ... and to give such groups ... or to the members thereof the sole right to obtain licenses for export to such markets' (cited in Alexander, 1977: 29).

In his analysis of the salt fish trade, Alexander concluded: 'With this Act, Newfoundland acquired a regulatory authority over its fisheries that was unequalled in North America and unsurpassed in Europe' (1977: 29–30). In comparison, the Norwegian Salt Fish Act allowed the government to authorize exporter associations. While this act in principle could have been used to gain control over the trade, the government remained at arm's length, in practice, and left matters to the merchants and their association (Christensen, 1986). In contrast, the Newfoundland government took active control over the industry.

An important reason for this difference pertains to the scale of the crisis. Although all the export-dependent fisheries across the North Atlantic suffered during the 1930s, Newfoundland was the worst off. Unlike the Maritimes, it lacked economic alternatives to the cod fisheries and the salt fish trade (Watt, 1963). Unlike Iceland and the Faeroe Islands, which shared Newfoundland's dependence on the fishery, its industry lagged seriously behind in technical and organizational development (Thormar, 1984). Compared with the major European producers, Newfoundland's salt fish was inferior in quality. In addition, Newfoundland was at a disadvantage during the return to barter trade in the 1930s, since most of the exports from the island went to Europe, while most of the imports came from North America (Alexander, 1977; Bates, 1944). Taken together, these factors explain the strong government involvement in the fisheries. The crisis not only represented a threat to fishermen and merchants, but to the economy of Newfoundland itself.

In addition, the political status of Newfoundland played an important role. With dominion status from 1855, Newfoundland did have its own political institutions. These institutions were never very strong, however. Except in St John's, local government structures had not developed (Innis, 1954). It was this political fragility, together with the economic dependence on one sector, the fishery, that led to the loss of self-governance in the face of the interwar economic depression. With the Commission of Government, traditional British paternalism towards the colony was played out with few restraining forces. Thus, the regulation of the salt fish trade was based on an explicit view that Newfoundlanders were unable to handle their own affairs and a very high confidence that the problems would be sorted out under British supervision.

In a comparative perspective, it is interesting to note that the model established in Newfoundland was exactly what the Norwegian fishermen had originally striven for, that is, centralization of the salt fish trade under government control. Although the Newfoundland merchants had been brought under government supervision, this did not mean that the fishermen's position automatically improved, however. The government's rationale for going into the salt fish business was to run it more efficiently, not to secure the fishermen's interests. Herein lies the most striking difference between Norway and Newfoundland: the fishermen's absence in the reform process, and, hence, legislation

The End of Commercialism 47

similar to the Trawler Act and the Raw Fish Act was not even considered in Newfoundland.

The Maritimes

Unlike the other fish-exporting nations in the North Atlantic, there were no attempts at government-assisted coordination of fish exports in the Maritime provinces of Canada during the 1930s. The main reason for this is to be found in the way the economic depression affected the region and its fishery. Compared with Newfoundland, Iceland, and the Faeroe Islands, the fishery was but a small fraction of the total Canadian economy. Hence, the fishery crisis was never a pressing concern for the federal government. The same could to some extent also be said about Norway. There, however, the regional dependence on the salt cod fishery, particularly in north Norway, was very large. In the Maritimes, both the fishery itself and the economy in general were more diversified, allowing both fishermen and merchants to seek out alternatives rather than mobilize for reform when the salt fish trade collapsed. In short, the Maritime fisheries were less vulnerable to economic depression. Having said that, it must be added that the depression was not without its effects, and it did trigger at least one reform, the trawler ban.

The Trawler Ban
From 1920 to 1929 the amount of fish salted in the Maritimes was almost halved, declining from 223 to 126 million pounds (Bates, 1944: 29). This was a direct consequence of the difficult situation in the international salt fish markets, and it forced the fishing industry of the Maritimes through drastic change. Bates (1944) has neatly summarized the development that pushed the Maritimers out of their traditional salt fish markets:

After the last war Norway and Iceland, for reasons of their own, improved their cures and quality and lowered their export costs and prices, and began the period of intensive international competition that was ultimately to drive Newfoundland fishermen below subsistence. Newfoundland had to sell salt fish, at whatever price would dispose of her annual catch. Canada then faced this sort of competition also in her markets, for Newfoundland had to strengthen her foothold in the western hemisphere as European markets became lost to her. (Bates, 1944: 30)

Hence, as Newfoundland lost access to its traditional European markets, it redirected its efforts towards the Caribbean area, the traditional Canadian market. As a result, the Maritimes had to seek other outlets for salt fish. Increasing attention was paid to the United States and the particular cures preferred there, salted boneless cod and pickled fish. But these markets were limited, and, as salt fish prices fell, more of the catch was diverted to the fresh fish trade. Thus, the rise of the fresh fish industry was closely connected to the decline of the salt fish industry: 'While steamships and railroads drove out the sailing ships and brought marked changes in the dried-fish industry, they gave rise on the other hand to the growth of the fresh fish trade' (Grant, 1934: 83).

The fresh fish trade in North America developed first in the New England area. As early as the 1880s this development started to affect the fisheries of the Maritimes, as fresh fish were shipped directly from Yarmouth and Digby, located in southwestern Nova Scotia, to Boston. However, the fresh fish trade in the Maritimes only blossomed when improved transportation facilities became available in the beginning of this century. After that the new industry developed fairly quickly. The percentage of the Canadian groundfish that was marketed in fresh and frozen form increased from 9 to 34 between 1920 and 1939. In the same period, the share of the catch that was salted declined from 80 to 54 per cent (Bates, 1944: 68).

The development of the fresh fish industry was intimately connected to the introduction of trawlers. Hence, the most important barrier against the development of the Canadian fresh fish industry were the legal restrictions on trawler technology. In the Maritimes, as in Norway, the fishermen were strongly opposed to trawlers. Trawlers were practically banned, for much the same reasons as in Norway.

The first restrictions on trawlers in Canada were introduced in 1908 when they were prohibited from fishing within the three-mile limit. The motive, exactly as in Norway, was to prevent the trawlers, mostly operating out of New England, from destroying Canadian fishermen's fixed gear. Despite these and other restrictions (Watt, 1963: 23-4), a small Canadian trawler fleet developed, and in 1926, eleven trawlers were in operation. At the end of the decade, however, restrictions were tightened. Through an amendment to the Fisheries Act in 1929, strict licensing measures were introduced. From 1931 the regulations included, first, a licence fee of $500; second, a rule that the applicants had to satisfy the Minister of Fisheries that a sufficient quantity of fish could not be obtained from hook-and-line fishermen to enable them to maintain their regular volume of business; and, third, a regulation that all new trawlers had to be built in Canada (Watt, 1963: 28). In 1933 the number of trawler licences was reduced by one-third. By 1939 there were only three trawlers remaining (Bates, 1944: 42).

The trawler restrictions were based on a coalition of quite diverse interests. As in Norway, the self-employed fishermen were in strong opposition to trawler technology, much in line with the 'traditional hostility of "hand-workers" to greater mechanization' (Bates, 1944: 41). Thus, the majority of the MacLean Commission, established to investigate the fisheries of the Maritime provinces, reported a substantial number of objections against steam trawlers from inshore fishermen:

(1) that they destroy the spawn of cod and haddock; (2) that they destroy the feeding grounds of fish, with disastrous result; (3) that they take large quantities of immature and unmarketable fish, the result of which, with intensity of fishing, will be the inevitable depletion of the fishing grounds; (4) that they are foreign-owned and foreign-manned; (5) that they destroy the gear of fishermen without making restitution; (6) that they market an inferior product, which in the end injures the industry by discouraging the consumption; (7) that they are responsible for over-production and the consequent 'glutting' of the market, thereby preventing the shore fisherman from disposing his catch, of superior quality, at a reasonable price; that because of the low prices offered, and the virtual control of the Canadian markets by the companies operating steam-trawlers, the shore fishermen are deprived of an adequate livelihood, with the resultant serious depopulation of the fishing villages in recent years; and that if steam-trawlers are allowed to continue to operate from Maritime Province ports, the fishing villages in these parts will soon be deserted. (Government of Canada, 1928: 92)

Of these objections, the commission concluded that the first six either were unfounded or beyond the power of the government of Canada alone to deal with. It was the seventh and final argument that the government found convincing and led it to conclude that 'steam-trawlers should in our judgement, without fear of the consequences, be prohibited from landing their fish and from obtaining supplies at Canadian ports, in order that the fishing population of the Maritime Provinces may be protected and retained' (1928: 102). When the Canadian government accepted this recommendation, the concern probably was a major one, as it was going to be for the Norwegian government eight years later. It must be added, though, that the inshore fishermen were not alone in their fight against the trawlers. First, the schooner owners, who feared that the trawlers would render their schooner capital obsolete, joined the fight. Second, a group of fish-processing firms also opposed trawlers. Stuart Bates explains this somewhat unexpected alliance as a result of purely strategic reasoning: 'A trawler license represented an asset of considerable value to the firm possessing it, and the competitors of such a firm, knowing that general opinion would not permit wholesale issuance of licenses, entered the opposition' (Bates, 1944: 41). In Nova Scotia, these groups together were able to enlist the provincial government in their resistance against the trawler technology. Together, they persuaded the federal government to adopt a restrictive line.

Despite the growing fresh and frozen fish trade in the Maritimes, the identity of the fishing industry at the end of the 1930s still rested in the commercialist salt fish trade. In Bates's words: 'The horizon of the fishing industry as a whole was essentially limited by the possibilities of the salt fish branch' (1944: 33). The strength of the commercialist tradition showed up in the trawler controversy, where it temporarily bridged the deep-set conflicts between the self-employed inshore fishermen, the offshore fishermen, the schooner owners, and the merchants. This is where we find the main difference between Norway and the Maritimes on the trawler issue. In Norway the resistance against the trawlers united the small-scale fishermen against commercialist merchants and modern capitalists alike. Since the fishermen's struggle was connected to market reforms, it would have lasting consequences in the postwar period. In the Maritimes, however, the resistance to the trawlers was built upon a coalition kept together by a common interest in the commercialist salt fish trade, which was in rapid decline.

The Struggle to Organize

We have seen how the world economic depression spawned fishery reforms in Norway and Newfoundland, as well as the Maritimes. Although these reforms were similar in some ways, they had significantly dissimilar results. First, government intervention in the fish export business differed. While exports were centralized and put under direct government supervision in Newfoundland, the Norwegian government allowed the exporters to continue traditional practices but under state protection. In the Maritimes the government did not intervene in fish exports. In contrast to Norway, the fishermen in the Maritimes were not strong or united enough to pressure the government to intervene. In contrast to Newfoundland, the economic significance of the salt fish trade to the Canadian economy was not great enough to induce the government to intervene without such pressure. Second, the trawler fishery was heavily restrained in the Maritimes, as well as in Norway. While the arguments for a trawler ban were virtually identical at the two sides of the North Atlantic, the social forces on which it was based were different. In the Maritimes it rested on a coalition of fishermen and merchants, temporarily brought together by the threats of industrialization. In Norway the trawler issue united small-scale fishermen against capitalists of commercialist, as well as industrialist, bent. In Newfoundland there was no similar movement. Here, the organizational and political weakness of the fishermen was almost matched by the entrepreneurial weakness of the business class, and the trawler issue did not arise. Third, the creation of fishermen's sales organizations with monopoly control over firsthand sales, the most conspicuous of the reform measures of the 1930s, only appeared in Norway. While merchants' exploitation of fishermen was also a well-known problem in Newfoundland and the Maritime provinces of Canada, no similar attempts at fundamental market reorganization occurred there.

Several factors are required to explain these differences. One is the relative importance of the fisheries in the total economy of the three nations in question and the availability of economic alternatives in the face of crisis. The weak organizational response in the Maritimes, for instance, is to a large extent explained by a diverse economic structure and the relative ease by which the fisheries could convert from salt fish to production of fresh and frozen fish. Another factor is the nature of the political system. In Norway the fishermen's struggle for reform was greatly aided by the general political shift towards the left in Norway during the 1930s. In Newfoundland, on the other hand, the fishermen's already weak access to political resources was even further limited with the end of responsible government from 1934. The most important explanatory factor, though, lies in the fishermen's capacity for collective action. It was the presence of a politically effective, common interest organization of fishermen, the Norwegian Fishermen's Association, that made it possible to overturn the traditional power relation between fisherman and merchant. Such organizations were present in neither the Maritimes nor in Newfoundland, and the traditional power relations in those regions remained virtually unchanged through the crisis.

In Norway the first local fishermen's organizations were formed at the end of the nineteenth century. From then on, they gradually grew in number and size, and from 1915 they began to join forces in larger regional organizations. The first and largest of these was the North Norwegian Fishermen's Union, formed in 1915. Seven other regional fishermen's associations were formed during the next ten years, covering the whole western and northern coastline. In 1926 all these associations joined together in one organization, the Fishermen's National Association (Fiskernes Faglige Landslag), which from 1930 was renamed the Norwegian Fishermen's Association (Norges Fiskarlag). This organization would gradually be recognized as the representative of all Norwegian fishermen. It did not, of course, have unanimous support from the start. The number of local branches increased from around 300 in 1929 to 900 in 1939. In the same period, the number of individual members grew from 6 to 20,000. Although this was only a fraction of the total number of fishermen, 120,000 in 1939, it had no rivals, and it covered the whole country and all major fisheries (Hallenstvedt and Dynna, 1976).

By contrast, the fishermen of the Maritimes did not manage to join forces. This does not mean that there were no efforts towards organized resistance among fishermen. But these efforts never were able to bridge the deep geographic and structural cleavages that criss-crossed the Maritime fisheries (Barrett, 1979; Clement, 1986). First, the diversity of the fishing industry in the region made it difficult for the actors within the sector to identify and act on common interests. While groundfish dominated and 'cod was king' in Newfoundland and north Norway, the Maritimes had important fisheries of lobster, pelagic fish, flatfish, and estuarial fish in addition to groundfish, all of which required different technologies and different social organization. Second, the various fisheries were not spread out evenly throughout the region. While Nova Scotian fishermen mainly directed their efforts towards the Atlantic, those of the other provinces had their most important fisheries in the Gulf of St Lawrence, the Bay of Fundy, and Labrador (Bates, 1944: 27). Third, the relative closeness of the Maritimes to New England provided an alternative market for fresh and frozen fish when the salt fish market collapsed. Fourth, the Maritime fisheries were also influenced by the ethnic and cultural cleavages of the region. The most important boundaries divided British-dominated Nova Scotia and the Francophone Quebec, but the Acadian enclaves that sprinkled the Maritimes were also an important factor in the fisheries. Fifth, the region covered four provinces. This was of particular importance in the fisheries, since the provinces differed in their constitutional powers within this sector. While Quebec retained jurisdiction over its sea fisheries, the rest of the Atlantic fisheries were under federal government control (Bates, 1944: 26). These factors represented insurmountable barriers against collective action, partly because they helped buffer the fisheries against the economic depression, partly because they left the fishermen with alternatives other than organized resistance, and partly because they made identification of common interests and goals among the fishermen difficult.

The fisheries of Newfoundland were much more homogeneous than those of the Maritime provinces of Canada. During the interwar period, there was no realistic alternative to the fishery and, within this sector, production of salted cod was the only option. As a result, Newfoundland fishermen were even more exposed to the shock of the economic depression than were their Norwegian counterparts, and, one might have thought, more prone to organized resistance. Such was not the case, however. This is perhaps surprising, since an organization that seemed to be a good starting point for collective resistance among the fishermen was formed during the first decade of the century.

The Fishermen's Protective Union (FPU) was formed in Newfoundland in 1908, and was a mix of a union, a cooperative, and a political party. The primary task of the FPU was to destroy the truck system. To break their dependence on the merchants, fishermen had to be able to buy and sell for cash. For this purpose, the FPU set up the Union Trading Company. In 1911, William Ford Coaker, the FPU's dominant leader stated: 'The Union cash stores will sell for cash, and it won't be many years before the credit system will disappear' (cited in Inglis, 1985: 57). By 1919 the company operated forty cash stores across Newfoundland. The FPU set up a Union Export Company, which became a major salt fish exporter. Further, the union published a daily newspaper in St John's. On the Bonavista Peninsula, the FPU set up a model community:

At its height, Port Union had fish plants, warehouses, a seal oil plant, a ship-building yard, a wood-working factory, a machine shop, a cooperage, a soft-drink bottling plant, a movie theatre, a church, fifty tenement houses for employees, and a Congress Hall for the Supreme Court of the Union. The Union Electric Light and Power Company made the Bonavista area one of the few rural districts to be supplied with electricity, and in Port Union itself, powered such futuristic innovations as modern mechanical fish-dryers and even elevators in the factories. (Inglis, 1985: 54)

With its 20,000 members, almost half of Newfoundland's fishermen, the FPU was also a considerable political factor. In the 1913 election the union formed an alliance with the Liberals. Eight of the nine FPU candidates were elected. During the First World War, a coalition government of Conservatives and Liberals was formed, and Coaker became a member of the government. After the war the FPU joined forces with the Liberal Reform party, which subsequently won the election. Coaker became deputy prime minister, as well as fisheries minister (Inglis, 1985: 56–7).

As a minister, William Coaker tried, unsuccessfully as it turned out, to restrain the merchants through government intervention in the export of fish. In 1920 he managed to have the House of Assembly adopt the 'Coaker regulations.' Among these, the Codfish Standardization Act established a commission with responsibility for regulating all aspects of catching, processing, culling, storing, and transportation of salt fish. The purpose was to re-establish the quality of the Newfoundland fish, which had rapidly deteriorated during the war. Further, the Salt Codfish Exportation Act established a board, chaired by a member of the government, with the mandate to license the export of salted codfish, and, for the purpose of ending distress selling, with the authority to set minimum prices and maximum volumes for any given market. These regulations, which Coaker saw as the beginning of a national marketing agency, were endorsed by the assembly. After a few months, they were in ruins, however, broken by the banks and a few powerful merchants and politicians. 'By an unexpected tightening of credit, the banks placed financial pressure on a number of firms which encouraged them to dispose of their fish rapidly and at prices below those established by the Exportation Board.' (Alexander, 1977: 23). As a result, the Newfoundland government repealed the Coaker regulations.

In addition to this failure, attempts to restrain the merchants' power through cooperative strategies within the FPU came to nought during the 1920s. Because of the FPU's extensive business involvement, the postwar depression weakened the organization. Gradually, Coaker's organization left its cooperative commitments and harked back to traditional commercialist principles. While Union stores had been founded on the principle of cash trade, Coaker in 1927 instructed his store managers: 'Never take [fish] if you have to pay for it.' The FPU was no longer an instrument in the fishermen's struggle against the merchants and the truck system. It had become a commercial empire over which the members had little more control than they had over traditional private companies (Inglis, 1985: 57). In this way the organization that might have turned the fishermen into powerful force in Newfoundland fisheries withered away. The experience with Coaker and the FPU itself became an obstacle to collective action among Newfoundland fishermen. For instance, when new organizational efforts among Newfoundland fishermen were instigated in the late 1960s, the fishermen were sceptical: 'We had a co-op before, you see. All our fathers was into it. It went broke ... people lost money – it was always an argument – some people thought others was getting more out of it than they should ... There was no way in the world that fishermen in Port au Choix was going to get involved in a co-op' (Port au Choix fisherman, cited in Inglis, 1985: 20).

Why did FPU fail? In north Norway an organizational effort along lines very similar to the FPU, the North Norwegian Fishery Union (NNFU), was established in 1915. This organization was, just like the FPU, part union and part coop, and had a strong political affiliation via its ties to the Norwegian labour movement (Hallenstvedt and Dynna, 1976: 31). In 1917 the organization had 6,500 members in 210 local branches. Like FPU, the NNFU tried to break the merchants' power over the fishermen, and for that purpose it established members' stores, a fish trading operation, a fishermen's savings bank, and a newspaper (Hallenstvedt and Dynna, 1976). As was the case with the FPU, the postwar depression threatened to destroy the organization. In contrast to its Newfoundland counterpart, however, the NNFU survived. Although the organizations' business ventures – the members' stores, the fish trading operation, the savings bank – all failed during the early 1920s, the NNFU in 1926 joined forces with the other regional fisheries associations in

Norway and subsequently became one of the most powerful actors on the Norwegian fishery scene.

The two organizations, the Fishermen's Protective Union of Newfoundland and the North Norwegian Fishermen's Union, were very similar. Still, the former failed and left Newfoundland fishermen virtually without a voice in the reform process of the 1930s, while the latter was the stepping stone that would bring Norwegian fishermen a very large say in the sector. The explanation for this resides in the constituencies of the two organizations, which, although similar in many respects, were different on key points.

Both the Fishermen's Protective Union and the North Norwegian Fishermen's Union were located in an environment where a large share of the households survived on a combination of fishing and other economic activities, primarily farming. This type of economic adaptation creates unfavourable conditions for collective mobilization. The social forces that otherwise could have provoked organized resistance became obscured. Occupational pluralism prevented the establishment of a clear-cut identity as an economic actor, and it thereby frustrated identification of common interests. Against this background, we would expect that the propensity to organize as fishermen would improve significantly if, in addition to the occupational pluralists, there existed a group of full-time, professional fishermen.

In Norway there was a substantial group of full-time fishermen in the interwar period. The Income Task Force (Government of Norway, 1937: 23) reported that the share of full-timers was 28 per cent in 1920 and 44 per cent in 1930. The Director of Fisheries, in his speech to the Norwegian Fishermen's Association in 1930, was a bit more conservative in his estimate, suggesting that 30,000 out of the total 100,000 Norwegian fishermen had fishing as their only occupation in 1930 (Hallenstvedt and Dynna, 1976: 62). Comparable estimates do not exist for Newfoundland. However, several factors suggest that there was a systematic difference here, and that the fishermen in Norway during the interwar period were more specialized than their counterparts in Newfoundland.

One such factor concerned differences in the seasonality of the cod fishery, which gave better conditions for specialization in Norway than in Newfoundland. In Newfoundland, a combination of the stock migration pattern and the ice and weather conditions made the summer the most important fishing season. Fishing and agriculture, therefore, had to be conducted simultaneously, creating tight time budgets and impeding development in both sectors. In Norway, on the other hand, the cod fishery occurred during the winter and early spring, and was finished by the time the growing season started. Within the confines of the traditional economic system, the seasonal pattern in Norway therefore allowed for greater development in fishing as well as in agriculture (Solhaug, 1976: 616).

In addition, there was a difference between Norway and Newfoundland in the fishermen's participation in fish curing. In Newfoundland, as late as the 1930s, the bulk of the cod was cured by the fishermen themselves. In Norway most of the fish was at this time sold fresh and then cured by specialized processors. An important reason was the development of a raw fish market, which happened in connection with the introduction of salt fish production during the late eighteenth century. This market grew significantly during the nineteenth century as salt fish took over from stockfish as the dominant trade product in Norwegian fisheries (Solhaug, 1976).

The establishment of a raw fish market was important partly because it allowed the fishermen to specialize and leave the curing of the fish to others, but also because it was a cash market. In Norway the existence of a raw fish market, therefore, helped break the truck system of credit and the fishermen's complete dependence on the merchant. Hence, by the interwar period, the truck system was no longer an important factor in Norwegian fisheries. The Income Task Force in 1937 observed that credit from the merchant had no practical significance (Government of Norway, 1937: 72). In comparison, the Amulree Commission condemned: 'the almost universal practice for each fisherman in the spring to approach either a local merchant or one of the large mercantile houses in St John's with a view to obtaining, on credit, sufficient supplies of gear, salt and provisions to enable him to conduct his fishery operations and to maintain himself and his family during the fishing season' (Great Britain, 1933: 102).

The creation of a cash market in the fisheries was profoundly important to the question of an interest organization. First, through the cycle of better prices, reinvestment and specialization, it allowed the development of a class of full-time, professional fishermen. Thus, while the seasonal pattern in Norway fisheries had allowed some specialization within the traditional mode of domestic commodity production, the cash market encouraged the fishermen to take the step into a petty capitalist mode of production. Second, the cash market helped break the merchants' dominant political and cultural position in coastal communities.

The truck system, on the other hand, closed the fisherman into a system where virtually all his contact with the world outside the local community went through and was controlled by the merchant. In addition, the lack of local government structures in Newfoundland made the merchant the local authority figure, and the natural mediator in all dealings with government authorities. This led to an overwhelming lack of alternative power centres and perspectives, which was further underscored by a political situation where both major parties were dominated by the mercantile class (Alexander, 1977). Hence, the Newfoundland portrayed by the Amulree Commission in 1933 still had much of the hierarchical and concentric structure of a feudal society, where all power strands ended up in the hands of the St John's merchants. By contrast, the coastal communities of Norway at this time had taken on a more modern social formation, characterized by a plurality of power centres and institutions.

In the fisheries, the establishment of the cash market was a key to this development. With this the fisherman could choose to whom he wanted to sell his fish and from whom he wanted to buy his supplies. Although the number of realistic alternatives might not have been great, the idea of having a choice, and with that, alternative sources of information that could be weighed against each other, was far-reaching. This shift of perspective was backed up institutionally, since the cash market was accompanied by the establishment of new financial institutions; first, the local saving banks and, later, the insurance companies (Christensen, 1995). The same development towards a plurality of institutions and perspectives also occurred on the political scene. Municipal government had been instituted in Norway during the 1830s. While the merchants for a long time would continue to dominate local politics, the establishment of these structures broke their political monopoly. From the turn of the century, the local government structures also formed the basis of real political alternatives, as the emerging labour movement started to win seats.

These two factors, the development of a petty capitalist class of fishermen and the development of a plurality of institutions and perspectives, distinguished the fishery sector of Norway from that of Newfoundland during the interwar period. The two were closely interconnected. Formal interest organizations are intimately tied to the ideas of individual choice, freedom of association, and institutional pluralism (Coleman, 1974). In contrast to traditional forms of organization, which find their place within a single power apex, ultimately divine in character, modern organizations are built from the ground up, and rest on the commitments of individual members. Thus, we can note that the two organizations, the Fishermen's Protective Union and the North Norwegian Fishermen's Union, had significantly different structures. In Norway the NNFU was formed in a bottom-up process. While the initiative came from within the labour movement, it was built on the basis of pre-existing local organizations (Hallenstvedt and Dynna, 1976). During the economic depression of the 1920s, the local branches remained largely intact, while the centralized business operations were abandoned. In Newfoundland the FPU was formed in a top-down process, very much in the paternalistic, colonial, tradition. The organization was centralized to an extreme, with all control resting in the hands of its creator,

William Coaker. In the face of the postwar depression, this meant that all efforts were directed towards saving Coaker's interests, while the losses were taken at the members' end. What had started in the disguise of a social movement was gradually revealed to be a traditional commercial empire (Inglis, 1985).

Conclusion

During the interwar period, the international economic depression caused an upheaval of political struggle and organizational reform in all the export-dependent fisheries of the North Atlantic. As we have shown, there were striking similarities in the responses to the depression in Norway, Newfoundland, and the Maritimes. On both sides of the North Atlantic the crises of the 1920s and 1930s marked the transition from a traditional commercialist to a modern capitalist economic order. To a large extent this occurred in the context of political struggles over the legal and organizational frameworks of the fisheries. These battles should not be understood primarily as confrontations between groups committed to commercialist principles on the one hand and agents of modernization on the other. Instead, they were struggles over different paths of modernization. As will be elaborated in the next chapter, the Norwegian fisheries did not wind up on the same path as the fisheries of Newfoundland and the Maritimes. In Norway the reform process of the interwar period acts formed the basis for the development of a petty capitalist system. While similar reforms occurred in the Maritimes and Newfoundland, these did not go as far, and they did not allow petty capitalist fishermen, or any other groups for that matter, to take command. In their absence, the fisheries were left open to large-scale industrial capital. To explain such differences, we have pointed out a number of factors that made it easier for fishermen in Norway than in Newfoundland and the Maritimes to form effective collective organizations.

2

The Rise of Industrial Capitalism

The commercial salt fish industry had been shaken to the core by the interwar economic depression, despite the attempts to protect it. Although salt fish continued to be an important commodity in the world fish trade during the postwar era, the commercialist principles on which it was based lost their dominant position. Which principles should replace them and form the basis for the fisheries of the modern era became the subject of intense political struggles. When the Second World War was over, the governments of fish-exporting nations across the North Atlantic had come to virtually identical conclusions: frozen fish markets represented the future, and they could only be accommodated through a Fordist model of large-scale, vertically integrated, and capital-intensive industrial technology. Substantial economic and political resources were mobilized to implement this model in the fishery. These efforts were important in substituting modern capitalism for traditional commercialism in the fishery. Still, the results were very different on the opposite sides of the North Atlantic.

In Canada the Fordist model gained a heavy presence in the fisheries in the postwar period. The Department of Fisheries and Oceans (DFO) estimated that the large trawlers owned by fish-processing companies caught about 70 per cent of the groundfish landed in the Atlantic region in 1976. At this time the twelve largest vertically integrated companies accounted for 80 per cent of the output of the fresh and frozen groundfish production and 45 per cent of the total fishery production of the region (Government of Canada, 1976: 25–7).

In contrast, the Fordist model did not become dominant in Norway. We only find one pure example of a vertically integrated firm, Nestle-Findus in Hammerfest. At its height this firm controlled no more than 7 per cent of Norwegian fish production. If we measure the success of the Fordist model by the use of trawler technology, the gap between Norway and Atlantic Canada was slightly narrower. By 1976, for instance, the trawler fleet accounted for 35 per cent of the catch in the groundfish sector (NOS, 1978), of which some were operated by independent fishermen. Still, the continuing and persistent dominance of the small-type, coastal fleet is a striking feature of Norwegian fisheries, which sets it apart from all other western fishery nations.

Fordist processing companies thus controlled more than two-thirds of the groundfish catch in Atlantic Canada, whereas they controlled less than one-third in Norway. This is an indication of fundamental differences in the sector systems of Canadian and Norwegian fisheries. In Canada most tasks within harvesting, processing, distribution, and marketing were coordinated within the command structure of large firms. Norwegian fisheries have relied on other principles of coordination, namely, collective organization and bargaining. Building on the reforms of the 1930s, a system evolved in which the mandated sales organizations (MSOs) within the harvesting, processing, and export subsectors, respectively, engaged in negotiations with each other within a framework of corporatist arrangements (Hallenstvedt, 1982; Holm, 1995). This system resulted from the wave of reform during the 1930s, which had a strong anti-industrial edge. An important motive behind the Trawler Act and the Raw Fish Act was thus to keep the Fordist model from gaining access to the fisheries. Instead, Norwegian fisheries were to remain the domain of owner-operated, small-type enterprises, with close ties to local communities. This suggests that, while the Atlantic Canadian fishery developed into modern-type industrial capitalism, Norwegian fishery, protected by social democracy and corporatist arrangements, held on to traditional technologies, structures, and values.

Some pieces, however, do not fit perfectly into this picture. One is the postwar development in the number of fishermen. If the interpretation suggested above is correct, and modern technology and organizational form gained access to the Canadian fisheries but were barred from Norwegian fisheries, we would expect a sharper decline in the number of fishermen in Canada than in Norway. As Figure 3 shows, however, this did not happen. Because of poor statistical records from Newfoundland, we do not know the exact number of fishermen in Atlantic Canada until after 1955. Drawing on the Walsh commission's estimates, the total number in the Maritimes and Newfoundland was probably in the area of 60,000 in 1945 (Government of Canada, 1953), In Norway the number of fishermen at this time was 118,000, almost twice as many as in Atlantic Canada. By 1975 the number of fishermen had dropped on both sides of the North Atlantic. But while 83,000 fishermen had left the Norwegian fisheries, only 24,000 had left the fisheries of the Atlantic provinces. During the next twenty-five years, the downward trend in Norway continued, leaving 28,000 fishermen in 1990. In Atlantic Canada the number of fishermen actually increased after 1975, reaching 68,000 in 1988. Over the whole period from

Figure 3. Number of fishermen in Canada and Norway, 1945–1990.

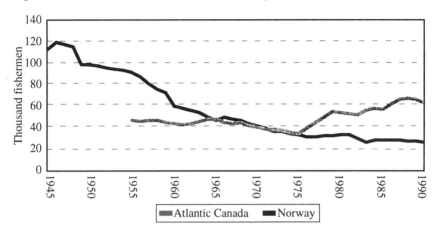

1945 to 1990 the number of fishermen in Atlantic Canada increased slightly, while 90,000 fishermen disappeared in Norway (NOS, 1995).

These findings are puzzling, since they suggest that the Atlantic Canadian fisheries, at the same time as being reorganized along Fordist lines, underwent less technological development and rationalization than the Norwegian fisheries, where the Fordist model was successfully resisted. In trying to make sense of this, which is the main task of this chapter, we suggest that the contrast between Norwegian and Canadian fisheries in the postwar period should not be interpreted as one between 'modern' and 'traditional' solutions. Instead, the fisheries at opposite sides of the North Atlantic became sites for two different models for modernization. In Norway the interwar period had seen the development of a well-organized group of professional fishermen. With support from the social democrats, they had won the most important battles over the hegemony of the sector. With the Trawler Act and the Raw Fish Act, the Fordist model was effectively resisted, and the fishery was modernized on the basis of the interests of the petty capitalist fishermen. As we shall see, the petty capitalist model had a double edge. On the one hand, it entailed fierce resistance against Fordist solutions. On the other, it undermined the traditional economic adaptation, as it put the fisherman-farmer under pressure to become either a professional farmer or a professional fisherman. In this way, the rejection of the Fordist model did not mean retreat to tradition, but a quick transformation from occupational pluralism and domestic commodity production to specialization and technological innovation within a petty capitalist framework. By contrast, the fishermen of Atlantic Canada had not been able to establish an organizational basis for collective resistance during the interwar period. The fishermen's absence from the political scene continued in the postwar period and allowed the Fordist model to assume a dominant role in the development of the industry. This model was introduced from above, however, and did not replace, but came in addition to the traditional economic adaptation of the fishing communities. As a result, the fisheries of Atlantic Canada took on a dual structure, characterized by an uneasy coexistence of large-scale Fordism and a small-scale domestic commodity production.

Fordism and the Fishery

The Second World War played a crucial role in the transition from commercialism towards industrial capitalism in the North Atlantic fisheries. One reason for this was the emergence of a new trade pattern during the war period, which turned out to be favourable for the diffusion of modern product types and capitalist production forms. While Britain's fishing fleet had been engaged in war activities, its fresh and frozen fish markets had been supplied by the Faeroe Islands, Iceland, and Newfoundland (Watt, 1963: 43). In the Maritimes, the war brought high prices for all fish products, but highest for fresh and frozen products in the U.S. market. In only four years, from 1939 to 1943, the share of groundfish marketed as fresh and frozen went up from 34 to 50 per cent, while the share of salted products declined from 54 to 38 per cent (Bates, 1944: 99). Alexander's (1977) observation regarding the role played by wartime experience in the case of Newfoundland is also applicable to the other north Atlantic fishery nations: 'Gaining entrance to Britain's wartime fresh fish market awakened interest in this sector of the industry, and if a fresh/frozen trade was to be pursued, then modernization of production was inescapable' (Alexander, 1977: 7).

The wartime production of fresh and frozen fish products also established industrial structures that were be extended and developed after the war. In Norway, for instance, the Germans had built up a system of frozen fish plants along the coast. This was part of a plan for a restructured fish trade when they had won the war, according to which trawlers and industrial freezing plants in Norway would supply the fish in the 'Grosse Reich' (Frionor, 1971: 14). In addition, the war influenced political structures and coalitions in ways that would be decisive for the direction of postwar fishery policy. In Canada, for instance, the war economy killed off most of the remainder of the schooner fishery (Bates, 1944) and, hence, weakened the coalition that had prevented the introduction of trawlers. More importantly, the war established new contacts between government and business elites. This turned out to be very important in creating a common view of the problems of the fishery and their solution in the postwar period. Peter C. Newman's description of how a Canadian establishment was created in Ottawa during the war applies equally well to what happened in London, where the Norwegian government was exiled:

It was the network and connections and interconnections between business and government, fathered by Clarence Decatur Howe, that became the Canadian Establishment ... They had come to Ottawa as individuals; they left as an elite ... When the dollar-a-year-men fanned out at the close of World War II, to run the nation they had helped to create, the attitudes, the working methods and the business ethic they took with them determined the country's economic and political course for the next three decades. (Newman, 1975; cited in Kimber, 1989: 76).

In Canadian fisheries, both Ralph Bell and C.J. Morrow, the creators of National Sea Products, were among these 'dollar-a-year men' (Kimber, 1989: 77). Across the Atlantic, the same type of elite network between government and business led to the 'London plan,' which shaped the Norwegian government's policy towards fisheries through much of the postwar period (Brox, 1989).

On both sides of the Atlantic, the plans for the fishery sector created by the business-government elite were heroic. The time had come for a fundamental reconstruction of the fisheries. Bates's analogy of the transition from horses and mail coaches to railroads captures the prevailing mood:

It then became apparent that future extension of transport could not be achieved by adding more mail coaches; something new had to be done: a railway had to be built and it had to be built whole with lines, stations, trains, and the like. It was not possible to do this gradually and cautiously; a whole railway had to be built or none at all: a revolutionary step had to be taken. Boldness was then, and still is, the keynote of expansion. (Bates, 1944: 125)

The plans that were laid during and after the war to industrialize the fisheries of the North Atlantic were bold indeed. They not only paid close attention to the need for an integrated approach that included market issues as well as plans for stabilizing the supply, they were closely linked to the broader issues of national economic and industrial policy. Although these plans were adapted to the individual nation's slightly different market situation, resource endowments, and location in the global economy, their overall thrusts were strikingly similar.

On the market side, the industrialists relied heavily on the promise of frozen fish. The traditional products, salted and dried fish, were, as a whole, sold in relatively poor, politically unstable, agricultural countries. This meant low prices as well as political and economic instability. In contrast, the consumer-packaged, frozen fillet was the product of the future. It would tap into the wealthy North American and European markets, and, since it did not require lengthy preparation, it would appeal to the modern housewife. Bates points out how the fisheries could follow in the footsteps of the agricultural industry, which in North America was:

> on a wave of expansion that filled the housewife's shelves with canned fruits, vegetables, meats, soups, and an almost infinite variety of other packaged food in varieties and forms that her grandmother never knew; an industry advancing with a barrage of propaganda that reminded the buyer by radio, newspaper, and shop displays, of the simplicity, the quality, the health-giving, labor-saving attributes of the new foods. (Bates, 1944: 62)

Bates's recipe was for the fish processor to take command of the whole production chain, pushing his control over quality and price as far towards the consumer as possible (1944: 115). If this was a daunting task in North America, it was of course more so in Europe in the postwar years. While the fish processors in the U.S. market could compete with other manufacturers for a share of the 'housewife's shelves,' those in Europe would have to help create those shelves, that is, they would have to build up a complete distribution system for refrigerated and frozen foods (Frionor, 1971). In any case, success in either market required resources much too large for the traditional fish merchant or processing company. The modern food market was a playing field for giants (Frionor, 1971; Kimber, 1989: 80).

To Bates, freezing technology was an important part of the solution to the quality problem. In addition, freezing would allow 'buffer' storage, and thereby prevent gluts and wide price variations (1944: 116). This concern for stability of operations, which resides at the centre of the Fordist model, also had consequences for the supply side. Only the trawler technology could secure a stable supply of raw materials, since 'the dragger and trawler can operate more continuously at sea and they can be made to fit better into the continuity of plant operations' (ibid.: 118). Traditionally, processing firms had solved the problem of variability on the supply side by increasing their normal labour force when fish were plentiful at the plant. This would not do in a modern frozen fish industry. Proper integration between catching and plant facilities, which meant large operations in a few centres, would facilitate constancy of operations and mechanization, and allow for the efficiency that was so important in building the new industry (ibid.: 106). Advantages of scale, stability, and mechanization, in short, mass production, were the key to the modernization of the fishery.

Industrialization in Two North Atlantic Settings

Norway: The Modernization Campaign

In Norway the end of the war marked the beginning of an intense effort to transform its backward and partly wartorn economy into a modern industrial one. The fishery problem was but one instance of the more general underdevelopment problem of the Norwegian economy. In short, too much production happened in labor-intensive primary production, too little in the capital-intensive industrial sectors. In north Norway the 'backwardness' was particularly prevalent. While the region had 12 per cent of the Norwegian population, it only produced 6 per cent of the national product. The reason, of course, was the large share of the labour force in fishery and agriculture, that is, 'the sectors that give a small return per employed person' (Grønås et al., 1948, cited in Brox, 1989: 71). The solution, in the fisheries as in the Norwegian economy as a whole, was industrialization and capitalization: 'A rapid and strong increase in the nation's productivity, which must be the foundation for an increase in the national income and a higher standard of living in the future, must to a large extent build on rationalization, modernization and continued development in the industrial sectors' (Government of Norway, 1947: 27).

From this perspective, primary industries took on a role as a pool from which labour could be tapped as the new manufacturing industries were constructed:

Because of the too high increase in the number of fishermen before the war, one must presume that this sector in the years ahead will be able to supply substantial amounts of labor to other sectors without reducing the size of the catch substantially ... There is, however, a limit to the amount of manpower one can draw from the fisheries before its operational methods must be changed. At the same time, the opportunities for replacing labor by way of investment in more capital intensive methods are fairly promising. To what extent such changes should be initiated, must at any time depend on the economic return that the labor released from the fishery can be expected to give in other sectors. From a fishery standpoint, there is no reason to restrain a development that leads to fewer people wanting to be employed in the fisheries. (Government of Norway, 1947: 25)

In other words, restructuring the fisheries would bring a double gain, both making the sector itself more efficient and productive and releasing labour for the new industries.

This analysis of the Norwegian economic problem had broad support in the political establishment. Thus, the first two decades after the war witnessed a massive campaign to transform the fisheries, along with the rest of Norway, into a modern industrial state. One example is Frionor, formed in 1946 as a centralized marketing company in the Norwegian frozen fish business. The Frionor model followed from the industrialization idea. Marketing of frozen fillets in Europe and the United States would require large players. Frionor, therefore, was granted a monopoly in the export of frozen fillets to all markets (Frionor, 1971). A second example is the establishment of Finotro in 1950, a company comprising seven large industrial filleting plants along the north Norwegian coast (Vik, 1982). A third example is the establishment of Findus in Hammerfest in 1951. An offspring of Freia, one of the few Norwegian industrial enterprises of international scope, Findus was a vertically integrated fish processor, complete with trawlers and retail outlets to market consumer-packaged frozen fish under its own label in Europe (Lien, 1975). All these projects were supported by the Labour government.

In addition, the industrialization campaign led the authorities to initiate institutional reforms. The Trawler Act had restricted expansion of the trawler fleet. In addition, the Ownership Act, adopted during the war, reserved the right to own fishing vessels to active fishermen. Both acts were contrary to the industrialization plan, which relied heavily on processing plants supplied by their own trawlers. As a consequence, the government set out to weaken the trawler legislation and subsidize the building of trawlers (Sagdahl, 1973; Mikalsen, 1977).

Counterforces
The industrialization campaign was met with heavy opposition from the fishermen and their organizations. Based on their successes in the 1930s, fishermen in the postwar period started out with grand ambitions. They now wanted to finish what they had started, creating a cooperative sector model with the fishermen in complete control, not only in harvesting but also in processing and exports (Sagdahl, 1973). This meant, of course, that the fishermen's perspective on the future of the fisheries clashed heavily with the government's view. Instead of Fordism and industrial mass production, the fishermen wanted to continue with the small-scale, petty capitalist approach of the prewar period. The scene was set for confrontation between the industrialists in business and government and the petty capitalist fishermen in the Fishermen's Association.

The establishment of Findus in Hammerfest clearly represented a victory for the industrialists. At its height in the 1960s, after it had been bought by the multinational food giant Nestlé (Rudeng, 1989: 348), it operated twelve trawlers and processed 15 to 20 thousand tonnes of groundfish annually. During this period it accounted for 3 per cent of the Norwegian groundfish landings and controlled 7 per cent of the production value in Norwegian fisheries (Throne-

Holst, 1966). Marketing frozen fish in Europe under its own label, it clearly represented the vertically integrated model the industrialists saw as the ideal.

Findus, however, was the only pure example of the Fordist model in Norwegian fisheries. The Finotro plants, which the planners in London had hoped would spearhead the restructuring of the north Norwegian fisheries, did not own trawlers until 1970. One reason for this was that the Finotro plants were partly owned, and to a large extent controlled, by the fishermen's organizations. Instead of profit centres, they saw the plants as landing stations where the fishermen could dispose of their catch, especially for the seasonal fisheries (Vik, 1982: 38). This was a role for the processing plant quite different from the one Stewart Bates had envisioned.

Nor did Frionor, the frozen fish marketing company, conform to the ideal of vertical integration and Fordism. Since it was granted a monopoly in the export of frozen fillets, it did become a large actor. Frionor controlled all Norwegian exports of frozen fillets from 1946. From 1951 this was relaxed to accommodate the establishment of Findus in Hammerfest. From 1967, Nordic Group, a common sales organization constructed along the same line as Frionor, was allowed in. These three actors remained in charge of the Norwegian frozen fillet exports until 1991, when all privileges were abandoned.

Constructed as a common sales organization and owned by the processing firms, however, Frionor was not much of a Fordist construction. Instead of the hierarchical command structure of a unitary company, Frionor had the complex decision structure of a cooperative. Formally, Frionor's role was to service the individual member processing firms, the primacy of which was reflected in their status as Frionor's owners. Compared with the Fordist version of vertical integration, the Frionor model was a very modest step in bridging the gap between processing and exports. Instead, it was in line with the cooperative approach to fish exports that the fishermen wanted. Thus, Johan J. Toft, then leader of the Fishermen's Association, expressed strong support for the organization at Frionor's twenty-fifth anniversary in 1971 (Frionor, 1971: 54).

The most intense confrontations in Norwegian fisheries during the postwar era occurred in connection with the introduction of processor-owned trawlers. Since the industrialization program required that the legal barriers against vertical integration be removed, or at least lowered, the Norwegian Parliament became an important arena in this conflict. Here, the fishermen were able to muster considerable support, particularly through the close ties between the Fishermen's Association and the Labour party.

Despite this, the restrictions against trawlers were softened through modifications of the Trawler Act in 1951, 1961, and 1971 (Sagdahl, 1982), and the restrictions against processing companies owning fishing vessels were relaxed through modifications of the Ownership Act in 1950, 1956, and 1961 (Mikalsen, 1982). During this process, the Fishermen's Association abandoned their absolute rejection of the trawler technology, and gradually came to accept trawlers as supplementary to the conventional fleet, as long as they were controlled by fishermen. This latter principle was an important basis for the revised legislation, at least at the symbolic level. In practice, however, the government was rather lenient in granting trawler licences to processors. Thus, the number of trawlers over 200 gross registered tons GRT increased from 29 in 1952 to 90 in 1971, most of which were owned by north Norwegian processing companies (Sagdahl, 1982). At this time, the trawlers accounted for about one-quarter of the groundfish catches in Norwegian fisheries (NOS, 1973).

The Petty Capitalist Sector System

Trawler technology and large-scale industrial companies did gain a foothold in Norwegian fisheries during the postwar period. They did not become dominant, however. The most important reasons for this are to be found in the legal and institutional order of the sector, which was committed to a petty capitalist model and gave the fishermen and their organizations the dominant power position. Figure 4 sketches the main features of the sector system in Norwegian fisheries during the postwar period until the 1980s. The confusing complexity of this organizational system, often noted as its main characteristic, should not distract us from its systematic effect on the distribution of power within the sector. In trying to make sense of the sector system, we begin by noting that it was based on a functional division of labour, dividing the fishing industry in three organizationally autonomous, but interdependent subsectors. This was a legacy of the interwar period, when the main production tasks were handled separately by economic units specializing in fishing, processing, and exports. This structure, which originated as a technical and economic solution to practical production problems, was held in place by the organizational and legal barriers created during the political struggles of the 1930s.

At the end of the Second World War, the Raw Fish Act was implemented throughout the commercial marine fisheries of Norway, resulting in twelve fishermen-controlled sales organizations (bottom left of Figure 4). On top of the sales organizations we find the fishermen's most important political organization, the Norwegian Fishermen's Association. It was not the only one, however. The figure also includes the Norwegian Seamen's Union, which had a considerable share of the trawler crews as members. Reflecting the relatively weak position of the Fordist model and the trawler technology in Norwegian fisheries, the Fishermen's Association remained dominant. This was not only because of the numerical strength of the petty capitalist fishermen. In addition, the Fisher-

Figure 4. The sector system in Norwegian fisheries during the postwar period: Economic organization.

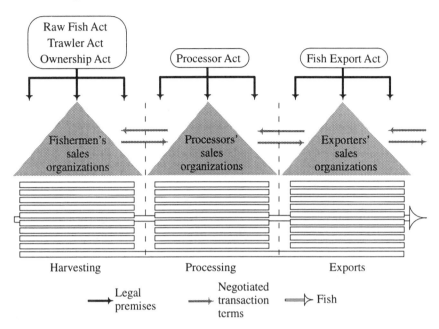

men's Association was financially supported by the fishermen's sales organizations. With authority from the Raw Fish Act they imposed a mandatory 1 per cent "organizational levy" on the catch. The main part of this money was channeled to the Fishermen's Association and formed the organization's most important source of income (Holm, 1995).

In the export subsector (located on the right-hand side of Figure 4), the situation was a bit more complicated. With a legal basis in the 1932 Salt Fish Act, export cartels had been established in some of the most important export branches, like salt fish and stockfish. During the Second World War, a different organizational form, the fish export council, was established in order to manage the tightly constricted trade of a war economy. The export councils did not replace, but were added on to the established export cartels. After the end of the war, both forms of organization were retained. The result was a twin structure, with fourteen fish export councils regulating the fish export *trade* through a system of licences, and fourteen trade organizations, organizing the *firms* engaged in fish exports.

In addition to the Raw Fish Act and the Fish Export Act, both of which originated in the 1930s, the postwar period saw the establishment of yet another legal device, the 1970 Processor Act. The act was built on the same principle as the Raw Fish Act, allowing the government to authorize processor organizations and give them monopoly rights in the trade of specified fish products. Only one organization was authorized on the basis of the act. This was Fiskeprodusentenes Fellessalg (the Fish Processors' Marketing Cooperative), which organized the producers of salted and dried fish in north Norway.

We are now in a position to give a more comprehensive explanation as to why the plan for Fordist restructuring of the fisheries did not succeed in Norway. It was not simply because the Trawler Act and the Ownership Act slowed down the creation of backward linkages. More importantly, vertical integration foundered on the established structure of the fish trade, as institutionalized by the Raw Fish Act and the Fish Export Act. At the same time as the government engaged in a struggle to eliminate the barriers against processor-owned trawlers, it assisted in the process of consolidating and extending the established system of sales organizations, which undermined vertical integration.

The most important reason for this inconsistency was the Labour government's commitment to the fishermen and the system of sales organizations. If the attempt to introduce trawlers was controversial, an attack on the Raw Fish Act would have been regarded as an act of treason. With this, however, we have only described the existing power structure, in which the fishermen were dominant enough to preclude a reconsideration of the Raw Fish Act. The basic question is how the fishermen had obtained such a position.

To understand this, it is important to have in mind that the organizational system outlined in Figure 4 had an impact on the amount and distribution of political resources within the fisheries. Although it was established as a trade system, it also transformed the sector as a political system. As a result of the organizational reforms of the 1930s, the fishermen's capacity for collective mobilization was vastly improved, while the processors and exporters were left hopelessly fragmented and politically impotent. The explanation for this resides in the path-dependent nature of the reform process, where the sequence of events within and between subsectors was very important.

As we have seen, the Norwegian Fishermen's Association was established in 1926, twelve years before the Raw Fish Act was adopted. Since a general interest organization existed before the fishermen's sales organizations entered the scene, the latter could stay out of sector politics and concentrate on the task they were authorized to do, namely, regulating firsthand sales of fish. In their turn, the sales organizations strengthened the association both organizationally and financially. In this way, the reform established a tidy division of labour, where a diverse set of economic organizations backed up one relatively strong political organization.

In the export segment, in contrast, no general interest organization existed before the adoption of the 1932 Salt Fish Act. This meant that the first systematic organizational effort among the merchants occurred in direct response to the international market crisis and was directly tied to the possibility of legal protection. This had important consequences for the organizational pattern that resulted. The exporter associations were cartels as well as general interest organizations. As cartels, they were defined according to specific market segments, and the fragmentation of the fish markets spilled over to the merchants' political organization. The exporters of salted and dried fish formed one association; the exporters of wet-salted fish another. Stockfish exporters organized separately from fresh fish exporters, and exporters of codfish did not want to associate with herring exporters, and so on. While the trade reform had created a unified political apparatus for the fishermen, the organizational mix of economic and political issues in exports left the merchants hopelessly fragmented and unable to define common interests (Hallenstvedt, 1990; Holm, 1995).

The Main Agreement

In addition to the three acts pertaining to the fish trade, Figure 4 also includes the Main Agreement, negotiated by the government and the Fishermen's Association. Valid from 1964, it set the conditions for state subsidies to the sector. In practical terms, it established the stage on which annual negotiations over the amount and distribution of state subsidies to the sector were conducted by representatives for the Fishermen's Association and the government. From 1964 to 1990, Nkr 15.4 billion was channeled to the fishery sector through the Main Agreement negotiations. This amounted to 19 per cent of the catch value during the same period (Holm, 1991).

The Fishermen's Association was the dominant organization within the sector, and it was the obvious counterpart for the government in the subsidy system. At the same time, the agreement consolidated the association's position within the sector. The government's explicit acknowledgment of the association through the agreement was important in itself. In addition, a considerable amount of money was channeled through the negotiations, and the fishermen had a real say over how it was distributed. To appreciate this point, it must be kept in mind here that the two parties to the agreement, the government and the Fishermen's Association, had very different ideals for the sector at the outset. In keeping with an industrialization perspective, the government had intended the Main Agreement as a program for structural rationalization, with investment support and buy-back programs as main instruments. Price subsidies were acceptable, but only as temporary measures until the sector had been rationalized and could manage on its own. To a large extent, it was this understanding that had been written into the Main Agreement (Government of Norway, 1963; Holm, 1991). However, the priority of rationalization over income stability was reversed during the practice of annual subsidy negotiations. For the 1964–90 period the bias towards the goal of income stability is illustrated by the fact that nearly 80 per cent of the subsidies were allocated to this purpose (Holm, 1991). In this way, the Main Agreement established a system where the state buffered the fishery sector from fluctuations in resources and markets. Instead of an instrument for rationalization and industrialization, as the government had intended, the Main Agreement had been turned into an economic safety net for the fishermen. Instead of reconstructing the sector, the agreement protected and reinforced the established order (Holm, 1991).

The fate of the Main Agreement illustrates our basic point: the dominance of fishermen in the Norwegian fishery sector. The fishermen had achieved this position on the basis of a number of different ideological, political, and structural factors, of which the history of struggle against the merchants and relatively favourable conditions for collective identification had been crucial. During the 1930s this had allowed them an organizational starting point from which their capacity for political action grew exponentially. This process was greatly aided by the organizational weakness of processors and exporters. During the postwar period the difference in political effectiveness between the two sides escalated. While the fishermen were able to fortify and improve their positions, the processors and exporters were not even allowed to participate on the most important sector arena, the Main Agreement negotiations.

We argue that the sector institutions created during the 1930s allowed the fishermen to take a dominant position within Norwegian fisheries. However, 'fisherman' is a label that covers a diverse set of actor types, including the domestic commodity producer who has fishing as one of many activities to make ends meet, the self-employed small-scale fisherman who owns the fishing vessel himself, the person crewing on larger vessels who must be regarded as a labourer whether he works for a wage or a share of the catch, and the capitalist boat-owner running his fleet from the company headquarters on land. It goes without saying that the interests of these groups are different. Which among these types of fishermen was predominant in Norwegian fisheries?

The reform movement in Norwegian fisheries had a strong populist bent. It had its basis in small-scale, coastal fishing and rural communities. The fishermen's primary enemy was the urban industrial capitalist. Their worst nightmare was losing their independence and becoming mere wage-earners in an offshore fishery. Thus, we can safely assume that neither the fisherman-capitalist nor his counterpart, the fisherman-worker played an important role in the Norwegian fisheries. The question is not settled, however, since there is a wide span separating the two remaining categories, the part-time fisherman of the multioccupational household and the petty capitalist fisherman. Both these groups have played an important role in Norwegian fisheries throughout the twentieth century. During this period, however, there was a definite shift from the former to the latter.

Brox (1989: Chapter 4) has pointed out that the Raw Fish Act, although it grew out of an explicitly anti-capitalist mood, formed the basis of a capitalist growth cycle in the fishery. The act not only resulted in increased price stability and, hence, allowed people to make commitments to a full-time career in the fishery. It also gave the fishermen a larger share of the sector's income and led to higher capital investment in the fishing fleet. In any case, the result was a crucial step away from the multioccupational adaptation of the commercialist system.

Another important step in this process occurred in connection with the government's industrialization program at the beginning of the postwar period. The part-timers in primary production were seen as the most unproductive element in the Norwegian economy. The industrialization of the fisheries had the dual purpose of creating efficiency in this sector and releasing manpower to other, more productive, branches of the economy. In the government's fishery policy, this analysis led to a strong bias towards full-time fishermen (Government of Norway, 1959). This was followed up in a number of ways, the most important of which was the Main Agreement. The preference for full-time fishermen here was expressed both symbolically in its preamble (Government of Norway, 1963) and practically in the choice of price support as the most important subsidy measure. A price subsidy favoured the full-time fisherman, both because it reduced income uncertainty and because it put a premium on quantity.

Brox (1966: Chapter 4) pointed out that the Fishermen's Association took the same position as the industrialists in the government on these issues:

The fishermen's own organizations play a crucial role in the effort to make the fishermen choose between being a full-time fisherman and not a fisherman at all, that is, destroying the possibility for the occupational pluralism that coastal people often find to be the most rewarding. In the same way as the Brofoss Committee, the Fishermen's Association primarily sees it as its task to secure the full-time fishermen a 'decent income.' (1966: 46–7)

To Brox, this was not only undesirable but also curious, since, according to

74   Section 1: Institutional Development

Brox, the farmer-fisherman at the beginning of the 1960s still was the dominant type of actor within the fisheries. While Brox probably inflated the importance of the farmer-fishermen in the postwar period (Drivenes, 1982), he was correct in pointing out the convergence between the government and the Fishermen's Association in their preference for the full-time fisherman. The organizational transformation of the fisheries that took place in the 1930s, and was consolidated in the postwar period, was formed on the basis of and further strengthened the fishermen's identity as petty capitalists.

The Maritimes and Newfoundland

On both sides of the Atlantic, by the end of the Second World War great expectations were tied to the Fordist solution to the fishery problem. In contrast to Norway, however, the plans for 'a revolution in the fish trade' (Bates 1944: 116) met with virtually no organized resistance in the Maritimes and Newfoundland. Nor was the development of a modern frozen fish industry hampered by institutions like the Norwegian Trawler Act, the Ownership Act, or the Raw Fish Act. In the Maritimes the only restriction of this sort was the trawler ban. The first move to weaken the ban occurred during the war, when subsidies were made available for the construction of trawlers and the conversion of schooners to trawler technology (Bates, 1944: 95). Trawler licensing restrictions were relaxed in 1949 and again in 1953 to allow foreign-built vessels. In contrast to Norway, where the Ownership Act made vertical integration difficult, the Canadian legislation encouraged vertical integration; trawler licences were issued only to companies or individuals 'affiliated with a plant capable of handling the catch.' (Watt, 1963: 40).

The paradigmatic example of the Fordist development in the Maritimes is National Sea Products. Established in 1945 by a merger of twenty-four fish companies in the region, the company soon became a dominant actor within Atlantic Canadian fisheries:

By 1977, the National Sea/Nickerson complex controlled 11 percent of all fish plants [in Nova Scotia], with a combined groundfish capacity exceeding 41 percent of the total in the province; employed 41 percent of the permanent fish plant workforce; purchased, according to 1971 figures, 34 percent of all fish landed by fishermen in the province; and, according to conservative estimates, had a market share in fish products exceeding 40 percent. (Barrett, 1984: 93–6)

In Newfoundland the interwar crisis had resulted in organizational structures that might have formed the basis for an alternative to the Fordist, free-enterprise model that ensued. The Newfoundland Fisheries Board, formed in 1936 and empowered to regulate 'all aspects of production, processing, culling, inspection, and distribution of fisheries products' (Alexander, 1977: 27), was, in 1947, complemented by the Newfoundland Associated Fish Exporters Limited (NAFEL). NAFEL was awarded the exclusive right to export salt fish from Newfoundland (Alexander, 1977: 34); this was a salt fish parallel to Frionor, the marketing cooperative in Norwegian frozen fish industry. Frionor remained a dominant actor on the Norwegian fishery scene until the 1990s; NAFEL, however, was destroyed during the 1950s. According to Alexander (1977: Chapter 6), this was the result of a combination of factors, including a declining European salt fish market and a severe exchange rate problem, but primarily because it did not have support in the political establishment. In 1953 NAFEL lost its monopoly right in bulk saltfish, the most important product. This was the beginning of the end of the organization, which was completed in 1959 when it lost the remainder of its trade privileges.

Newfoundland's union with Canada in 1949 was an important factor behind the loss of support for NAFEL. The federal government regarded salt fish as an obsolete product and preferred competitive selling and free enterprise to trade monopolies and cooperation (Alexander, 1977). Confederation with Canada brought the branch-plant form of organization (Laxer, 1991) into Newfoundland with full force. In 1950 the fishing industry in Newfoundland had been wholly owned and controlled by provincial interests; twenty years later half of all catching and processing was controlled by non-resident firms (Alexander, 1977: 141)

It would be wrong to portray this as a development completely without counterforces. One of the most important examples was the attempt in the early 1960s to reconstruct the fishery along cooperative lines by forging an alliance with agriculture. Although agrarian interests in general have not had much impact on the Canadian polity (Laxer, 1991), the cooperative structures in the agricultural sector remained as a legacy of the wheat farmers' protest movement. Thus, Premier Smallwood in Newfoundland launched an initiative for a national fisheries program built explicitly on the cooperative model in agriculture. The idea was to show that the fishery was structurally equivalent to agriculture and that it consequently needed the same type of solutions. The proposal for a national fisheries program included subsidies to reduce price and harvest uncertainties, improved credit facilities to fishermen, greater expenditures on the development of infrastructure, and the establishment of a common marketing agency (Alexander, 1977: 145–7). This program was built explicitly on the Canadian Wheat Board model. However, there were clear parallels to the Norwegian institutional arrangements in the fisheries, where the Raw Fish Act and the Main Agreement reduced price uncertainty and guaranteed income levels, the fishermen were provided credit through the State Fisheries Bank (Handegård, 1982), and massive investment in regional infrastructure took place through the regional development programs (Hersoug and Leonardsen, 1976). Except for common marketing, which was rudimentary, Norwegian fisheries might have served as a template for the Canadian program. The plan was rejected by the Canadian federal government, however, which defined the main problem of the fisheries as 'a limited resource and an excessive number of people depending on it' (Government of Canada, 1964; cited in Alexander, 1977: 148).

The main development strategy in the fisheries of Atlantic Canada, thus, was one of Fordist development through encouragement of private enterprise. In the Maritimes the industrialization policy was relatively straightforward, since the fisheries only constituted a small part of the total economy, and the people displaced by capital in the fisheries were assumed to be easily absorbed in other sectors. Besides, as Bates (1944: 115) pointed out, there was room for expansion in the fisheries, as the potential of groundfish stocks was not fully utilized, and, in any case, Canadian fishermen took less than 7 per cent of total codfish catches off the eastern shores (ibid.: 106). Still, Bates was careful to recommend a dual strategy, where the primary effort of creating large-scale, vertically integrated frozen-fish plants in a few main fishing ports was complemented by measures aimed at the inshore areas. Developing these areas, which were dominated by a multioccupational household economy that typically integrated fishing, farming, and forestry pursuits, was a complicated task: 'From a social point of view, the rehabilitation of these areas is an inter-administrative task, involving as it does farming, forestry, fishing, public health, education, electrification, roads, and other 'amenity' capital; consequently the administrative task varies in each district according to its relative assets in arable land, timber, fish and shellfish' (ibid.: 123).

If this type of adaptation represented problems in the Maritimes, it posed many more difficulties in Newfoundland. Here, the multioccupational economy Bates found in the Maritime coastal areas dominated the whole island, outside of the Avalon Peninsula. Further, the whole issue of modernization was complicated by the lack of alternatives to the fishery. While Norway and the Maritimes could try for an efficient fishery, Newfoundland needed a large one (Alexander, 1977: 159). This issue became less pressing with Confederation, since union with Canada lifted the barriers to the emigration of surplus labour, and created some hope of expansion in other industries. Thus, the 1951 Walsh Commission embraced the industrialization program:

A characteristic of modern industry, as contrasted with primitive industry, is a high ratio of capital to labor, in other words, a relatively large investment per man. For the greater part of the Newfoundland fishing industry ... this ratio is still very low. What is needed, therefore, is an increase in the size of fishing craft, in the efficiency of gear and in the extent and variety of processing facilities. With an increase in the fixed 'plant' comes a greater urgency to eliminate variability in the supply of raw material. This involves concentration of the fishing fleet serving the processing plant and consequently the building up of fishing communities on a substantial scale. For the fishermen, the effect of development of this kind would be vastly to increase the output per man – as is necessary, in order to raise incomes to the level required to retain men in the industry. (Government of Canada, 1953: 24)

In this way, union with Canada opened Newfoundland fisheries to the Fordist model and accelerated the transition from salt fish to frozen fish production. At the level of public policy, this was most clearly expressed in the resettlement programs of 1953 and 1965. Under the first program, the purpose of which was to assist people to move to larger centres where public services were easier to provide, some 115 communities containing 8,000 people were resettled (Wadel, 1969: 32). Under the second program, which ran until 1970, some $7.4 million was spent moving 3,240 families (Crowley et al., 1993: 358). There was an economic bonus on relocation to 'designated fishery growth areas,' and the relocation program was closely tied to the modernization of the fishery (Wadel, 1969: 33–4).

A policy of industrialization was pursued systematically in the Newfoundland context, but not done without second thoughts. The Walsh Commission, established to promote more effective utilization of the fishery resources in the Newfoundland fishing industry, was uncomfortable about the social consequences of unchecked industrialization. Thus, the commission raised doubts about the rate at which the surplus fishing population could be absorbed in alternative employment. In a footnote it pointed out that if the province's normal codfish catch were to be taken with trawlers and long-liners, it would not employ more than 2,000 fishermen, roughly 10 per cent of the fishermen at the time (Government of Canada, 1953: 24). The complexity of the fishery problem, and the commission's reservations as to the virtues of industrialization and frozen fish production, became explicit towards the end of the report:

It is clear to the members of this Committee that the standard of living of the inshore fisherman who cures his own fish is generally below the desirable level. However important, therefore, the salt fishery may be in the economy or the trade of Newfoundland or Canada, [the Committee] cannot say anything that may halt the exodus from it unless immediate and effective steps can be taken to assure to the fishermen a standard of living, including amenities and services, regarded as normal for rural areas. Nor is the problem capable of a ready solution by merely allowing the exodus to continue. Many, particularly married people, have their roots firmly in the settlements which have for years been established along the coast and on the islands nearby. They have not the skills to produce employment in other industries, if avenues of employment were open, and they have not the means to move with their equipment to other parts of the coast to engage in the fishery near the processing plants. Indeed, existing plants have the capacity to produce as fillets only a fraction of the salt fish production, even if a market were available for such a quantity of fillets. (Government of Canada, 1953: 102)

The solution to the fishery problem that the Walsh Commission hinted at lay in transfer payments from the federal government. Indeed, access to Canada's social security benefits had been a crucial issue in the 'Confederation battle':

Newfoundland had nothing at the time [of Confederation] that could be conceivably be compared with the social services of Canada. There were no family allowances. Old age pensions were paid to persons of 70 years and over on the basis of a means test and at the rate of $75.00 per year. There was no unemployment insurance. To a people whose social services were still fifty years behind the times and whose subsistence economy could allow life to be endured on catch incomes often less than the family allowance would produce, the attractions [of Confederation] were seductive. (Perlin, 1959: 54; cited in Wadel, 1969: 26)

Social security benefits became available in Newfoundland immediately after Confederation. Their importance to the fisheries, however, was drastically increased from 1957 when fishermen were made eligible for unemployment insurance (UI). Until then, no support had been provided for the self-employed, such as fishermen, on the grounds that they could manipulate their periods of unemployment. The UI program for fishermen treated the fish buyers as employers. The fishermen collected one stamp for each of fifteen weeks in which delivery was in excess of a specified quantity, and they were were entitled to draw benefits from December to mid-May. While there were many other sources of federal and provincial assistance to the fishermen, this has been the most important (Crowley et al., 1993).

It is crucial to understand that the UI benefits in the fishery had effects very far from, maybe even contrary to, the spirit of the industrialization program. Wadel's (1969: Chapter 3) discussion makes this very clear. He began by showing that the transfer payments, of which the UI benefits were the most important, accounted for a very large share, maybe as much as 40 per cent of cash incomes, of the household income in outport Newfoundland. Then he showed why, for at least three different reasons, the UI benefits undermined the government's industrialization program. First, the UI benefits tended to shorten the fishing season, since it took away the incentives to continue fishing once the maximum number of stamps had been collected (Wadel, 1969: 51). This was contrary to the industrialization program, which called for a longer fishing season and more fishing effort (Government of Canada, 1953). Second, the UI benefit scheme encouraged the fishermen to salt his catch instead of selling it to filleting plants, since the added work was important in collecting the maximum number of stamps (Wadel, 1969: 51). This was also contrary to the industrialization program, which called for more production of fresh and frozen fish and less salting. Third, UI income tended to be more important for outport than town households, both because it accounted for larger share of the cash incomes in rural areas, and because rural households tended to receive larger transfer payments (1969: 53). In this way, the UI scheme discouraged movement from outport to centre, prevented fishermen from finding alternative employment, and slowed down the transition from the traditional salt fishery to the modern fresh and frozen fish industry. Thus, it was hardly a coincidence that the number of inshore fishermen in Newfoundland had increased by 6,000 in the ten years after the introduction of UI benefits for fishermen in 1957. Wadel therefore agreed with Copes's (1964) conclusion, but without much of the latter's regret, that the UI benefits worked in opposition to the government's industrialization policy (Wadel, 1969: 51).

The late 1950s was a period with very low prices in international fish markets. This affected the fisheries in Norway as well as in Canada and induced policy changes in both places. In Norway 1957 was the first year of price subsidization in the form that later would be codified in the Main Agreement. In Canada the same market depression was an important factor behind the extension of UI benefits to fishermen. The contrast between the two responses is revealing. Whereas the price subsidies rewarded quantity and thereby professional, full-time fishermen, the UI system catered to the part-time fishermen and the occupational pluralist. This difference simply reflected the underlying class structures, with petty capitalist fishermen dominating Norwegian fisheries, while domestic commodity producers were much more important in Canada. In 1957 this difference was further strengthened with the institutionalization of contrasting subsidy mechanisms.

In Norway fishermen and the government had clashed over the issue of industrialization and vertical integration, and the Fordist model was successfully resisted. By contrast, 'industrialization and the elimination of the "cottage style industry"' (Alexander, 1977: 159) became the accepted development model for the Canadian fishery. The implementation of the first half of this development model, establishment of Fordist firms, progressed virtually without resistance. The second part, however, the elimination of the cottage-style industry, was not followed up in practical policy. Instead, the authorities encouraged industrialization along Fordist lines at the same time as they supported occupational pluralism and subsistence fishing, particularly through the UI scheme. To explain this inconsistency requires an understanding of the political basis of the decision to extend UI benefits to the fishermen.

The key to this outcome was the part-time fishermen's numerical strength as voters, particularly in the province of Newfoundland. Copes (1972a) demonstrated that the fishermen were the most important voter group in Newfoundland and formed the basis for Smallwood's long period as premier after confederation in 1949. Smallwood did that by making the fishermen's cause his own. Hence, Matthews (1993: 54–60) concluded that the provincial resettlement programs formed the exception and that the Newfoundland government during the postwar period systematically sided with the part-time fishermen in opposition to the industrialization policy pursued by the federal government.

One should not exaggerate the contrast between the provincial and the federal level. Both were characterized by considerable ambiguity in case of the fisheries policy. There can be no doubt that Smallwood was a firm believer in the promise of industrialization, and he followed up on that belief in practical policy (Copes, 1972b). Similarly, while the federal government pursued industrialization, it was far from immune to the problems and needs of the outport fishermen. Thus, the extension of the UI system to fishermen in 1957, while sponsored by the Newfoundland government and opposed by the federal Department of Fisheries, could not have been accomplished if the Canadian government had been set firmly against it.

Compared with Norway, where the fishermen were influential through their numerical strength as voters, as well as their organizational strength, the Canadian fishermen could primarily influence public policy as voters, particularly in Newfoundland. Newfoundland fishermen were not completely unorganized, however. The Newfoundland Fishermen's Federation had been established in 1951 on the initiative of Premier Smallwood. In the paternalistic tradition dating from the time of Coaker and the Fishermen's Protective Union, the federation was based on top-down rather than bottom-up control. 'Most fishermen seem to regard the Federation as an extension of the Provincial Government insofar as it can't do anything without the permission of the Government' (Wadel, 1969: 145).

The relative importance of the 'numerical' as opposed to the 'corporative' channel (Rokkan, 1966) meant a bias in favour of the part-time fishermen, simply because they represented more votes. This bias was reinforced by Smallwood's strategic manoeuvring. According to Copes (1972a), Smallwood did not maintain his position only by way of policy initiatives designed to attract the fishermen's vote. In addition, he systematically pursued a fisheries policy that would support a large number of fishermen, and he did not stop short of tinkering with the electoral system so as to give the rural constituencies, and thereby the fishermen, a disproportionately high representation in the House of Assembly. In this way, the strategic role of the fishermen's vote in Newfoundland gave the interests of outport fishermen a much heavier representation than one would expect from their organizational resources and numerical strength.

In concluding his analysis, Wadel stressed the need for a dualist policy in Newfoundland, stimulating development in urban and rural areas simultaneously (1969: Chapter 7). In the midst of the controversies over the relocation programs, one can understand that Wadel paid most attention to the strong industrialization rhetoric of the time. Looking back at this period, however, it seems that a dualist policy was exactly what was pursued. On the one hand, the lack of organized resistance from petty capitaist fishermen made it easy to pursue industrialization and Fordist development. Somewhat paradoxically, however, the weakness of petty capitalist interests forced the authorities to take on responsibility for the part-time fishermen in the outports. The result was a divided sector, characterized by uneasy coexistence of large and small actors, modern and traditional technology, and industrial and commercialist organizational forms.

Large-scale Industrial Capitalism versus Petty Capitalism in the Fisheries

From very similar starting points at the end of the Second World War, Canadian and Norwegian fisheries developed in different directions. In Norway the fishermen were able to resist attempts at Fordist reorganization of the sector. Instead, the institutional transformation that had commenced during the 1930s was completed, resulting in a petty capitalist sector system that gave the fishermen a dominant position. In Canada the industrialization campaign met with little organized resistance and was carried further than in Norway. At the same time, the domestic commodity form of production survived, in part because no alternatives existed, in part because it was protected by the UI system. The result was a dualist structure, where Fordist and traditional structures coexisted.

This difference in sector structures helps explain the different developments in the number of fishermen in Atlantic Canada and Norway in the postwar period that was pointed out in the introduction to this chapter. The transition from domestic commodity production to a full-time, petty capitalist form of production proved relatively easy within the Norwegian sector system. First, it represented a relatively small step that, unlike becoming employees in Fordist firms, did not require relocation or drastic shifts in lifestyles. Second, supporting systems like the Raw Fish Act and the Main Agreement encouraged such transitions, and allowed them to take place in small steps over a long period of time. Third, the fishermen's own organizations encouraged this development. The result was a relatively smooth transformation of the sector. At the same time, there was a strong increase in efficiency on the basis of a small-scale and decentralized production structure and an equally strong decrease in the number of fishermen.

In contrast, the industrialization strategy of Canada led to a more constrained modernization process. On the one hand, Fordist firms did become dominant, and allowed a large number of people to leave domestic commodity production for wage work. On the other hand, many people within the fishing communities did not get this opportunity or did not want it. With the Canadian UI system, which, in sharp contrast to the Norwegian subsidy system, supported traditional forms of domestic commodity production, the transition to modern forms of production was restrained. The result was a lower rate of rationalization and a slower reduction in the number of fishermen than in Norway.

Another difference between the fisheries of Norway and Canada pertains to the characteristic responses to fluctuations in resources. While both sectors have been heavily exposed to such fluctuations during the postwar period, they seem to have created more and deeper crises in Canada than in Norway. Bill Morrow, the chairman of National Sea Products, once suggested that there was a seven-year boom and bust cycle in the east coast fisheries (Kimber, 1989: 288). More generally, the history of Canadian fisheries in this century is one of recurrent crisis; of government commissions and task forces following each other in rapid succession; of a never-ending need for large-scale government bail-outs (Lamson and Hansen, 1984). The history of Norwegian fisheries also can be written as a history of crises, but they have been fewer in number and less drastic in consequences than the crises in Canada.

While not denying the possibility that Canadian fisheries may have been exposed to more severe fluctuations than Norwegian because of differences in ecological factors and market structures, we concentrate here on differences in the buffering capacity of the two sector systems.

First, differences in size between the economic actors in Norwegian and

Canadian fisheries are significant in this context. In Canada the large-scale, vertically integrated firms play a much larger role than in Norway (Apostle and Jentoft, 1991). An important argument in favour of the Fordist model had been its capacity to deal with the fluctuations of the fisheries. On the resource side, the trawlers would secure a stable supply of fish, since they could go out and find the fish in all but extreme weather conditions. On the market side, integration into the retail sector, coupled with systematic marketing efforts, would stabilize demand. The remaining variability would be buffered by way of freezing. In this way, the large capital cost of the industrial model would be offset by a large and stable throughput. This argument turned out to be misconstrued. In practice, neither the seasonality of supplies nor market fluctuations could be contained. Instead of efficient mass production and low unit costs, the freezing industry developed massive overcapacity. The introduction of the Fordist model exacerbated rather than solved the instability problem, since it replaced traditional flexibility with the rigidity of the large-scale firm. Thus, for example, Apostle and Barrett (1992) have shown that in the Nova Scotian fishing industry, the intermediate-sized firms have been much more successful in coping with resource and market fluctuations than the large-scale vertically integrated firms. One important reason for this success is the smaller firms' greater flexibility, which remains a crucial requirement in unstable sectors like the fishery. If nothing else, the rejection of the Fordist forms in Norway meant that some of the capital costs and inflexibility of the industrial model were avoided.

Another important difference between Norwegian and Canadian fisheries pertained to the systems for managing fluctuations at the sectoral level. In Norway the Main Agreement negotiations represented such a system. Here, representatives from the sector and the government met annually to negotiate the amount and distribution of subsidies that, given the prevailing resource and market situation, would be necessary to stabilize incomes. In Canada, no comparable mechanism for buffering the sector from fluctuations existed. Instead, the organizational apparatus required to handle the situation has had to be constructed from scratch every time market prices have dropped significantly, or the fisheries have failed. But the lack of routinization also meant that assistance from the government, while occurring often, could only be legitimized if it was seen as a response to an exceptional event, in other words, a crisis. Hence, the Canadian fishery was more exposed to crises than the Norwegian because it lacked permanent mechanisms for buffering the sector from fluctuations in the external environment.

The economic difference between these two ways of handling fluctuations may not necessarily have been large. In both Norway and Canada the transfers of government money to the fisheries have been substantial over the years (Crowley et al., 1993; Holm, 1991). However, when such transfers were negotiated within permanent organizational structures, as in Norway, it lent orderliness and normality to the situation. In contrast, the downturns of Canadian fisheries could only be handled if they could be recast as 'crises,' with the intense political and social mobilization that high levels of conflict and polarization that implied.

3

The Resource Management Revolution and Market-Based Responses

As we have seen in the previous chapters, the fishing industries on both sides of the Atlantic have always had to cope with problems caused by fluctuations in both resources and markets. This has led to the development of various strategies and institutions to buffer the industry from whatever the current problem was. In the first third of the twentieth century, most of the problems originated on the market side. With technological improvements and changes in the industrial form of the industry, however, harvesting capacity has increased enormously, making resource problems more dominant in the latter part of the century.

This shift in the source of problems facing the fisheries also has had an institutional counterpart. In this chapter, we explain how the institutionalization of resource management within the fishery came about. In our terminology, institutionalization is the process by which social power is allocated to the protection of some value or concern (Stinchcombe, 1968). The mechanism by which social power is allocated to this value is primarily organizational: money has been redirected to the task; organizations have been constructed to attend to the host of practical work involved; negotiation systems have been set up to engage the interested parties and have the necessary decisions made. As we shall see below, the institutionalization of resource management has led to a fundamental reconstruction of Norwegian and Canadian fishery systems. A very large share of the organizational resources within the sector have been redirected from other concerns to management-related issues. In the following we describe the process that made successful institutionalization of resource management possible. The key to this process was the construction of a scientific model of fishing that not only gave practical guidelines for management, but also supplied compelling reasons why science-based fisheries management had to be a state responsibility.

The Resource Management Model

Modern fisheries resource management was premised on the possibility of creating a theoretical model representing the fishery based on biological relationships. The model made two interconnected assumptions. The first was the idea that human activity is a primary determinant of the state of fish stocks. This idea is fairly recent. In the nineteenth century, the prevailing view was that human activities would have no effect on marine ecosystems, one way or the other. As T.H. Huxley put it in 1883, 'The cod fishery, the herring fishery, the pilchard fishery, the mackerel fishery, and probably all the great sea fisheries, are inexhaustible: that is to say that nothing we do seriously affects the number of fish. And any attempt to regulate these fisheries seems consequently, from the nature of the case, to be useless' (quoted in Gordon, 1954: 11).

Although the usefulness of regulating the fisheries was the subject of heated debates (Burkenroad, 1953; Smith, 1994), the management idea gradually won acceptance. Two events were of particular importance here. One was the regulation of the Pacific halibut fishery during the 1930s. The stock, which had been depleted, seemed to respond positively to the fixed-catch limit imposed under an agreement between Canada and the United States (Thompson, 1950). Another experience of paradigmatic importance for the development of modern fisheries management was the perceived effects on the North Sea demersal fishery of the two world wars, or, in the scientists' words, the two great fishing experiments (Smith, 1994). The larger catches in the immediate postwar years suggested that lower fishing efforts, in this case imposed by the war conditions, would produce larger yields (Beverton and Holt, 1957: 24). Together with the Pacific halibut case, the war 'experiments' were accepted as evidence that human activities could negatively affect marine fish stocks, and, equally important, that fish stocks would respond positively to restrictions on fishing efforts (Smith, 1994).

The second assumption on which fisheries management rested concerned the feasibility of quantifying the human impact on fish stocks. Even if human activities could affect marine ecosystems, this was of little significance as long as the impact was unpredictable and uncertain. Although the scientific theory of fisheries management has many contributors, the 1957 volume by Beverton and Holt was the seminal work in the literature (Larkin, 1977: 3). Fish stocks were viewed as systems, the state of which was determined by the relationship between reproduction, growth, natural mortality, and fishing mortality. A key assumption of the model was that fish stocks are reasonably stable and behave predictably under moderate levels of exploitation. 'One essential aspect of this synthesis is the recognition of a fish population or community of populations as a self-maintaining *open system*, exchanging materials with the environment and usually tending towards a steady state (Beverton and Holt, 1957: 23; emphasis in original).

Today, Beverton and Holt's model may seem simplistic with its single-minded focus on fishing effort and its implicit rejection of ecological interactions and environmental changes as major determinants of the status of fish stocks. Their model nevertheless was a symbolic representation of fish stocks that was more than adequate at this stage in the development of fisheries management. The major obstacle for management at the time was not the model's lack of sophistication, but that the model treated fishermen as an external factor. This would not have been a problem if fishermen had automatically followed the scientists' advice, reducing their efforts to the level where the yield would be optimal. Such was not the case, however. As Beverton complained in 1952, 'It is ... clear that factors exist in a fishery without external regulation, particularly the fundamental competitive element in fishing, which make it difficult if not impossible for the industries to regulate themselves to obtain a more favourable balance (Beverton, 1952: 1; quoted in Smith, 1994: 323).'

Thus, fisheries management required a model of the behaviour of fishermen as well as a model of how fishing would affect fish stocks. Although fishery science developed in direct response to social and economic problems, fishery biologists had been hesitant to include human behaviour in their models. The completion of the fisheries management model was therefore undertaken with crucial assistance from the economics profession. The result was the classic bioeconomic model, which was essentially developed by Gordon (1953, 1954).

The bioeconomic models came with specific recommendations of how the fisheries should be regulated. They were also very important in persuading governments that management could yield social gains and should therefore be a state responsibility. According to these models, a fishery left to itself will always be suboptimal. The combination of free access and individual rationality leads to overexploitation and low levels of income for fishermen. Achieving optimal social benefits in the fishery thus requires external intervention. This point was later forcefully expressed by Garrett Hardin (1968). To avoid 'the tragedy of the commons,' fish, like all common property resources, must be managed. Left on their own, rational actors are 'locked into a system' that compels them to increase their effort, even when they know that they are on a path that leads towards depletion of resources. In practice, the tragedy could be avoided by allocating state power behind explicit management, as prescribed by the model. Resource management thus builds on a partnership between science and the state. Science must establish the facts: How large are the stocks? Which level of effort will give the optimum return? The state must, besides funding science, regulate the fishermen's activities and prevent them from destroying their economic basis.

In the fisheries however, a main problem for the establishment of a resource management system prior to 1977 was exactly the lack of state power. Within Mare Liberum the state did not control access to international fishing grounds. The anarchy of international relations barred the state from taking on the role as responsible resource manager and forced it into the role as self-interested guardian of domestic industry. Thus, in the fisheries the tragedy of the commons played itself out on two levels: among the fishermen on the fishing grounds, and among the states around the negotiation tables in international fisheries organizations. The resource management reform was thus directly premised on a change in the international regime, such that management powers would be invested in the state. The resource management model, and hence the new oceans regime which was based on it, were intimately tied to the promise of science to provide knowledge and the state to provide power (Dahmani, 1987; Sanger, 1987). In this way, the institutionalization of resource management forged virtually unbreakable bonds between science and the state.

The Establishment of Sovereign Rights

The resource management regime was the start of a new role for the government in the fisheries, not only in Canada and Norway, but in other countries as well. The introduction of extended fisheries jurisdiction gave the state a much larger say in how the fishing industry was run. The fisheries became heavily regulated. Rules were imposed regarding sizes of boats, types of gear, fishing days, and many other aspects of fishing. The resource management model made it clear that fishermen, left to their own devices, would not be able to maintain a sustainable fishery. Worse, the government would lose valuable resource rents from the fishery, and fishermen would lose income. It seemed obvious at the time that the government needed to step in to save the fishermen from themselves, from behaviour that was rational on an individual level, but would lead to collective disaster on a communal level. There was great faith in the ability of science and government control to design a well-regulated fishery that would no longer need government intervention or financial support during downturns in either markets or stocks.

Although fisheries had been subjected to various regulations earlier, the regulations had applied only to fish stocks that remained well within the established national jurisdiction, or, if they ventured outside, for some reason or other did not attract international competition. Thus, the absence of state authority outside the territorial sea limits was the main impediment to implementing resource management regimes prior to the establishment of the new oceans regime. The change of regime did more than extend the reach of the state, however. The ratification of the United Nations Convention on the Law of the Sea (UNCLOS) also supplied a detailed management model, complete with objectives, basic organizational designs, regulatory instruments, scientific expertise, and legitimating accounts. In this way, the new oceans regime not only identified the management goal, it also came with a methodology for achieving it.

The new oceans regime made it possible to manage fish stocks by turning them into state property. But the mere declaration of a 200-mile exclusive economic zone (EEZ) was only the first step towards this end. In addition to formal rights, ownership must be based on actual control over the object in question (Agarwal, 1994). If state property rights in fish were to become more than an empty principle, practical mechanisms to determine which fish belonged to which nation state had to be established. The new oceans regime is based on territorial claims, the limits of which only occasionally correspond to stock abundances. To implement the management model, such territorial rights had to be translated into property rights in fish. For fish stocks remaining completely inside the 200-mile zone of one state, establishing and enforcing state ownership is easy. It suffices to prevent foreign fishing vessels from entering the fishing grounds in question. Things get more complicated for fish stocks that are shared by two or more states. In that case, ownership rights can be established as shares of a stock, for instance, on the basis of some rules of thumb as to how much of the stock that 'belongs' to the territory of each state in question.

The function of the institutional system that developed as a result of the new oceans regime was to construct and fortify state property rights in fish. The state's territorial rights were to be transformed into enforceable quota allocations for the industry. In practice, this process takes a great deal of work. Scientific surveys must be undertaken; biological data must be collected and analysed; agreements must be negotiated among the nation states involved; acceptance of the regulations has to be secured from industry representatives and other interested parties; the fishery must be monitored, the quotas must be enforced, and transgressions punished. All this requires considerable resources.

Limiting Access

The 200-mile EEZs, established by Canada and Norway as well as other nations in 1977, only meant that the state took formal responsibility for resource management in the fisheries. To be able to act as a resource manager in practice required much more. The first step in this process was, as we saw, the establishment of state property rights in fish. This was accomplished by limiting access by foreigners. The next, and much slower, more difficult, and conflictual step concerned limiting access domestically.

To a large extent, domestic access controls could be built directly on the basis of the institutional apparatus that had been constructed to enforce state property rights against foreigners. Enforcing a national quota would, at a minimum, require a system for shutting down the domestic fishery when the quota had been caught. External and internal closure were thus two sides of the same coin. In practice, it took a long time until the national quota limitations became effective. In the Barents Sea cod fishery, for instance, the first total allowable catch (TAC) was established in 1975, before the regime change. At this time, there was little concern for the status of the cod stock. The quota was generous and did not lead to substantial restrictions on fishing efforts (Sagdahl, 1992: 28). During the next few years, however, the new management regime was put to the test when the International Committee for the Exploration of the Sea (ICES) started worrying that the cod stock was in decline and recommended sharp quota cuts. Gradually, the quota limitations imposed real restrictions on the fishery. At first such restrictions were made binding only for the offshore trawler fishery. No serious attempts at restricting the coastal fishermen's efforts were made. The coast fishermen were allowed to continue fishing even after the Norwegian total quota had been caught. In 1981, for instance, this resulted in an overfishing of 100,000 tonnes. Until the end of the 1980s the most significant restrictions on the coastal fishery were periodic closures. This measure was first applied in 1980, when there was a one-week stoppage of the Lofoten fishery. From 1983 a maximum vessel quota was set in the coastal fishery. This was fairly large (400 tonnes), however, and had no consequences for most vessels (Sagdahl, 1992). Only with the drastic quota cuts at the end of the 1980s were the coastal vessels subject to a strict quota regime, and they had to stop fishing when the quota was caught.

The same process of gradual tightening of quota restrictions took place in Atlantic Canada. From 1977 the allocation of the TACs within the Canadian sector was dealt with within the framework of the annual Groundfish Management Plan. Here, the TACs were shared out among different groups of fishermen, identified by criteria such as size of vessel, type of gear, and geographic location. The allocation process was acrimonious from the start, since the sum of claims from the groups involved exceeded the quantity of fish available. As in Norway the regulations were different for the offshore and inshore fleets. The offshore sector was put under a quota regime, implying strict catch restrictions. In contrast, the inshore fleet was regulated under an allowance system which did not impose serious limitations on effort, since, first, the allowances were calculated on the basis of expected average catch, and, second, fishing was not stopped when the limit was exceeded (Parsons, 1993b: 126). As in the Norwegian case, this led to TAC overruns, and the inshore fleet was put under a quota system in 1981.

As this indicates, it has been more difficult to implement quota restrictions in inshore than in offshore fisheries. There are several reasons for this. The underdeveloped formal administrative structures in the inshore fishery created communication problems for bureaucratic management agencies. Together with the larger number of inshore vessels, this meant that the management and enforcement costs within this sector tended to be much higher than in the offshore sector. In addition, there was a long-standing tradition of open access to the fisheries. In the inshore sector, this tradition tended to be interpreted as part of the cultural heritage of coastal people, and therefore as a positive and inalienable right. Representatives for the inshore sector also pointed out that the major expansion in the fisheries occurred in the offshore fleet. This fleet segment was therefore responsible for the overexploitation, it was argued, and should pay the price. All these factors help explain the asymmetry between offshore and inshore regulations.

The gradual development of the national quota from theory into practice paralleled the development of an increasingly finely tuned system for allocating fishing rights among domestic fishermen. In Atlantic Canada, for instance, the implementation of a quota regime in the inshore sector in 1981 immediately led to the establishment of 'sector management.' Under this system, vessels under 65 feet were regulated within three separate geographic areas: Newfoundland and Labrador, Gulf of St Lawrence, and the Scotian Shelf–Georges Bank. Movements between areas were restricted. In 1983 the development of a stable allocation pattern continued, as the enterprise allocation (EA) system introduced a long-term division of TACs between offshore and inshore fleets (Parsons, 1993b: 137–9). Under this system the vertically integrated processing companies were granted individual quotas for their offshore fleets. That year an individual quota system was also tried out in the western Newfoundland inshore trawl fishery. Around 100 inshore vessels fishing cod and shrimp were put under a three-year individual quota (IQ) program. The system was based on non-transferable vessel quotas, calculated on the basis of average landings within fleet categories, and vessel size. It was regarded as a success, and extended in 1988 to the midshore dragger fleet throughout the Atlantic region, and then again in 1989 to include the Gulf of St Lawrence small groundfish dragger fleet. In the meantime, IQs had been introduced in the offshore lobster fishery from 1985, in the offshore scallop fishery from 1986, and in the offshore clam, northern shrimp, swordfish, and tuna fisheries from 1987. By 1991, when the groundfish dragger fleet under 65 feet in the Scotia–Fundy region also was included, the IQ system comprised all but the smallest vessels in the fixed-gear fleet (Parsons, 1993b).

The introduction of the IQ regime in Atlantic Canada was a step-by-step process that went on for almost ten years. In Norway the same process happened in two large jumps. The first occurred in 1978 when individual vessel quotas were introduced in the offshore trawler fleet. The second occurred in 1990 when an IQ regime was introduced in the coastal fleet as a direct response to the resource crisis. This system was two-tiered. On the one hand, the vessels that in one of the three previous years had landed more than a certain amount of cod, the exact requirement increasing with the vessel size, received an individual vessel quota. On the other hand, the vessels that had not landed enough to qualify for a vessel quota, could fish on a group quota. In practice, the quantity each of these vessels could fish was restricted partly by the competition with the other vessels in the group for a share before the whole quota was taken, partly by a maximum quota for each vessel. This quota was 2.5 tonnes for vessels below 12 metres and 3.5 tonnes for vessels longer than 12 metres. Of around 9,000 vessels that came under consideration, 3,300 were awarded a vessel quota, and the rest fished competitively under the group quota.

Quota regulations place restrictions on how much fish can be taken out of a stock. This is not the only way to constrain fishing effort, however. A fishery can also be regulated by controlling the number of fishermen and vessels that are allowed to participate. In practice, this can be done by way of licensing schemes. In both Norway and Canada the development of increasingly elaborate quota arrangements, as described above, paralleled and was reinforced by the development of increasingly elaborate licensing arrangements.

In Norway the Ownership Act, originally designed to exclude all but active fishermen from owning fishing vessels, was thus modified in 1967, allowing the ministry to stop the building of new vessels or to refuse access to new vessels into the fishery (Mikalsen, 1987). This change in the act must be understood against the background of the development in the herring fishery, where it was fairly obvious that the harvesting capacity was out of all proportion to the size of the stock. When the Participation Act replaced the Ownership Act in 1972, the idea of using the licensing mechanism to control access and prevent overcapacity was further strengthened. From then on, all fishing vessels over 50 feet had to be registered and approved by the Ministry of Fisheries. In accordance with the act, licensing systems were established for purse seine, trawl, and danish seine. The resource management role of the licensing system was most obvious in the case of purse seining, where the adjustment of harvesting capacity to stock sizes was given as an explicit purpose for the introduction of licensing.

In the Atlantic Canadian fisheries, limited entry was first introduced in 1967, in the inshore lobster fishery. Four years later, in 1971, it was extended to offshore lobster and scallop fisheries as well as to herring purse seiners. In the Atlantic groundfish fishery, movement towards limited entry began in 1973, when the Minister of Fisheries announced a new fishing fleet development policy for Canada's Atlantic coast, aiming to 'match fleet size to fish stocks by instituting a more selective subsidy program for vessel construction and establishing a new license control program' (Government of Canada, 1973; cited in Parsons, 1993b: 176). In accordance with this program, a freeze was introduced on offshore trawl licences from 1973–4. From then on, new vessels were allowed only as replacements of existing vessels. In the inshore sector, barriers to entry were gradually raised starting in 1976. Until the end of the 1970s, however, the licensing system in the Atlantic groundfish fisheries developed in a fragmented and piecemeal manner, and was described as 'complicating and confusing and not easily understood by those in the fishing industry or even those administrating the system' (Levelton, 1979, cited in Parsons, 1993b: 176). After a comprehensive review of the existing system, a unified licensing framework was introduced. By 1982 all Atlantic fisheries had been placed under limited entry.

In both Norway and Canada the primary focus of licensing policy has been on access requirements for vessels: their size, which stocks they could fish, which gear they could apply. But licensing arrangements have also regulated the access of fishermen. Who had legitimate claims to go fishing, who had not? Who should count as a bona fide fisherman? Along with the development of increasingly elaborate regulations of the fishery itself, the criteria that must be met in order to qualify as a fisherman have been tightened. In Norway the most important legal mechanism is the Fishermen Registry (Fiskarmanntallet), first established by the 1908 Accident Insurance Act for Fishermen. In the registry, two types of fishermen are defined according to level of activity. The part-timers are registered on Sheet A, the bona fide fishermen on Sheet B (Ørebech, 1984).

Originally, the Fishermen Registry was just that: a mechanism for registering those who actually were fishing. Gradually it has become an important factor in regulating access to the fishery. One step was taken with the Ownership Act, where status as an active fisherman was required to become the owner of a fishing vessel. Another important step was taken during the industrialization campaign of the 1950s and 1960s, when subsidies and other economic measures were made contingent on the fisherman's status, as established by the registry. In line with this trend, the Ministry of Fisheries in 1992 proposed to establish a new Sheet C in the fisherman registry. Only C-registered fishermen would be allowed to own fishing vessels. To qualify, a fisherman would have to be engaged in fishing for at least three of the previous five years (Government of Norway, 1991–1992: 91). In addition, the ministry suggested that formal education would be a relevant criterion for registration on Sheet C. During the debate in Parliament on the proposal, the Fisheries Committee stated, 'It is important that those who are permitted to acquire a fishing boat first have education relevant to the profession of fishing, preferably both theoretical and practical' (Government of Norway, 1994: 13). Although no formal requirements to this end have been adopted, it is about to be a practical possibility through the establishment of a fishing course ('the blue line') in high school. The same tendency is apparent in Atlantic Canada, for instance, in the Cashin Task Force's idea of a more 'professional' fisherman (Cashin, 1993: 67–72). As in Norway, professionalization in practice means that formal training and exams could count as entry tickets to the fishery.

Regulation and Enforcement

The implementation of the resource management model in Norway and Canada took the form of an increasingly complex system of rules and regulations. Once the fish quotas have been determined according to scientific procedures, with modifications during the domestic and international negotiation process, they are distributed in specific patterns among groups of certified fishermen. Modern fisheries management thus builds on detailed systems of rules defining how much fish can be taken, when and where fishing is allowed, which type of vessel and gear are allowed, and who count as fishermen. For resource management to work, such rules have to be effective guides in the practical realities of the fishery. To a certain extent this can be ensured by explicit enforcement. Taking on the role as resource manager not only required the state to set up a system that could produce and allocate fishing rights. Substantial societal resources also had to be allocated to the enforcement of these regulations.

On both sides of the North Atlantic, the decision to strengthen fisheries surveillance and control and the decision to establish the EEZ came as parts of the same package. If the claims to exclusive state jurisdiction to the vast ocean grounds were to be effective, they had to be policed. In Norway the government thus proposed to establish a Coast Guard based in northern Norway under the administration of the Ministry of Defence (Government of Norway, 1976). The proposal was followed up in 1980, when the Coast Guard was established by government decree. Three ice-classed, helicopter-carrying vessels were built. In addition, the Coast Guard took over one vessel from the old Fishing Patrol and chartered four others (Hønneland, 1993: 64). In Canada the preparations for extended fisheries jurisdiction began in 1976, when the government set forth a five-year plan for increased surveillance. During this period an additional $4 million was allocated to offshore surveillance annually. This would more than double aircraft and ship time in offshore surveillance and allow foreign vessels to be boarded at least four times and domestic vessels at least two times a year (Parsons, 1993b: 630). Unlike Norway, however, enforcement was performed by the boats and officers of the Department of Fisheries and Oceans (DFO), not the Canadian Coast Guard's. As a result of Program Review, a budget-cutting process in Canada, the Canadian Coast Guard became part of the Department of Fisheries and Oceans on 1 April 1995, although it took another year to work out the details of the transfer from Transport Canada to DFO.

Just as we have seen for the other management instruments, enforcement tasks have changed over time. At the outset, the main task of the Coast Guard was to secure national territories and quotas from foreign trespassers. As the management system gradually extended its reach into domestic fisheries, the focus of the enforcement agencies also shifted inward. In Canada, for instance, the observer program, which started up in 1977, at first targeted mainly foreign vessels. Over time, however, it was extended to include also the Canadian offshore fleet. In 1987–8, the program in principle covered all vessels larger than 45 feet, but concentrated on vessels above 100 feet. With the fisheries crisis of the 1990s, resources have been provided to achieve 100 per cent coverage on the northern cod fishery. In addition inshore/nearshore patrols have been established in the Atlantic region and dockside monitoring programs have been established to ensure compliance within the fisheries regulated under individual vessel quota (IVG) systems.

In Norway the introduction of the IVQ regime in 1990 drastically increased enforcement needs. From then on it was necessary to keep track of actual catches as compared with quota allocations for each individual vessel. To handle this vastly more demanding task, the enforcement system was strengthened on three points. First, the Coast Guard has received more resources. Second, the Control Section of the Fisheries Directorate, which traditionally has been responsible for product quality control, is now heavily involved in quota controls, particularly in connection with landing. Third, the fishermen's sales organizations are engaged in quota control. Their system for registering landings by individual vessels, originally developed to control prices and other transaction terms, could easily be converted into a system for quota control (Hallenstvedt, 1993). In addition, and paralleling the development in Canada, nearshore/inshore fishery patrols were established from 1996.

Transforming Exogenous Constraints into Endogenous Values

Explicit surveillance and control, coupled with threats of or actual legal sanctions, go a long way in ensuring compliance. Such measures are quite costly, however, and they will be most efficiently applied if they can be concentrated on a small number of offenders. For any law or regulation it is therefore a huge advantage if those who are subject to it believe that it is necessary and just, and in accordance with their own interests. In the case of fisheries regulations in Norway as well as in Canada, this has been a key problem. Resource regulations have to a large extent been imposed from above. For many people in coastal communities, these restrictions are regarded not only as an illegitimate expropriation of their birthright. They are also perceived as outrageously unfair, since the majority of small-scale fishermen have been forced to pay for the sins of overexploitation committed by the few industrial offshore enterprises. It is thus no doubt that large groups within the industry regard the regulations as harassment, and they will break them if they can get away with it.

Large groups, particularly in the traditional coastal fisheries, have perceived resource management as an assault on their interests and way of life. However, some important groups have welcomed the change. These are primarily those who can meet the new demands and make the transition to fit into the new, more professional category of fishermen. For such groups, the resource management regime is a huge advantage. When national quota allocations have been made socially binding, they can be shared among the fishermen as tangible and stable commodities. Instead of the chance of a lucky catch, the fisherman can make his plans on the basis of a pre-established quota allocation. To the professional fisherman, the resource regulations therefore not only represent constraints but also rights and assets. The active support of this group is an important source of stability and legitimacy for the resource management institution. It is furthermore clear that the balance between hostile and supportive sector groups will tip in the latter's direction once the resource management regime is in place. This is not only because the professional fishermen, who are favoured by resource management, are generally more effective political actors than the part-time, multioccupational fishermen. In addition, the resource management reform tends to exclude the part-timers from the fisheries, and hence they gradually lose their access to sector-political arenas.

Another mechanism that helps improve the legitimacy of resource management is the domestic system of industry consultation in management decision-making. On both sides of the North Atlantic the fishermen are represented in the management process. In Norway the Fishermen's Association has five of fourteen members on the Regulatory Council. In Canada a much larger number of fishermen representatives are involved in the more pluralist Atlantic Advisory System. In any case, the fishermen have a voice when regulations are decided on. As Weber (1978) pointed out, the involved parties' acceptance of decisions made in such systems hinges, not on the degree to which the decision accommodate their interests, but on a pre-established endorsement of the procedures by which they are made. Since the fishermen, via their organizations and representatives, are involved in making the rules, they are themselves partly responsible for and bound by these rules.

Resource Management and Modernization

By the 1980s resource management regimes were in place in both Canada and Norway that were supposed to prevent overexploitation of fish stocks by setting quotas based on scientific evidence regarding sustainable harvesting levels and limited access to the fisheries. Through science, it was argued, natural decreases in stock levels could be detected early, in time for preventive action. Instead of letting stocks decrease even further from fishing, quotas could be cut in the hope of making the downturn smaller and thus helping the stocks to recover faster. At the beginning of the 1990s, however, scarcely fifteen years after the new oceans regime had been established, the fisheries of Norway and Atlantic Canada were in deep crises. The new regime had been legitimized on the promise of rational management and sustainable fisheries. This promise burst with these crises. The scientific estimates proved not to be as reliable as resource managers would have liked to believe. There was also an underestimation on the part of the resource managers of the political pressure to prevent quota cuts being made because of their impact on the incomes of fishermen. In the end, resource management had resulted in the same outcome as the earlier open-access regime, a collapse of fish stocks. This time, though, the decline in catches was controlled by institutional arrangements.

The responses to the resource crises in Norwegian and Canadian fisheries at the beginning of the 1990s, thus, can be seen as the success rather than the failure of resource management, as they were the first comprehensive demonstration of the tremendous amount of social power at its command. In contrast to earlier situations when fish stocks were threatened, the warning signs were taken seriously and acted on without hesitation. Drastic quota cuts, and even complete shutdowns, were implemented in spite of the massive economic and social hardships that followed. To the extent such hardships were deemed unacceptable, it was not the management regulations that gave way, but the constraints on government budgets. The idea of long-term stock management had become institutionalized. When the management requirements during the resource crises of the 1990s clashed with the centuries-old tradition of open access to fishing grounds, it was the latter that folded.

The resource management regime led to the creation of new institutions and the revamping of old institutions in the fisheries. Concomitantly, changes were also happening on the market side that would also affect the insitutional structure of the fisheries. In particular, there was a large shift in patterns of trade in fish products as a result of the new oceans regime and, later, there were changes in the structure of the world fish markets themselves that required a new institutional response. Thus, beginning in the late 1970s, there were both internal and external pressures that would have an enormous effect on the insitutional structures in Norway and Canada that had been built up since the 1930s.

Market-Based Responses

The introduction of extended fisheries jurisdication, almost by definition, brought about a change in the patterns of trade in fish products. In a very brief period, a large proportion of fish that had been 'domestic,' because it was sold on the domestic market by the home distant water fishing fleet, suddenly became traded fish as coastal nations moved quickly to establish their 200-mile EEZs and removed the distant water fleets from their waters. Coastal nations with relatively abundant fish stocks became major suppliers of fish on the world market. Countries that had traditionally fished around the world became importers of fish. There was thus an enormous increase in measured trade in fish products. For example, between 1978 and 1990 international trade in fish products grew by nearly 76 per cent (FAO 1994).

The imposition of the 200-mile EEZs changed the structure of world trade in fish products. Japan and the Soviet Union, the two nations with the largest distant water fleets, could no longer fish freely for their domestic markets, except in limited areas of the oceans. They had to rely on imports or joint ventures instead. European nations that had large offshore fleets to supply their domestic markets, such as Germany, the United Kingdom, and Spain, lost access to traditional fishing grounds and had to rely increasingly on imports to meet domestic demand. Countries such as Canada, Iceland, and Norway, which already had traditions of exporting fish products, became major players in supplying the world's demand for fish. For example, prior to 1992 when a fishing moratorium was imposed on many of the cod stocks off the east coast of Canada, Canada was the leading exporter of fish products, although it was only around sixteenth in the world in terms of landings. Norway dropped in terms of landings from fourth in 1977 to thirteenth in 1990, but was fourth in terms of exports. This primarily reflects the growth of Norway's salmon aquaculture industry.

Other coastal countries that had not played a large role in world markets before the establishment of extended fisheries jurisdiction, such as Thailand, New Zealand, and South American countries, have meanwhile become important suppliers of fish products. Extended fisheries jurisdiction thus led to a build-up of export sectors in the coastal nations.

Over time, as countries developed their export sectors, the interdependence of fish markets increased. Events in one regional market began to affect other regional markets as buyers looked around the world for sources of fish. Traditional species and products began to face competition from new species and products. These changes put pressure on existing institutions and marketing channels. These pressures had a particularly strong impact in Norway, where the institutions connected to the fisheries were more formally arranged. But even in Canada, traditional marketing channels were modified as a result of the new competition. Ties between buyers and sellers that had existed for several generations for the small companies began to unravel, while vertically integrated companies began to look for new markets and partners.

As a result, there was a shift in how firms operated in the market. Specialization of activities was no longer within a given plant, but within a given firm, and it might be spread all over the globe. Raw materials were not necessarily supplied by the local fishing fleets. For countries such as Canada and Norway, where an abundance of natural resources has provided the basis for much of their trade, this shift in the world economy has meant profound changes for the industries based on these natural resources. In particular, the creation of a global fish market has meant a weakening of the links between harvesting and processing, especially for the large, vertically integrated firms (Jønsson, n.d., p. 4), as it became cheaper for companies to source product elsewhere in the world, rather than harvest it locally with their own vessels, or to buy from the inshore fleet. The stock crises in the Barents Sea in 1990 and in the Northwest Atlantic in 1992 merely reinforced this structural change. With few or no domestic supplies of fish, the large processing companies, and even smaller companies, began to buy raw fish and intermediate inputs from around the world, rather than relying on domestic supplies. The result has been that the traditional North Atlantic suppliers of groundfish, particularly cod, have lost ground as the dominant suppliers of fish, especially in the commodity market. This would probably have occurred even without the stock crises, but it was exacerbated by the lack of fish on the part of the Norwegian, Canadian, and Icelandic suppliers, successively, in the early 1990s.

In the following, we describe in more detail how both of these processes, resource management and market changes, played out in the Norwegian and Canadian fisheries. While there are similarities between the two countries, the differences are more striking. Two factors were of particular importance. The first is the differences in institutional systems that were in place in each country before changes in the market started to make themselves felt. Because of the more complex corporatist system built up around the fish trade in Norway, the consequences of globalization meant a more drastic institutional transformation process there than in Canada. The second factor had to do with the deeper resource crisis in Atlantic Canada starting in the early 1990s. This forced the authorities, as well as industry actors, into much tougher choices regarding access to and allocation of fish resources than those faced in Norway.

Norway

In Norway during the 1980s and 1990s pressures were brought to bear on the carefully constructed corporatist institutional system that had been developed prior to the Second World War. Compared with Canada, as we shall see below, the changes that resulted were more visible and dramatic in Norway, since they started from a quite complex institutional system. On the other hand, the Norwegian fisheries did not come under such extreme pressure as in Canada, since Norway escaped relatively lightly from the resource crises that swept over the North Atlantic during the 1990s. Thus, while the qualitative gap from the established institutional order was larger in Norway, the pace and energy at which the new order was established was considerably less.

Chapter 2 described the sector system that dominated the Norwegian fishing industry in the postwar period. The reform process of the 1930s had been a political battle over the organization of fish trade. Because of the free-rider problem, reorganization of the fish trade had to be done through law. The reform process therefore centred on the construction and implementation of a set of acts regulating the fish trade. Three such acts were adopted, one for each of the main groups of actors within the fisheries. The 1938 Raw Fish Act regulated firsthand sales; the 1955 Fish Export Act organized the export trade; the 1970 Processor Act pertained to sales from processors to exporters. The three acts were variations on the same theme, as they all allowed the government to establish sales organizations with monopoly rights within narrowly defined market segments. Above we showed how the three acts fit into the sector system. The fishermen, exporters, and processors had all been allowed to establish mandated sales organizations (MSOs). The organizational pattern that resulted was different in the three subsectors, however. The reform process also had very different consequences with respect to the political effectiveness of the three groups. The MSO reform propelled the fishermen to a dominant position within the sector, but it left the merchants fragmented and weak.

The MSO system was dominated by the fishermen, and despite an impressive array of organizations, processors and exporters remained divided and weak. Although the government played a crucial role in the construction of the

MSO system, its control over the sector should not be exaggerated. On the contrary, the MSO system can be characterized as a co-management system, where the government went a long way in delegating tasks and responsibilities to sector groups, in particular to the fishermen. The MSO reform vastly improved the fishermen's capacity for political mobilization. In Chapter 2 we saw how the fishermen were able to subvert the industrialization project of the postwar period, and how they even managed to turn the Main Agreement into one of their sector-political strongholds. The MSO reform, then, not only gave the fishermen control over the fish trade, but it also provided an organizational basis for consolidating and improving their dominant position within the sector.

The Emerging Sector System

This system was partly dismantled, in the 1980s and 1990s, not as the result of a coordinated reform process, but, rather, as the result of several apparently independent episodes. First, the legal system regulating processing and exports was revised, reflecting a new reality in the way firms organized both fish processing and exports. A new Fish Export Act, adopted in 1990, cancelled the system of authorized exporter associations and replaced the eleven export councils with one. Firms engaged in exporting fish still need to be approved by the export council. The minimum requirements to qualify as an exporter are low, however, and once approved, a firm can export any kind of fish products. The new export council's responsibilities include collection and analysis of trade statistics, market information, generic marketing, administration of export stipends, and advising the Ministry of Fisheries on questions concerning fish exports. The new export legislation, together with the cancellation of the Processor Act in 1993, cleared away the barriers against common interest organizations among processors and exporters. As a direct result, the multitude of branch associations joined together in one single organization, the Federation of Norwegian Fishing Industry. All in all, the result is a much simpler organizational structure, giving processors and exporters more freedom to respond to market fluctuations and, concurrently, considerably more clout in sector politics.

Second, the legal system pertaining to the raw fish market was revised in several important ways. In the 1992 modification of the Raw Fish Act, the sales organizations lost their right to authorize and control fish buyers. That same year the general ban against foreign vessels landing fish in Norway was lifted through a modification of the Fishing Limit Act. Until then, such landings could be allowed by the ministry, but usually only if the sales organizations approved. Finally, the practice whereby the sales organizations charged a 1 per cent levy on all catches on behalf of the Fishermen's Association was banned. This was the association's major source of income, and a new funding system based on member fees had to be devised. Although the main principle of the Raw Fish Act, the sales organizations' monopoly rights in firsthand trade, is still in place, these changes clearly diluted the sales organizations' powers and severed the ties between the sales organizations and the Fishermen's Association.

Third, the Main Agreement has been relegated to obscurity. This happened with the adoption of the Free Trade Agreement on Fish within the European Free Trade Association (EFTA). The agreement, concluded in 1989, abolished all the tariffs and quantitative import restrictions on fish and fish products within the EFTA region from 1 July 1990. Its consequences for state subsidies in the fisheries are more important in this context, however. All subsidies not directed towards structural measures were cancelled by 31 December 1994. Income support, the main instrument within the Main Agreement system, was no longer allowed. Hence, roughly 70 per cent of the aid of the 1989 subsidy agreement would have been illegal according to the EFTA agreement. The agreement on a European Economic Area (EEA), reached between the European Union (EU) and EFTA in 1992, reconfirmed these principles.

These changes weakened the fishermen's power position within the sector. The strength of the Fishermen's Association is waning, both because the ties to the sales organizations have been severed and because the subsidy negotiations have lost importance. In addition, the association is strained by the internal conflicts of quota allocations, which it has been forced to take on in the Regulatory Council (see below). Meanwhile, the processors and exporters are gaining power in the nexus between the Fish Export Council and the Federation of Norwegian Fishing Industry. The legal mechanisms that prevented them from joining forces, the old Fish Export Act and the Processor Act, are gone. The arrangements that subordinated them in relation to the fishermen, the Main Agreement and the Raw Fish Act, have been weakened.

Thus, within just a few years, the MSO sector system, which had withstood the large-scale and systematic attack from the industrialists during the interwar period, crumbled. How can this be explained? We shall describe the forces behind the demise of the MSOs, concentrating first on external, then internal, factors.

The Demise of the MSO Sector System

External Pressures
An important reason for the destruction of the MSO system was a declining faith in the corporatist organizational principles on which it was built. This was not a phenomenon exclusive to the fishery. On the contrary, it was but one

manifestation of a general trend that during the 1980s swept through most western capitalist nations. Deregulation and privatization are at the centre of this trend; the state should be rolled back and the market principles given more space. The virtues of the market model are argued in the language of efficiency and rationality. 'Rolling back the state' is necessary because government bureaucracy is overloaded with tasks and responsibilities. The tendency towards globalization, which has stripped the state of important policy instruments, makes this even more urgent. At the same time, it is assumed that the remaining tasks can be carried out more efficiently by substituting market incentive structures for bureaucratic controls.

The present wave of reform parallels closely, albeit with opposite thrust, that surging through Western Europe during the 1930s. At that time, the economic crisis undermined the faith in free market arrangements. In Norwegian fisheries, as we have seen, this triggered a reform process based on the interest organization model, a reform that eventually grew into a full-fledged corporatist sector system. In much the same way, the present reform in Norwegian fisheries reflects a general crisis of legitimacy for the corporatist model. The market model, placing efficiency in the centre, has replaced the corporatist model as the guiding principle.

The fishery reform of the 1980s was also closely connected to the Norwegian adaptation to potential membership in the European Community. In the event that Norway became a member of the EU, and even in case of alternative ties, the Norwegian system of institutions simply was not acceptable. The EU institutions within the fisheries, based largely on free-market principles, supplied a fairly concrete model for the reform. Reference to the EU system was made at several points during the reform, notably when the Processor Act was canceled, and the subsidy system was changed. However, the modification of the Raw Fish Act also complies with the EU system. Furthermore, even though this part of the reform was accomplished before Norwegian membership in the EU was a realistic option, the new Fish Export Act clearly pre-empted adjustments that would have been required in any case.

At a more basic level of legitimacy, the reform seems to bring the institutional arrangements of the fishery more in line with standard expectations of how business and sector politics should be organized. An idea basic to corporatism is that there is a close relation between common concerns and the interests of sector groups. In the Norwegian fisheries this idea is reflected in the close ties between economic and political organizations. In the MSO system, the common interest of all fishermen was assumed. Although there was room for disagreement on details, you could not leave the organization or work for your interests through alternative organizations. Furthermore, the sector and the authorities also were assumed to share a common interest. The government delegated important tasks to sector groups, under the assumption that this will work out for the common good. The reform replaced this belief system with one that is considerably more distrustful towards special interests. Now it is assumed that interest organizations as a rule will work for their own goals and will, unless they are kept in check by the authorities, undermine the common good. Furthermore, there is no longer any need for government to support an interest organization. The basic market mechanism will do its work here. If the interests are important they will find ways to set up organizations and make themselves heard, without the aid of government. In any case, should the government influence the structure of interest organization? While the obvious answer to this question in the immediate postwar period had been 'yes,' by the 1980s it had become equally obvious that it was 'no.'

Internal Pressures
Several forces were converging to create pressure on the MSO system from the outside. These probably would not have been enough to set off major institutional reforms if the externally generated charges against the system could not have been tied to internal problems and interests. An important reason for the fishermen's ability to withstand the industrialization campaign in the postwar period had been exactly the internal conflicts among and weakness of the industrialists' clients within the sector: the fish processors and exporters. In contrast, the external pressures on the fisheries in the 1980s and 1990s catered to important internal groups and offered solutions to important problems in the fishery, problems that went unresolved within the MSO system.

Throughout the postwar period, a process of structural change was in progress. In part, this process was a strand of broader technological and economic developments in the economy. In part, it was spurred by specific events in the sector, in particular, the adoption of the Raw Fish Act and the Main Agreement. The MSO system thus improved conditions for investment and technological development in the fishing industry and led to an impressive structural development within the sector during the postwar period. Gradually, the number of fishermen and fish plant workers declined and the primary production units became fewer, but larger and more capital intensive. At the same time, both at the production and sector levels, the economic units were uncoupled from households and local communities and embedded in formal administrative structures.

These developments were initiated by the creation of the MSO system. At the same time, they gradually undermined the structural basis on which the MSO system was built. In processing and exports, for instance, the organizational system built on the situation in the interwar era, when there had been a high degree of specialization in branches, and, within each branch, processing and exports were conducted by separate firms. In the postwar period there were persistent tendencies towards horizontal integration across branches and vertical integration between processing and exports. Thus, the organizational structures established and protected by the Processor Act and the Fish Export Act imposed barriers and an organizational structure that were increasingly irrelevant at the firm level. A similar process of development occurred in the harvesting sector, where institutions like the Raw Fish Act and the Main Agreement catered to the petty-capitalist fisherman. From the beginning, these arrangements helped the traditional multioccupational fisherman, the fisher-farmer, into becoming a professional, petty-capitalist fisherman. When resource management turned, in theory at least, unstable fish stocks into predictable and controllable property, some of these petty capitalists were in a position to transform themselves into real capitalists. To do that, however, they also had to break free from the established institutional framework.

In this way, the introduction of a resource management model contributed to a long-term process of structural change that gradually transformed the MSO system from a supportive to a restraining construct for more and more actors, particularly in light of the changes in the international economic system. In addition, however, resource management played a much more direct role in the weakening of the MSO system. With the institutionalization of resource management, the gravitational point was moved from the axis between the Fishermen's Association and the government, to the outside of the industry, in the axis between science and the state. The weakening of established institutions happened not only because scarce organizational attention and resources were withdrawn from them and allocated to new concerns. In addition, there were sharp contradictions between the MSO system and institutionalized resource management. Even in a world of unlimited organizational resources, the former could not have been upheld along with the latter. For instance, within the MSO, the fishermen system was perceived as the weak party in need of protection; under the resource management system, it was portrayed as a destructive force that had to be disciplined. It was this redefinition of the fundamental situation of the fisheries that gradually undermined the legitimacy of the MSO system, particularly at the harvesting level.

The 1980 skirmish over the Main Agreement was an early sign of what was about to happen. Bjørn Brochmann, an official in the Ministry of Fisheries, provoked a storm of indignation and protest when he attacked the Main Agreement. From the perspective of bioeconomics, he argued that the fishery subsidies, rather than improving the efficiency and profitability of the fisheries, which were their formal purpose, contributed to overcapacity and thereby undermined that very goal. The logic was as follows: in the fishery, both market prices and resources will fluctuate. In good years, some of the profits will be converted into larger vessels and more efficient gear. In a poor year, however, the subsidies will allow the fishermen to survive economically. Thus, the Main Agreement had set off the fishery on a one-way street, heading relentlessly towards growing overcapacity and dependence on subsidies (Brochmann, 1980, 1981).

What Brochmann did was simply to apply the logic of the resource management model to an institution created within the MSO system. From that viewpoint, the Main Agreement in particular, but the entire MSO system more generally, was not only inefficient and irrational, but counterproductive. As a result of these viewpoints, Brochmann became immensely unpopular in the fishery sector. He was thoroughly battered, and left the sector a few years later. But his point of view did not disappear. Gradually, the logic of resource management undermined the established order. When the EU question was raised towards the end of the 1980s, the integrity of the MSO system was already lost, and virtually no resources could be mobilized in its defence.

Canada

The situation in Canada was much different from that in Norway. First, Canada did not have a similar elaborate institutional structure as Norway, so that the introduction of resource management did not lead to major institutional reforms. Second, there was no institutionalized division between processors and exporters, as in Norway. Most processors did their own exporting, although some small processors did sell through other processors. Third, because Canadian processors had operated in a more *laissez-faire* market situation and were, especially in the offshore sector, more vertically integrated than their Norwegian counterparts, the world movement towards trade liberalization and global markets did not represent as large a change for the Canadians as for the Norwegians.

Nevertheless, the resource crisis of the 1990s was more severe for Canada than for Norway, and it had come on the heels of two other 'crises,' including a financial crisis in the offshore sector in the early 1980s. These crises had put a large drain on the public purse in Canada, one that seemed to be never-ending. With increasing attention paid to deficit reduction and reducing government expenditures, and with science partially discredited by the apparently sudden and devasting collapse of groundfish stocks, the federal government began to pull back from its interventionist role in the fishery. In addition, the passage of the new Oceans Act in 1996 formalized a change in the policy direction of the federal Department of Fisheries and Oceans, from fisheries to the broader notion of oceans, a process that had been under way in the department for some time.

As we saw in Chapter 2, the Atlantic Canada fishery developed along two lines: an offshore sector, characterized by vertically integrated firms and large trawlers, and an inshore sector, characterized largely by owner-operators, selling fish both to independent processors and even to the offshore companies. The fishery, of course, was more complicated than this, as there was a generally lucrative shell fishery, and the inshore fishery was further split by type of gear and size of boat. In general, however, there was a division between the vertically integrated offshore sector and the smaller, owner-operated inshore sector.

Indeed, such a division was reinforced by the fisheries management regime itself. In the 1970s Minister of Fisheries Roméo Leblanc gave priority to the inshore sector in the allocations of the overall total allowable catch (TAC), and the inshore allocation was rebuilt until it was roughly 50 per cent of the TAC for all groundfish species. By Northwest Atlantic Fisheries Organization (NAFO) sector and species, the allocations were not 50–50 between inshore and offshore, but on an overall basis, the TAC was allocated roughly evenly. It should also be noted that the inshore covered a wide spectrum of fishing gear and size of boats that makes the inshore sector less homogeneous than the policy would imply.

A factor behind this was the perceived need to provide increased employment opportunities in areas of Canada with traditionally high unemployment rates. This division lasted until the mid-1980s, with only a few exceptions, such as northern cod, where the offshore was given the bulk of the TAC. The resource management regime as established in Atlantic Canada, then, reinforced the bifurcation of the groundfish fleet, between large, vertically integrated firms and smaller, owner-operated boats.

The setting of annual groundfish management plans, dividing the TAC between the inshore and offshore sectors, and then dividing up the inshore's share by stock, area, fleet, and type of gear, led to major battles over access to various stocks. The process by which stocks were allocated among the different fleet sectors is discussed in more detail in the next section.

Allocation became very political, with federal Cabinet ministers and provincial ministers brought into the fray. Unlike in Norway, where the fishermen spoke with a single voice in their negotiations with the government, in Canada, the groups were splintered, and only the offshore sector, because of the relatively small number of companies, was able to provide anything like a united front.

Much more than was the case in Norway, the internal and external factors affecting the fishing industry in Canada in the 1980s and 1990s were intertwined, as the Canadian industry had fewer buffers between it and external factors. Regional politics also meant an interventionist approach to downturns in the fishing sector. The political aspect meant, however, that fluctuations in markets and stocks had to become "crises" before there was any policy response. Hence, the lack of institutionalized measures to deal with fluctuations meant the Canadian policy response was more often ad hoc than the Norwegian case.

The Crises

The Canadian fishing industry faced two crises after the imposition of the 200–mile limit. The first, in the early 1980s, affected mainly the offshore sector and was the result of overexpansion in the face of historically high interest rates, high fuel prices, and falling demand, as a result of the worldwide recession induced by oil price increases. The second, in the early 1990s, was a more general crisis, induced by the unexpected collapse of the northern cod stocks, followed by stock collapses elsewhere in the Atlantic.

The federal government's response to the financial crisis of the offshore sector was quite interventionist, compared with other industries, or even with the inshore sector, which also had some financial difficulties. Rather than letting the offshore processing companies go bankrupt, a massive effort was undertaken by the government to save them. Although it was never explicitly mentioned in the public documents, a major reason for the intervention was to keep a bank that had lent heavily to the offshore companies from going under. A Task Force on the Atlantic Fisheries was set up under the direction of Michael Kirby with the mandate to recommend 'how to achieve and maintain a viable Atlantic fishing industry, with due consideration for the overall economic and social development of the Atlantic provinces' (Kirby, 1982: 363).

The Task Force on the Atlantic Fisheries produced a report and issued a number of recommendations. The report claimed to offer a 'restructuring' of the Atlantic fisheries and a new policy direction. In the end, however, there was little restructuring, except for the merger of some of the offshore companies. Seven vertically integrated companies were rolled into two: National Sea Products, based primarily in Nova Scotia, and Fishery Products International, based primarily in Newfoundland. Instead, the main result of the task force was a refinancing package for the offshore sector. The federal government bought shares in the two new companies and paid off their loans to the bank, although there was also private equity involved, particularly in the case of National Sea Products. Thus, the main rationale for the Task Force proved not to be a restructuring of the fisheries, but preservation of the offshore companies. Indeed, Kirby was very explicit on this point, after the fact: 'So we should understand ... that restructuring means the orderly reorganization and refinancing of a group of insolvent or nearly insolvent companies with the objective of creating new enterprises which have a good chance of long-term viability' (Kirby, 1984: 2). In effect, the resolution to the crisis proved to be one aimed at saving the Fordist solution the government had promoted in the 1950s and 1960s, rather than an actual change in the industrial structure of the fisheries.

This was quite different from Kirby's position before the report was published. Indeed, it is quite different from what is in the report itself, which argued that fundamental changes in structure were needed for a long-term solution to the recurrent crises of the fishing industry. That the solution to the 1980s crisis in the fisheries turned out to be one that preserved the Fordist organization of the fishery was not entirely the fault of the Task Force on the Atlantic Fisheries. Like every other fisheries task force before it, it had not been able to resolve the contradictory goals of establishing the fishery as a viable industry and maximizing employment and maintaining fishery-dependent communities without consideration of the costs (Shrank, et al., 1992).

There were two major sources of political pressure to prevent the offshore processing companies from closing. First, there was pressure from provincial governments, which emphasized the short-term effects on employment and which were adamantly opposed to any plant closures or other changes that might result in fewer people being employed in the fishing industry (Kirby, 1984). Second, there was a federal election in the middle of the refinancing negotiations. The Liberals were replaced by the Conservatives, who had a more *laissez-faire*, market-oriented view of industry and who were committed to decreasing the role of government in business. Both events led to the government retreating from its interventionist stance once it was clear that the refinancing of the offshore companies was not going to falter.

The parliamentary form of the federal government has meant it is often quite easy for overall national interests to be ignored in the face of regional pressures. This has occurred particularly when the fisheries minister has come from the Atlantic provinces. He thus has, in essence, a conflict of interest between his obligations to his constituency and his obligations as Minsiter of Fisheries.

In summary, then, the financial crisis of the early 1980s resulted in continued government support for the Fordist solution. The offshore sector was reorganized into two main companies, National Sea Products and Fishery Products International, along with about sixteen smaller companies that owned offshore vessels. There was little change in the inshore sector, and virtually no change in composition of the fleet. No effort had been made to reduce capacity, either in harvesting or processing. Indeed, one of the avowed objectives of the refinancing had been to prevent plant closures. Except that there now existed two vertically integrated companies rather than seven and enterprise allocations had been introduced in the offshore, little had changed.

Although the task force was credited with solving the 'current' crisis in the fisheries, in some ways, it was the markets that resolved the problems of the fishing sector in the 1980s. As the world began to move out of the recession in 1983, the markets picked up. Prices were high; landings were up, reflecting, so it was thought, the successful implementation of a well-designed resource management scheme. It was felt that the promise of extended fisheries jurisdiction and a scientifically based regime of resource management was finally being realized and that government intervention to save the offshore companies was a sound investment.

The fishing industry seemed well on its way to a viable position. In 1985 the Conservative government, which had committed itself to reducing the size of government, cut the budgets of a number of departments, including the Department of Fisheries and Oceans. In 1986, as a result, there was a significant reorganization of DFO, including the closure of its marketing branch and the Marine Ecology Laboratory in Halifax. Since the fishing sector finally appeared to be able to stand on its own feet, it was felt there was no longer a need for the same level or type of government services as had once been proved. There was also a shift in philosophy on the part of the government, from a somewhat paternalistic view that 'government knows best' to more willingness to let the 'markets' sort things out with minimal government intervention.

This was not to last, however. The second crisis, the collapse of northern codfish stocks, started in 1989, resulting in significant cuts in the offshore quotas, then to inshore quotas. By 1992 the situation was such that the scientists felt the only hope for northern cod stocks was a moratorium, which was put in place in July 1992. Originally, it was to be for two years, but it soon became obvious that it would have to be in place longer. It is quite likely it may still be in place beyond the turn of the century. In addition, other stocks were found to be in decline, and there were more quota cuts and fishery closures.

At the same time, the increasing globalization of the fish markets and changes in the supply situation of the United States meant that the usual market response to decreased Canadian supply, increased prices, did not materialize. The 'Americanization' of the Alaskan pollock stocks in 1989 meant that vast quantities of Alaskan pollock began to appear on the U.S. market, at prices considerably lower than those for cod and haddock. Farmed catfish also began entering the U.S. market in sizable quantities (Barnett, 1990). Further, other countries, such as New Zealand, Thailand, and Korea had entered the U.S.

market. Cod began facing competition from species it had not had to face before. The Canadian fishing industry for the first time was beginning to feel the effects of an increasingly global seafood market. Canadian cod was simply and easily replaced with alternative sources of white fish.

This time the government barely intervened in the industry. In the 1970s and early 1980s the main political objective of fisheries policies had been to keep processing plants open so as to maximize employment opportunities. In part, this policy was designed to give access to unemployment insurance to as many people as possible. Unemployment insurance was considered to be a part of the regular annual income stream of many fishermen and plant workers, particularly in areas where the fishery was highly seasonal. Now, at the end of the 1980s, mainly because of fiscal pressures, the federal government was no longer willing or, indeed, able to keep plants open at any cost. This reflected, in part, a shift in government philosophy, which relied more on market-based solutions to problems. Part of the shift in policy was also practical. The crisis of the 1980s was primarily financial, so it could be 'solved' through the infusion of money to the financially strapped companies. The crisis of the 1990s was more fundamental. It was not simply a question simply of helping companies with high debt loads to refinance so that they could stay open and provide jobs. With no supply of raw material, it would have been futile to insist plants stay open.

Assistance packages were put in place, but they were targetted at individuals, not companies. National Sea Products shrank to one plant. Fishery Products International, on the other hand, had foreseen the crisis, and positioned itself to source fish from other countries, thus becoming part of the global seafood market.

The crisis in the early 1980s had been financial, primarily affecting only the offshore segment of the fishing sector. In part because different constituencies had different agendas, it had been caused by an overly optimistic expansion that the political process had failed to check. In the end, short-run regional interests won out against long-term national interests. The crisis of the early 1990s was different. Once again, it was a stock crisis. This time, however, it was not simply a decrease in landings, but a complete disappearance of the stocks. Unlike the Norwegian case, where the stocks rebuilt within a couple of years, the collapse of northern cod and other groundfish stocks has no foreseeable end.

In the earlier crisis the offshore processors had had some clout with government, but by the stock crisis of the 1990s they were essentially ignored. In both crises, the fishermen had little power to influence policy, except to push for assistance packages. The government's response to both crises had offered few new solutions. What measures were proposed seemed to be driven more by a need to cut budgets than a coherent long-run strategy for building 'a fishery of the future.'

The impact on institutions was less obvious in the Canadian case than in the Norwegian case, mainly because there were fewer formal institutional structures to affect. Internal pressures from government budget cuts, together with an apparent failure of the resource management regime, led the Department of Fisheries and Oceans to pull back from active management of the fisheries. DFO had suffered significant budget cuts in 1995 and 1996 and found its mandate shifting from 'fisheries' to 'oceans.' DFO introduced individual quota systems, both transferable and non-transferable, and started 'partnership' fisheries. New allocation institutions, such as the Fisheries Resource Conservation Council (1993), were established. The inshore sector found itself moving from groundfish to shellfish. While National Sea Products had shrunk, Fishery Products International had expanded, not in terms of plants in Canada, but in terms of increasing its access to fish supplies from around the world. The increasing internationalization of seafood markets required new ways of operating in the market, and not everyone was able to adapt.

Drawing Things Together

In this chapter, we have looked at the process by which resource management was institutionalized and transformed from theory into reality. Until the new oceans regime allowed state power to be allocated behind the resource management model, the regulatory measures in the large ocean fisheries of Norway and Canada were primarily for the protection of fishermen, not fish.

In Norway, as we have seen, a comprehensive institutional system for the protection the fishermen's interests was constructed. In spite of the government-sponsored industrialization campaign in the postwar period, the weak and vulnerable fisherman in need of protection formed the image around which the whole sector system revolved. This image was built into the key institutions of the sector. The Fish Export Act protected the fisherman from unpredictable fish markets; the Raw Fish Act protected him from greedy fish buyers, the Trawler and Ownership Acts protected him from capitalist takeover, and the Main Agreement protected him from economic uncertainty in general.

In Canada the more unrestrained Fordist philosophy of the postwar period led to a removal of fishery regulations that had been established to protect the fishermen, notably the trawler restrictions. As we saw in Chapter 2, however, the image of the weak and vulnerable fisherman was institutionalized there too, particularly in Newfoundland, primarily in the guise of the 'social fishery' of the outports, as protected by the unemployment insurance system and the Newfoundland provincial government.

Ultimately, all attempts at regulation and management in the fishery are 'for the benefit of man, not fish' (Burkenroad, 1953). This does not mean that all management regimes are equal in purpose and consequence, however. The contrast between the traditional sector systems in Norway and Canada and the new resource management regimes is revealing. Where the former sought to protect the fishermen, the latter were designed to protect the fish. This was not only a shift of priorities that has rendered the fishermen's interests a secondary concern. The institutionalization of the resource management model has established the ideas of open-access and self-seeking actors, in short, the tragedy of the commons, as the authoritative account of the basic problem of the fishery. The fishermen have been recast from heroes to villains in the drama of the fisheries and must consequently be restrained rather than protected.

In both Norway and Canada the resource management revolution converged with the forces of globalization to create dramatic changes in the fisheries in the early 1990s. Figures 5 and 6 show the changing balance among the major actor groups in Norwegian and Canadian fisheries throughout the century.

In Norway the fisheries in the twentieth century have gone through two large transformations. The first was the MSO reform of the 1930s, by which the MSO system replaced the truck system. Under the truck system, the merchants had been the dominant sector group, exploiting the fishermen through their control over credits and fish markets. The state was not heavily involved in the fisheries, which were traditional, household based, and embedded in local communities.

The MSO reform turned the traditional power relation around and propelled the fishermen into a dominant position. This happened 'through law and organization,' as Hallenstvedt (1982) put it. The fishermen gained the upper hand because they were lent power by the state. As we have seen, the MSO system was set up to protect fishermen from the uncertainties of the market, from merchant exploitation, and from capitalist takeover. The image of the weak and vulnerable fisherman in need of protection was built into the sector structure, through core institutions like the 1936 Trawler Act, the 1938 Raw Fish Act, the 1954 Ownership Act, and the 1964 Main Agreement. The economic actor that was defined and protected in this system was a petty-capitalist fisherman.

As argued in Chapters 1 and 2, the MSO system was set up in sharp opposition to large-scale industrial capitalism. But it also enticed the fishermen out of their traditional subsistence economy, where fishing had been one element in a household-based economy. Despite this, the small-capitalist fishermen came

Figure 5. Sector reforms in Norwegian fisheries.

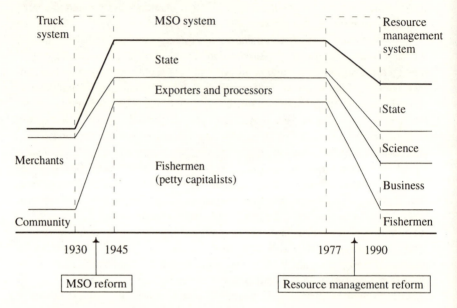

from and still remained embedded in local communities. The MSO system's commitment to community values is apparent in many ways. One illustration is the Main Agreement, which in its later phase was legitimized, to a large extent, on the grounds of regional policy, that is, the needs of the coastal communities (Holm, 1991; Jentoft and Mikalsen, 1987). Another example is the Raw Fish Act and the collectivist principles by which the raw fish market was organized. The community logic of the sales system is apparent in the way the sales organizations disciplined the processors. Under the MSO system, the processors were never allowed to develop into strong and autonomous firms based on capitalist principles. Instead, processing plants were regarded as landing stations, places where the fishermen could dispose of their catch. Since the fishing communities were scattered on a long coastline, and the fleet consisted of a large number of small and not very mobile vessels, the sales organizations – through their rights under the Raw Fish Act to set prices and authorize and control fishbuyers – sought to maintain decentralized landing facilities. In practice this meant that virtually every fishing community had its own processing plant, and most processors remained fairly small. Rather than a production facility for transforming raw fish into marketable products, the processing subsector within the MSO system was defined as a service facility for the fishermen.

Figure 6. Sector reforms in Canadian fisheries.

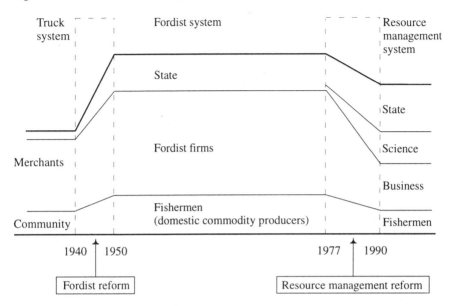

All this has changed dramatically since 1977 and the establishment of the 200-mile limit. Under the resource management system, fishermen were gradually redefined from heroes to villains. Whereas the MSO system protected fishermen, the resource management system protects the fish. This means that the state no longer trusts co-management principles of organization and that the practice of delegating power to the fishermen has been abandoned. Under resource management fishermen must be strictly disciplined. To this end, scientific institutions, bureaucratic organizations, and enforcement agencies have moved in to control the relationship between the fisherman and the fish. This relationship has been made explicit, formal, and controllable through quota arrangements, licences, rules of access, and resource rights; mechanisms that tend to disembed fishermen from the local communities and re-embed them in state and market structures.

While resource management has submitted fishermen to strict controls, processors and exporters have been let loose, escaping from fifty years of subservience within the MSO system. Playing on the rhetoric of globalization, European integration, and the importance of competitive advantage, processors and exporters are trying to reconstruct their sector from a service facility for the fishermen into a modern, high-tech, value-added food industry. This project has branched out in many directions. One example is the quality management reform that swept over the Norwegian fishing industry during the 1990s. Inspired by the Total Quality Management (TQM) philosophy, this reform represents a large-scale attempt to upgrade the formal management and control structures of the fish-processing firms, turning their focus from the sea towards the markets. Another example is the continued attacks from the fish-processing industry and their organizations on the relics of the MSO system. While these interests in the postwar period had a good case for complaints, they now have the necessary power base to make them count. A third example is the state subsidies that are channeled to encourage structural change in the fisheries. In the 1990s these subsidies are no longer channeled through the Main Agreement, which was developed precisely for this purpose. Instead, the structural measures are controlled by the newly established Regional Investment Fund, which is heavily committed to the large-scale industrial model (Holm and Mazany, 1995). In this way, the industrialization project, which failed when it confronted the MSO system in the postwar period, reappeared during the 1990s.

Figure 6 shows the development of the fisheries in Atlantic Canada in the same period. The starting point, the merchant-dominated truck system at the beginning of the twentieth century, was virtually identical with the situation at the western shores of the North Atlantic. While the Canadian fisheries have been exposed to much of the same problems and reform attempts, the outcomes have not been the same.

Many of the events of the 1920s and 1930s, that in Norway led to the MSO reform also appeared in the Maritimes and Newfoundland. In contrast to Norway, however, they did not lead to lasting organizational changes. Lacking a solid organizational basis for resistance, the fisheries were left open for the industrialization campaign during and after the Second World War. Hence, where the postwar period in Norwegian fisheries was dominated by an MSO system with the fishermen holding key positions, the result in Canada was a Fordist system with large-scale vertically integrated companies in the lead. The shift in power structure was greater in Norway, nevertheless, the breakthrough of Fordism in Canada did represent a discontinuity, transforming a commercialist order into an industrial capitalist one. In contrast to the total domination of the merchants within the truck system, the power of the capitalists within the Fordist era was somewhat curtailed. In part this was because of the tight integration between capital and the state, in part because the state also was committed to the protection of the traditional, community-based small-scale fishery, particularly through the unemployment insurance system. In contrast to Norway, where the MSO reform built on and reinforced a petty capitalist fishery, the result in Atlantic Canada was a dualist structure, where large-scale Fordist structures and traditional small-scale operations coexisted uneasily.

The difference in the power balance between Norwegian and Canadian fisheries during the postwar period also meant that the outcome of the reform processes of the 1980s and 1990s differed. In both Norway and Canada this reform was created as the introduction of resource management converged with the tendency towards globalization. In both settings the main focus of the government intervention in the fishery sector shifted from trade and industrial structures to resource management issues. While the state, aided by science, built up a heavy presence on the fishing grounds, it left the land-based part of the fisheries to take care of itself. In Norway this process led to a shift in the power balance between fishermen and the fish-processing industry, restraining the former and leaving the latter more scope in which to manoeuvre. In Canada, where industrial capitalism already had a privileged place, introduction of resource management meant new constraints also for the Fordist firms. Figures 5 and 6 also indicate the importance of the fisheries in society at large. Both the MSO reform in Norway and the Fordist reform in Canada propelled the fisheries into greater significance. This happened both because these reforms improved the technical efficiency and productivity of the fisheries, and because they led to greater political and economic integration of the sector in society. The first attempts at modernization followed somewhat different tracks in Norway and Canada, nevertheless, they both relied on a dual definition of fisheries – as a way of life as well as a means to income and profits. Without restraints on resource extraction, this led to overexpansion and depletion of fish stocks. Thus, the resource management revolution has not only led to restraints on resource utilization, but also to a lowered importance of the fisheries and of their status in the society as a whole.

The resource management reform and the globalization tendency suggest that the way ahead is towards completion of the modernization process, transforming the fisheries into a small but rational economic sector, disembedded from local communities and traditional coastal cultures. It is not, however, obvious that this would be the best way to utilize the fishery resources. Nor is it the inevitable outcome of a preset course of development. First, while the resource management model comes with a bias towards closure and further rationalization, it also comes with wide opportunities for political choice. This theme will be pursued in Section 2 of this book. Second, it is not obvious which type of industrial model stands the best chance in the globalized fish markets. The large-scale Fordist firms no doubt will retain their strong position in the commodity market for frozen white fish, but globalization has opened up possibilities for different types of actors and organizational models. In the high-quality luxury segments, for instance, local networks and flexible specialization may be an attractive solution. This theme will be pursued in Section 3. Finally, there is the question of whether the nexus of modern-type institutions and techniques of control will be able to contain the fluctuations, uncertainties, and chaos of the fisheries. In the Conclusion, we will return to this question.

SECTION 2
RESOURCE REGIMES – CO-MANAGING THE COMMONS?
THE POLITICS OF FISHERIES MANAGEMENT IN ATLANTIC
CANADA AND NORWAY

Introduction: Resources Regimes

The ineffectiveness and outright failures of management revealed in the crises of the 1980s and 1990s triggered an international debate on the proper role of government, science, and user groups in the management of marine resources. There is now widespread concern that national jurisdiction and centralized decision-making based on biological data and bioeconomic models may be neither capable nor sufficient for the conservation and enhancement of marine resources. The ensuing debate, to which social scientists have contributed (Jentoft and Kristoffersen, 1989; Lamson and Hanson, 1984; MacInnes and Davis, 1990; McCay and Jentoft, 1996), is part of a broader critique of explanatory models such as the tragedy of the commons. Models that are used to support reliance on centralized, science-based regulatory regimes to counter the effects of competitive, open-access use of common property resources often mask or weaken the capacity of decentralized, user-based systems to manage natural resources (McCay and Acheson 1987; Matthews 1993; Ostrom 1990).

The present section contributes by addressing how the efficacy of regulatory measures is influenced by political conditions and institutional design: by interest group politics, administrative structures, and the organizational framework within which management decisions are made and implemented. It focuses on fisheries management as the management of people rather than of fish and, therefore, as a political and social process where the scope and content of regulatory policies are shaped by the interaction of scientific advice, government initiatives, and demands by user groups. In this context 'fisheries management' refers to structures and activities pertaining directly to the choice and implementation of regulatory instruments – be they licences, quotas, closed seasons, or limits on technology. 'Management' includes the setting of objectives and policies, the provision of knowledge, and the articulation of interests, as well as the process of allocating licences and quotas.

We consider three main themes in our comparison of fisheries management institutions in Norway and Atlantic Canada. First is the question of what institutions exist, how they have evolved, and how they work. The focus is on management decision-making, institutional dynamics, and the politics of institutional design. Second, we examine recent trends in rethinking management policies and schemes, delineating, in particular, how the issue of market-based approaches is being handled in both countries. Third, we focus on the relationship between regime characteristics and management programs by discussing how institutions and procedures may have constrained or facilitated certain policies and decisions. Particular attention is paid to the ways in which the institutional structures of regulatory policy-making may affect the adoption of new ideas and proposals, in particular 'privatization' through the introduction of individual transferable quotas (ITQs) and 'regionalization' through the delegation of decision-making power.

Crisis and Reappraisal

Recent developments, as well as the historical experiences of both Norway and Atlantic Canada, outlined in earlier chapters, point to fisheries management as the management of crises – and to the challenge for any management regime of tackling today's problems in ways that avoid or minimize future upheavals. Experiences with crises highlight three challenges to fisheries management institutions.

First, there is the 'classic' question of how to deal with overcapacity and continuous pressures on the resource. This is a matter of perpetual dispute in both Norway and Atlantic Canada. Whenever the stocks fail there is always the question of what went wrong and why. The recent crises have sharpened the conflict between inshore and offshore and breathed new life into the perennial debate about appropriate harvesting technologies and the future structure of the fishing industry.

Second, there is the question of the role of science. With hindsight it seems that scientists on both sides of the Atlantic erred considerably in their stock assessments during the 1980s. In Atlantic Canada the state of the stocks was overrated, and quotas were set accordingly. For this and other reasons the collapse of the stock of northern cod was both a crisis and a surprise (Steele et al., 1992; Finlayson, 1994; cf. Hutchings and Myers, 1994; Myers et al., 1996; Walters and Maguire, 1996). In Norway, on the other hand, stock estimates may have been too pessimistic, and quota cuts too severe, considering the speed with which the resource recovered. In this sense, both crises dramatically suggest shortcomings of current scientific concepts and procedures

Introduction   123

and thereby threaten the political legitimacy of contemporary management regimes.

Third, the recent crises pose a challenge to the ways in which management decisions are made. By increasing the ecological awareness of public interest groups, for example, the crises have generated pressures, not just for a shift in harvesting technologies, but also for extending the management policy community beyond those traditionally involved. Demands on both sides of the Atlantic for representation on management councils and committees suggest that there may be a better balance to be struck in fisheries management between the economic interests of fishers and the broader environmental concerns of the general public. Management institutions, in other words, are being challenged by groups that find participation too exclusive and 'corporatist.'

To the extent that the crises are seen as signs of the ineffectiveness of regulatory regimes, they have weakened the authority of science and government and contributed to a stronger focus on the social and political nature of fisheries management. It is now widely agreed that management is not purely a technical process. Rather it is about choosing among alternatives on the basis of political pressures, ambiguous rules, conflicting preferences, and incomplete scientific, anecdotal, and other information. The complexity and relative inaccessibility of the marine biological environment generate an enormous information problem that undermines any pretence of scientific certainty. Add to this the observation that scientists work within a social environment and that their judgments may not be immune to political pressures (Finlayson, 1994), and one sees the origin of recent critiques of reliance on dynamic models of standard fish population and quota-based regulations in fisheries management (Wilson and Kleban, 1992; Wilson et al., 1990). For these and other reasons, policy-makers on both sides of the Atlantic are re-examining basic assumptions about the structure and operation of their regulatory regimes.

The Research Problem: Comparing Regulatory Regimes

A regulatory regime is a system of governance established to make and implement collective decisions within a specific area or issue (Young, 1994). It includes not only the structures, rules, and procedures that govern the course of regulatory decision-making but also the values, goals, and principles that underpin regulation in a particular place and time (Krasner, 1983; Elkin, 1986). One factor that controls the nature of regulatory regimes is where the power and responsibility to make regulatory decisions is located within the state (Francis, 1993). Are they vested exclusively in central government agencies, shared between central and regional institutions, or delegated to subnational units?

Federal and unitary states differ sharply from each other, with important implications. For example, jurisdictional disputes are more common in federal systems, such as Canada, but unitary states like Norway risk losing advantages of decentralization (Sabatier, 1977).

Second is the question of how regulatory tasks and responsibilities are divided between the state and the private sector. Participation by user groups is an important aspect of most management regimes (McCay and Jentoft, 1996). Do users of the resource participate actively in the framing of regulatory structures and standards? Are those who participate truly representative of and accountable to those with both interests and rights?

At the core of our research lies the assumption that management schemes, in order to work, must enjoy at least some support among those being managed, and perhaps some confidence among the general public. There is, in other words, a trade-off to be made between bioeconomic imperatives and political considerations and between centralized control, democratic participation, and market dynamics. We therefore address how legitimacy and compliance are generated through institutional design and through underlying tensions between (a) centralized jurisdiction and regional control, (b) participation by user groups and the representation of broader, non-industrial concerns, and (c) the pursuit of economic efficiency through marketlike solutions and pressures for retaining the crucial role of the state.

Decision-making: The Role of Central Government

The balance between central authority and regional autonomy is often contested in modern democracies. The Canadian state, for instance, is said to be characterized by a certain 'institutionalized ambivalence' (Tuohy, 1992) which is apparent in the uneasy and contested relationship between federal control and provincial jurisdiction. In Norway the relationship between central government and subnational institutions is more stable and better defined. Local government is not anchored in the Norwegian constitution, and the scope and content of regional and local autonomy are largely decided by the state. There is, furthermore, a fairly strong consensus about the legitimate domains of national and local institutions, and, consequently, little jurisdictional rivalry between these two levels of government.

In Canada managing the fisheries is a federal responsibility. Management decisions and policies are made within relatively centralized administrative structures in which bioeconomic perspectives shape the policy process (Bannister, 1989; Phyne, 1990). Scientists and ministers are key players in resource management, with provincial authorities taking a more marginal position.

Attempts at soliciting representative inputs from user groups have only been moderately successful, and there are some misgivings about the lack of user-group cohesion and the fragmented nature of government–industry interaction.

In Norwegian fisheries management centralized control is modified by corporatist structures where well-organized interest groups, in collaboration with government officials, play a prominent role. There is considerable emphasis on user-group participation, and corporatist arrangements are utilized to solicit representative inputs from the industry (through peak organizations) about both scientific and allocative issues (Hallenstvedt, 1982; Jentoft and Mikalsen, 1987; Mikalsen and Sagdahl, 1982).

Changes in international circumstances (for example, the Law of the Sea Convention), however, as well as in national concerns (for example, resource crises and economic downturns) have made policy-makers look more carefully at the presumed strengths of their different institutional structures. In Norway fisheries resource managers have been exploring the possibility of introducing a more marketlike approach to management, while at the same time retaining both the formal authority of central government and the strong position of user groups (Government of Norway, 1991–2). Meanwhile, Canadian policy-makers contemplate ways to generate more industry involvement in fisheries management, but continue to emphasize the need for executive discretion and the crucial role of disinterested expertise and professional judgment (Haché 1989; Government of Canada, 1990, 1993).

Our aim is to elaborate these differences by describing and contrasting the two regimes on several dimensions. In what ways do they actually differ with respect to institutional structures and reform strategies? What is at the root of these differences, and what are their implications, if any, for the scope and direction of management policies? Since management decisions are usually justified by stock assessments, the role of science (and scientists) is crucial. How, within our two regimes, is scientific advice balanced against the (often) short-run interests of the industry and the broader objectives of government?

Decentralization: The Role of Regional Agencies

The role of central government has become a contentious issue in fisheries management. Demands for 'regionalization,' that is, more decentralization of decision-making, seem to have have gained momentum in both Norway and Atlantic Canada. Such demands can express territorial responses to competition for scarce resources (as is the case in both countries), but they can also be part of a strategy for adjusting management policies to the needs of coastal communities or to the specifics of ecological systems. Regionalization, furthermore, pertains both to policy, in that it can signify a 'principle' used in the allocation of resource rights, and to structure, in that it can mean the delegation of decision-making authority to regionally based agencies or groups.

Considering the basic institutional differences between a federal state (Canada) and a unitary system (Norway), and given the pressures, in both countries, for some form of regionalization, we will examine whether there in fact are trends towards more decentralization in fisheries management. At issue here are the pressures for addressing regional peculiarities and local needs while at the same time accommodating national priorities and concerns. How is this dilemma reflected in institutional structures and management policies? What structures do exist for incorporating information about territorial diversity in what appear to be fairly centralized systems? Are provincial and regional authorities, or other agencies at the sub-national level, key players or marginal actors in fisheries management?

Co-management: How User Groups Participate

The involvement of user groups in public policy-making is a conspicuous phenomenon in modern democracies. On this point, Norway displays a well-integrated corporatist system of interest associations, providing for stable and highly formalized relationships between government and interest groups within most sectors of the economy (Christensen and Egeberg 1994; Kvavik, 1976; Rokkan, 1966). In Canada the fragmentation of interest associations seems to have inhibited the development of a corporate system of the kind found in most European countries (Pross, 1992: 236–7). There are, of course, arrangements of functional representation in Canadian politics, but these 'subgovernments' seem to lack the cohesion and authority of their Norwegian counterparts. It is more pluralist than corporatist (Pross, 1992).

Given these differences, and the obvious need for securing the compliance and support of those affected by government decisions, the question of how user groups participate in fisheries management becomes interesting. For instance, fisheries management in both Norway and Atlantic Canada is fairly centralized yet allows for considerable participation from user groups. What we do not know, however, is why such participation, in Norway, takes place through formal consultative arrangements at the national level, while Atlantic Canada displays a complex, and much more loosely coupled, network of advisory committees at the regional and local level. What are the reasons for this, what are the rules and procedures governing such participation, and how does it influence the legitimacy of decisions and policies? Will the particular patterns of government–industry relations in fisheries management reflect the more fundamental differences between a pluralist system (Canada) and a corporatist regime (Norway)?

Groups other than fish harvesters and processors may claim a voice in management. Consumers have legitimate interests; both price and quality may be affected by regulatory measures. Local and regional authorities take a keen interest in the effects of management policies on employment opportunities and the welfare of local communities. Environmentalists and animal rights groups claim to speak for the fish, birds, and marine mammals of ecosystems. Ethnic groups demand acknowledgment of historical rights and the importance of preserving traditional ways of life. In both countries, demands from such quarters for representation on management committees and councils raise the question of balance between the immediate and legitimate interests of user groups in the fishing industry and the more long-term and diffuse concerns of the general public. On this point we will examine the pressures for the redesign of management institutions so as to make participation less exclusive and 'corporatist.' What exactly are those demands, from whom do they come, and what have been the reactions from government and the fishing industry?

'Privatization' and Property Rights: A New Role for the State?

Managing the fisheries through individual quotas (IQs) is a recent and rapidly spreading approach. Whether we choose the terms *rights-based regimes* (Crowley and Palsson, 1992) or *privatization schemes*, there is little doubt that IQs have some of the earmarks of private property: the creation of more or less exclusive rights to harvest particular stocks or species. A crucial issue is whether these rights should be freely tradable, that is, as individual transferable quotas (ITQs). ITQs are now found in Australia, Canada, Iceland, New Zealand, the United States, and other countries, but not in Norway.

Debates about IQs and ITQs reflect larger questions about the proper role of government in managing the economy. In both Norway and Atlantic Canada it has become commonplace to assert that market forces should play a larger role in determining the distribution of rights and responsibilities in the fisheries. In both places resource managers have promoted measures to deregulate management and 'privatize' the rights to harvest resources. Within industry opinions have been far more divided and cautious, if not outright negative. On this point we will look more closely at the background and objectives of IQ proposals: why are they introduced, how are they decided upon and eventually implemented, and how do they affect other characteristics of the present regimes?

On the face of it, the concept of transferable rights (ITQs) seems to have found more favour within particular segments of Atlantic Canada's fisheries than in Norway. To explain this, we turn to variations in political practices, institutional structures, and 'contextual' pressures. What should be kept in mind, however, is that proposals to deregulate and 'privatize' have been challenged by most of the groups affected, especially in Norway where the 'T' of transferability has yet to be officially recognized. This is, however, an ongoing process whose course and eventual outcome may provide interesting and important clues as to relationships of power and influence within the fishing sector, as well as to the more general conditions for comprehensive regulatory reform.

A Note on the Comparative Analysis of Management Regimes

Our study is a series of focused comparisons (Harrop, 1992: 8) of the management institutions of a single sector, fishing, across two countries. With renewed interest in the study of institutions in most social science disciplines, including economics (North, 1990; Williamson, 1985), sociology (Powell and Dimaggio, 1991), and political science (March and Olsen, 1989; Thelen and Steinmo, 1992; Young, 1994), there is little agreement as to what should count as an institution (Thelen and Steinmo, 1992).

Recognizing the proliferation of existing definitions and uses of the concept of institution, our usage is closer to that of political science than sociology or economics. It is reflected in the following: 'the formal rules, compliance procedures, and customary practices that structure the relationship between individuals in the polity and economy' (Hall, 1992: 96); or 'sets of rules of the game or codes of conduct that serve to define social practices, assign roles to participants of these practices, and guide the interactions among the occupants of these roles' (Young, 1994: 3).

Institutions are thus the basic organizational structures and principles of modern democracies. At this level, an institutional analysis of fisheries management involves recognition of the embeddedness of particular procedures and practices, and functionally specific institutions, within broader institutional arrangements (Young, 1994). We are led to ask: What difference, if any, does it make that one of the countries in question is a federal system with a somewhat uneasy balance between central government and provincial authorities, whereas the other is a unitary state with fewer conflicts over issues of power and jurisdiction? We explore how such differences at the 'macropolitical' level may affect the characteristics of the regulatory regimes in question, from the division of management responsibilities between government agencies, interest organizations, and individual actors to the more specific rules and procedures that shape the course and outcomes of management decision-making.

Introduction 129

At a 'meso-level,' institutions include political parties, interest groups, and policy communities. For instance, some countries have a multitude of political parties, others are basically two-party systems; some exhibit cohesive networks of industry-specific associations, while others are characterized by organizational fragmentation and little collective action. Some have established highly formalized arrangements of functional representation in public policy-making, as in corporatist systems, while others rely on centralized control or pressure-group politics. Intermediate institutions of this kind provide arenas, or 'channels,' for the coordination and articulation of group interests in the political process. Differences in such 'meso-level' institutions may affect the coordination and articulation of economic interests in the fisheries and, in turn, the access and influence of user groups in management decision-making.

At the 'micro-level,' management regimes may differ as to the division of tasks and responsibilities among administrative levels and units, in the character of formal legislation underpinning regulatory schemes, in the amount of autonomy and discretion granted to bureaucratic agencies, and in practical arrangements for monitoring the implementation and enforcement of political decisions. At this level, we assume that formal procedures, administrative regulations, management objectives, and other structural features of specific fisheries regimes affect the shape of the regulatory process as well as the scope and content of management policies and decisions.

Thus far we have emphasized the likelihood that structures, processes, and outcomes within a given policy area will be a function of the constitutional and political characteristics of the nation in which they are developed. Comparing the fisheries management regime of federal Canada with that of unitary Norway one would, according to this line of reasoning, expect to find significant variations in regulatory structures and policies, if only because some of the basic structures of politics are different. However, it is also possible that the structures, processes, and policies of fisheries management are the result of factors unique to the issue area itself. According to the 'policy sector hypothesis' (Vogel, 1986: 195), the nature of the problems to be solved limits and shapes the solutions. From this perspective, one would anticipate that the comparative analysis would show that fisheries management institutions of Canada and Norway have striking similarities, despite major differences in political culture. Putting these alternatives to a rigorous test is beyond the scope of the present study. Rather, they represent competing perspectives that will permeate our analysis throughout and serve as benchmarks in the summary chapter.

In Chapter 4 we depict how management decisions are currently made, examine the politics of regulatory decision-making, and discuss some of the problems and dilemmas of user-group participation – be it 'pluralist' or 'corporatist.' In Chapter 5 we outline current management policies and examine the objectives behind them. Chapter 6 is devoted to an analysis of recent attempts at introducing new regulatory practices – with particular emphasis on the ways in which the broader institutional context may affect the prospects of policy change. In Chapter 7 we examine the pressures and proposals for institutional reform in fisheries management, linking recent demands for structural change to the perceived deficiencies of current arrangements. We conclude Part 2 of the book with a summary of findings and arguments and with an attempt at a more systematic and pointed comparison of the fisheries management regimes of Atlantic Canada and Norway.

4

Managing the Fisheries: Procedures and Politics

The introduction of exclusive economic zones (EEZs) in 1977 signaled the shift of most responsibility for fisheries management from international fisheries organizations to national institutions. In both Norway and Atlantic Canada, management came to be based on a process of consultation between government, science, and industry: in Norway chiefly through formal consultative arrangements at the national level; in Atlantic Canada through a fairly complicated, and more loosely coupled, network of advisory committees at the regional and local level. These two systems, different in structure as well as procedures, have adopted the same general approach to management. They have thus come to struggle with similar problems of legitimacy, justification, and efficiency, generating similar pressures for institutional change and policy innovation.

Management by 'Inclusion': Norway

Three generalizations arise from an examination of Norwegian fisheries management (Hoel, Jentoft, and Mikalsen, 1991): political and administrative decision-making is centralized, it is corporatist in that organized interests play a strong role in policy-making, and yet there are strong internal differences and organizational cleavages within the fishing industries involved. Policy-making and implementation at least formally are the exclusive domain of central government, notably the Ministry of Fisheries in Oslo. Another key institution is the Directorate of Fisheries in Bergen, essentially a 'professional' or staff institution whose main role is to provide expertise and advice to the Minister of Fisheries. It is one of the oldest institutions within the fisheries bureaucracy (established in 1900), and it is clearly more influential than its formal advisory role suggests. Its directorship is considered one of the most powerful positions in fisheries matters, and it 'sits' on top of an administrative structure that includes regional and local offices such as the Fisheries Extension Service and the Agency for Enforcement and Control (of which more below). Counting its regional branches, the directorate employs almost 500 people. Its annual budget is around Nkr 200 million (1994), compared with an annual budget of Nkr 42 million for the ministry, which has around 80 employees. The directorate is responsible for assessing the 'technical' merits of various solutions – and for putting forth proposals for concrete measures. The ministry's task is to assess the political feasibility of the options available and eventually decide what steps to take. The ministry is also in charge of negotiations with other nations. The directorate participates at this level too, but only in an advisory capacity.

There is little decentralized decision-making. The ministry has an extension service that serves regional and local industries and provides advice to the directorate and ministry. However, regional directors and local offices lack any real decision-making power, except in the allocation of 'recruitment quotas,' which are fish quotas set aside by the ministry to encourage younger people to enter fishing.

Recruitment quotas are not necessarily available every year. That is up to the ministry to decide. For example, they were available in 1991 and 1992, but not in 1993. Demand has always greatly exceeded supply. In the county of Tromsø only about 5 per cent of the applicants have been successful; in the county of Nordland in 1992 there were, 169 applications for 24 quotas. The criteria for allocation are not very precise – a few general guidelines set by the ministry – and they give plenty of room for local discretion. Age is important, of course, and in Tromsø the great majority of those awarded recruitment quotas were between 25 and 40 years old.

In response to demands for more 'regionalization' of fisheries management, county councillors and officials are now usually consulted before new regulations are enforced. They, in turn, consult the regional director and his staff, providing an opportunity for feedback to the ministry via the county administration. There is, at least in some regions, a trend towards closer if informal cooperation between the Regional Director of Fisheries on the one hand and county politicians and officials on the other.

The practice of consulting the regions has been driven by the ministry. The directorate has been against it, possibly because it encroaches on its advisory power. It should also be noted that the regional director and his or her staff are involved in managing the fisheries in several of the fjords along the Norwegian coast. This involvement takes place through the Advisory Committees for Local Regulations (Jentoft and Mikalsen, 1994), where the relevant gear groups are well represented. These committees, however, are advisers (to the Director of Fisheries) rather than decision-makers in their own right, and they cannot be considered important players within the management system.

In a sense balancing the highly centralized nature of this system is the strong role played by organized interests, as is true in other sectors of the Norwegian economy. Interest groups are either directly represented in government through a multitude of councils and committees or delegated public authority within particular policy fields (Hallenstvedt, 1982; Smith, 1979).

Fisheries policy-making is thus characterized by a highly formalized system of consultation and negotiation at the national level. On the other hand, it is influenced by internal differences and organizational fragmentation. Cleavages are *regional*, as between north Norway and the south, *economic*, as between processors and fishermen, and *technological*, as between offshore and inshore fisheries. These cleavages are often mutually reinforcing, and they go a long way in explaining the plethora of organizations in Norway's fisheries. Hallenstvedt (1982) identified well over eighty different organizations and associations in the fisheries, local branches of nationwide associations included. On the national level, fisheries management thus becomes an exercise in the quasi-resolution of conflict through a time-consuming process of consultation and bargaining.

The nature and outcomes of this process, to which we will return, also depend on the outcome of bilateral negotiations between Norway and other fisheries nations. The creation of the 200-mile EEZ in 1977 led to a major change in the conduct of international fisheries negotiations: from multilateral bargaining within the framework of the North-East Atlantic Fisheries Commission (NEAFC) to bilateral negotiations between coastal states sharing an exclusive right to particular stocks. Relevant is the fact that 80 per cent of the Norwegian catch comes from stocks shared with other countries, notably Russia (formerly the Soviet Union) and the European Union (EU).

Formal Organization

The first step in the regulatory process takes place at the international level – within the framework of the International Council for the Exploration of the Sea (ICES) and its Advisory Committee on Fisheries Management (ACFM). The ACFM receives scientific data from the relevant organizations of the member countries and gives its recommendations as to the overall quota, or total allowable catch (TAC), for the stocks in question. The next step, on the Norwegian side, is working out strategies for bilateral negotiations with Russia and the European Union. This takes place within a special committee (Sjøgrenseutvalget) which includes representatives from the Fishermen's Association, the

Ministry of Fisheries, the Foreign Office, the Ministry of Defence (Coast Guard), the Norwegian Seaman's Association, the Marine Science Institute, the Plant Workers Union, and the National Association of Fish Processors – an appropriate introduction to the diversity of interests involved. On the basis of consultations among these groups, the final negotiating strategy is set down by the Minister of Fisheries.

Determining the overall quotas (TACs) for particular stocks takes place through the bilateral negotiations already mentioned: with Russia on the stocks in the Barents Sea, with the European Union on the fisheries in the North Sea, and with Poland and Sweden on the fisheries in Skagerrak. Collaboration with Russia takes place within the Mixed Norwegian–Russian Fisheries Commission, where the agenda revolves around the setting of TACs for shared stocks, and deciding each country's share of the TAC and the allocation of quotas to so-called third countries. Shared stocks are cod, capelin, and haddock. Since the creation of the 200-mile EEZ in 1977, there has been a set rule for the sharing of quotas: for cod and haddock, a 50–50 split; for capelin, 60 per cent to Norway and 40 per cent to Russia. Cooperation with the European Union takes place through a broad delegation, usually chaired by an official from the EU Ministry of Fisheries.

The resulting TACs are the starting point for the consultative process at the national level. The most important (and difficult) issue on the agenda is allocating Norway's share among different groups of fishermen or 'segments' of the fleet. This is the primary task of the Regulatory Council (Reguleringsradet), an advisory body to the ministry which includes representatives from numerous organizations and institutions (see below). The council's agenda is prepared by the Directorate for Fisheries, which also works out a proposal for regulatory measures during the coming year, along with detailed specifications of its distributional implications. A few weeks after disclosure of the proposal, the council meets to discuss and decide on the options outlined. Reportedly, there is, pressure for consensus and unanimity, but a vote is always taken when there is obvious disagreement, which is often the case. The role and working of the council will be discussed more fully below. The final decision as to management strategies lies with the Ministry of Fisheries: a decision that may or may not require the approval of the Cabinet. Figure 7 gives an outline of the formal procedures.

Providing the Data: The Scientific Process

Norway and Canada follow the same procedures for stock assessment. They

Figure 7. Management procedure in Norway.

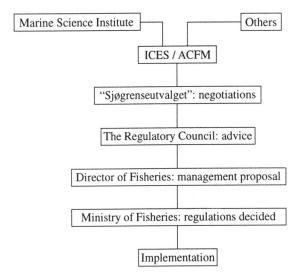

rely on government-sponsored surveys at sea and on commercial catch data for information on biomass, fishing mortality, and population dynamics. In the Barents Sea, Norway's stock assessments are based on fall and winter surveys. Acoustics and trawling are used in parallel, and the data are organized for time series analysis. There are plans for conducting a more comprehensive survey in the autumn, which is necessary because now there are a lot of older fish in the cod population. These are not covered during the main winter survey because they migrate towards the coast (Lofoten in particular) – and into areas where it is difficult to operate research vessels. There is some collaboration with the Russian researchers, who conduct surveys in October and December.

Data from commercial fisheries are collected by the sales organizations, and the information is processed by the Directorate of Fisheries. Where sales slips are not considered adequate sources of information, as is the case for the offshore fleet of fresh-fish trawlers, vessel log data are also used. This system has generally produced consistent data. However, there are problems with discarding and underreporting. Given overcapacity in almost every fleet, there is considerable incentive to circumvent regulations. Nevertheless, bureaucrats and policy-makers generally believe that the enforcement system now in place is more effective than similar systems in other European nations. For example, discards have been made illegal, even though there are not enough inspectors to cover the entire fleet. A system for closing areas with an abundance of small fish has also been instituted. This is considered one of the most successful regulatory measures of the 1980s, well worth the estimated Nkr 20 million it costs to enforce it. In addition, area closures have increased cooperation between government and scientists in promoting more selective gear. For example, they have developed an 'exclusion grid' (Nordmore grate) for the shrimp fishery that is very useful in diminishing the by-catch of some cod.

Because Norway shares jurisdiction for 80 per cent of its stock with other nations, international scientific cooperation is crucial. The major forum is ICES and its Advisory Committee for Fisheries Management (ACFM), which was created in the 1970s to deal with fisheries issues. The ACFM has established subcommittees for pelagic and demersal fisheries which meet twice a year, in the spring and fall. The most important meeting for Norway occurs in the fall, when advice on the TAC for cod is generated.

Today the ACFM subcommittees tend to be responsible for covering particular areas rather than selected stocks or species. The idea behind this change is multispecies management, a worthy but elusive goal. There are elements of multispecies management in determining capelin TACs, but in areas such as the North Sea such a strategy is close to impossible because of the large number of species, actors, and interests, as well as the difficulties in developing a satisfactory multispecies population assessment model.

The advisory process is currently based on a three-fold typology of fish populations. First, there are stocks considered beyond safe biological limits, because of either a small spawning stock or a combination of low recruitment and intensive harvesting. Such stocks are usually the subject of concrete science-based advice. Second, there are stocks on which very little information is available, as is the case for some redfish. Because assessment is very difficult, advice is usually provided in the form of precautions. This is frequently equivalent to saying that no further expansion of the fishery should occur. One person interviewed noted that maintaining recent catch levels is a typical recommendation because managers require 'something' to justify their decisions. Third, there are species that are considered to be in no immediate danger from overexploitation. In these instances, managers' choices can be influenced by socioeconomic considerations, the weight of which must be decided at the political or managerial level.

According to a scientist interviewed for this research, there is currently, especially within the European Union, a trend towards emphasizing socioeconomic considerations, not least in situations where politicians are advocating an increase in TACs. Biologists do not deny that there may be a case for this, but the problem is the lack of solid evidence as to the socioeconomic effects of manipulating the TAC. In the opinion of our informant, Norwegian economists appear to be utilizing old-fashioned biological models in their work. He commented, with regard to a recent discussion with an academic economist, 'I know very little economics, and I wish your economists would admit to knowing as little about biology.'

Recognition of a dramatic decline of the Norwegian-Arctic cod stock in the late 1980s led to sharp reduction in fish quotas, and hence 'crisis,' between 1989 and 1992. What is remarkable about the recent fishery crises in both Norway and Atlantic Canada is that resource managers had the social power to sharply restrict fishing. In Norway, however, the quota reductions, which may have helped restore the cod stocks, would not have been possible without the cooperation of the Fishermen's Association. The downturn in the cod fishery around the Lofoten Islands in 1987 probably made fishermen painfully aware of the need for tight regulations. Nonetheless, with quota cuts, relationships between the scientific community and the industry have been strained.

Scientists have been criticized either for inaccurate predictions or because their assessments are at odds with 'local' knowledge or the perceptions of fishermen, as is also true in Canada. Norwegian fishery scientists evince disdain for anecdotal information as a basis for making stock assessments or constructing regulatory measures. Like their Canadian counterparts, they tend to have more faith in their sampling procedures and in the knowledge generated by the offshore sectors of the Norwegian fleet, than in information provided by fishermen, particularly small-boat inshore fishers.

Norwegian scientists will admit that fishermen do have knowledge, but they will often contest the scope and quality of that knowledge. According to one scientist, small-scale fishermen, for example, may think that because there is an abundance of cod in the local fjord, the stocks must be more plentiful than biologists suggest. Having been challenged on their estimates by the owner of an offshore vessel some years ago, the Marine Science Institute hired his vessel for surveys and brought him and his crew along. The vessel owner, much to his surprise, realized that there were in fact large areas with few signs of cod – even smaller ones – and he promptly changed his opinion on the issue.

On the other hand, Norwegian fishery scientists generally keep in touch with the industry by attending the annual meetings of fishermen's associations, at the local, regional, and national levels. They are often invited to give talks within their respective fields of expertise.

Seminars for elected representatives from fishermen's associations have also been held – to educate and inform about research methods and procedures. In the 1980s several trawlers with crew were put at the scientists' disposal, and they spent a full week 'mapping out' the Barents Sea, offering the fishermen a unique look into the nature of scientific work. The event was later called the 'cod adventure' – and repeated four years in a row.

The Marine Science Institute and Directorate of Fisheries in Bergen see themselves as providing scientific and technical information and management alternatives for the regulatory council, as well as expert advice on enforcement. Representatives are careful to abstain from the explicitly political discussion about allocations that the regulatory council must make. For example, as of 1996 the director of the Marine Science Institute, who sits on the regulatory council, consistently abstains from voting on allocative issues as a matter of principle. Some scientists have suggested that the institute should not even be represented in the council, because its agenda is dominated by issues that are not scientific in nature, and hence outside the specialized competence of marine biologists. The issue has been raised with the ministry, but it obviously prefers to have the scientists represented as full members rather than as observers. Were the scientists to give up their representation, the 'balance of power' in the Regulatory Council would tip even further towards the industry.

The Consultative Process: The Regulatory Council

The regulatory council, set up in 1983, is part of a long-standing tradition of consultation in Norwegian fisheries management. It replaced two other 'management' committees set up in the 1970s, the licensing committee and the regulatory committee, and represents a continuation of a policy of consultation initiated with the appointment of members to the Trawler Council in the early 1950s. The latter included representatives from industry and government, and it was supposed to be consulted before trawler permits were issued. The basic idea behind these and similar institutions is to provide arenas for the exchange of information and advice and for the clarification of problems and positions. From the government's point of view, such institutions strengthen the legitimacy of management decisions; on the industry side, they are instrumental in the articulation of interests and demands, that is, in wielding influence over management policies.

The regulatory council was established through an amendment of the Salt Water Fishing Act, the most comprehensive of the six different pieces of legislation relevant to fisheries management passed in the early 1980s. The mandate of the council is broader than that of its predecessors: it includes the right to be consulted on decisions pertaining to the setting of TACs (for stocks managed exclusively by Norway), as well as on decisions concerning the allocation of quotas among different groups and regions.

The composition of the council is as follows: Norwegian Fishermen's Association (five members), National Association of Fish Processors (two), Norwegian Seamen's Association (one), Norwegian Union of Plant Workers (one), Directorate for the Management of Natural Resources (one), Directorate of Fisheries (two), Marine Science Institute (one), Saami Parliament (one). Note the strong representation from the Fishermen's Association (five of fourteen), and the fact that the fish-harvesting industry as such commands a clear majority (nine of fourteen). The latter point, however, is of little consequence, as the industry seldom operates as a political coalition because of internal conflicts.

Another feature of the council is the inclusion of organizations and institutions 'representing' the environment, that is, the Directorate for the Management of Natural Resources and, since 1990, the Norwegian Environmental Conservation Society (as an observer only). This broadening of representation does not, as yet, seem to have had any major impact on management policies, and it may thus have more symbolic significance than real effect. It is important, however, in that it reflects a growing awareness of the need to broaden the premises of management decisions beyond the (economic) interests of those directly affected. In addition, it represents a positive response to demands from environmental groups and an acknowledgment that fisheries management is becoming a matter of national concern.

The granting of observer status to the Environmental Conservation Society, however, met with vociferous opposition from individual members and local associations of the fishermen's Union: 'We've had enough, the union should now reconsider its representation in the council,' one of the union's representatives was quoted as saying. Others voiced concern about 'the broadening of participation from organizations and groups based on [irrational] sentiments.' There was, at the time, fear that pressure groups outside the industry would come to dominate management policy-making, and several local and regional associations passed resolutions to the effect that this had to be stopped. See, for instance, *Fiskaren*, 30 November 1990, and *Sunnmorsposten*, 5 December 1990. At present, however, this does not seem to be an issue at all – possibly because of the 'responsible' behaviour of the new observers.

At the Council's meetings there are three types of participants: full members such as the representatives of industry and the Saami Parliament; observers like those representing the environment, and 'others,' mostly bureaucrats and experts. Meetings are not public. Only members of the first group are allowed to vote. Observers can take part in the discussion but they cannot vote. The rest are expected to talk only when called upon to do so, for instance, to clarify particular issues. It is cast as 'wrong' for members of the ministry, for instance, to actively shape recommendations, but they are important participants, frequently called upon to speak, especially to clarify factual points and questions of policy. The procedures are fairly loose, but more formalization may follow the recent increase in the number of participants and interest in the council's work from the news media.

The official definition of the council is as a 'professional,' non-political body, rather than an arena for regulatory policy-making. However, a close look at the composition of the council should suffice to convince any observer that this is not a body of neutral, disinterested experts, and it was probably never intended to be. Rather, it is a mixture of the purely professional and overtly political: a body of knowledgeable people with interests at stake. According to members interviewed, this is also reflected in its deliberations.

On the one hand, there is relatively little controversy about fisheries science. It does happen, of course, that biological estimates are challenged, but the scientific recommendations are generally accepted. On the other hand, the paucity of scientists represented suggests a certain ambiguity about the role and functioning of the council. In most circumstances, it is difficult to draw a straight line between professional advice and political preference, even more so when the livelihoods of those represented are at stake. Even scientists, when arguing in the council, seek to strike a balance between biological necessity and political feasibility. The representatives of the Fishermen's Association are directly instructed and closely supervised by their national executive; this is a sign of the overtly political nature of the council's work.

The representatives of the Fisherman's Union tend to have little leeway for compromise. They usually meet with a fixed mandate in the sense that they are bound by instructions from the union's national executive. This, however, must be seen in relation to the fact that the union is a fragile coalition of various groups with conflicting interests on management issues. Union policy tends to be carefully worked out compromises, or 'minimal solutions,' that seldom give much room for manoeuvre (Hallenstvedt 1982; Hoel, Jentoft, and Mikalsen, 1991).

It is, of course, also interesting to note that in order to reach a decision, a vote often has to be taken. Several members of the council have, in interviews, pointed to voting as a fairly dominant form of decision-making. Compromise and consensus are, in other words, not as prevalent as the official perception and status of the council would lead one to believe.

Preparing the Agenda: The Role of the Directorate

The groundwork of the regulatory council is largely done by the directorate, with its economics department and legal division as key players. There are no formal procedures for preparing the agenda. Usually it starts with an 'in-house' meeting to discuss issues to be faced by the council, problems, and controversies that are likely to occur, and who shall be responsible for preparing particular issues. A preliminary agenda for the council's meeting is set up and distributed to its members for approval. The directorate also advises on how to manage particular fisheries, including a proposal of how the Norwegian TAC should be divided among different sectors of the fleet. There are also informal attempts to 'sound out' the fishing industry, through informal contacts with the associations.

Officials spend much time attending fishermen's associations up and down the coast to find out 'what the fishermen like or do not like,' as one official put it. In the end, the associations will have considerable influence on how the TAC is shared. A central concern, since 1989, has been a 'quota ladder,' allocating percentages of the TAC of cod between the inshore and offshore sectors. The idea for this came from the Norwegian Fishermen's Association.

The chief task of the directorate is defined as making certain that the quotas are not overfished – rather than to deal with allocation. Nonetheless, it does involve itself in allocation, particularly when fishermen's solutions are viewed as much too complicated to implement. When proposing or demanding a particular solution, fishermen tend to emphasize justice and fairness. They are not, in general, concerned with the costs of regulation or the feasibility of enforcement. The directorate may also intervene to impose its views of fairness when it appears that some interests are marginalized. In the words of one official, 'We leave the sharing as far as we can to the fishermen, but we know that some of these decisions are taken at 3 o'clock in the morning. We know there are groups that are not present at the table in the fishermen's associations, so we have to make sure that these are treated fairly.'

The regulatory council usually meets twice a year, in June and December. Issues such as the allocation of the TAC are dealt with in December, when the advice on TACs from ICES and the proposal for next year's management plan from the directorate are available. Management issues are generally dealt with on a species-by-species basis. From time to time there are meetings devoted to particular stocks, but there are no special sessions to analyse and discuss the current state of stock assessment science and other affairs, as has been the case in Atlantic Canada. 'I don't think we ever had a special meeting just for the biologists to tell fairy tales,' as one official, who happened to be an economist, put it.

The Regulatory Council: Powerful but Irresponsible?

The regulatory council debates and gives advice; final decisions reside with the

Ministry of Fisheries. Our impression, from interviews and documents, is that the ministry usually follows the advice of the council, or in case of a vote, the position of the majority. This suggests that the council, or even a small majority of its members, may wield more influence than its advisory status implies. To what extent then, are we dealing with an institution with considerable power but no responsibility? The ministry, of course, is directly answerable to Parliament, and eventually to the public. Not so with the council, even though its position has become problematic. Although the Ministry of Fisheries has borne most of the brunt for the fisheries crisis, the council has been criticized for failing to give sound advice. Interest group representation, some argue, is tantamount to letting the fox into the henhouse, or, in Norwegian terms, letting the goat into the oatbag.

A member of the regulatory council – the representative of the fisheries branch of the Norwegian Food and Allied Workers Union – made the following observation: 'One of the main causes of the current resource situation is the lack of responsibility among the actors in the industry; among legislators, bureaucrats, and fishermen. The distance between user groups and policy-makers has been so wide that the individual participants have been able to neglect their responsibility. At the same time user groups have wielded substantial influence over regulatory decisions. In this way, resource management has become an object of logrolling' (source: letter of 27 August 1990 to Landsdelsutvalget for Nord-Norge og Namdalen).

In its decisions the ministry seems to follow the proposals put forward by the council. In case of disagreement and a split vote, it may choose to follow the minority, but a compromise is usually found. On this point, the ministry may experience a conflict between its role as 'defender' of the stocks and its role as 'arbitrator' among conflicting interests. As the second task will be more immediate and pressing, solutions that will satisfy all groups involved take higher priority. A leading example is increasing the TAC. This strategy also helps solve the problem of how to reduce transfer payments to the industry. Larger quotas mean higher incomes, which in turn reduce the need for government subsidies. Another reason for increasing the TAC could be the need to keep one's share of the market by preventing an interruption of supplies.

In the past the TAC has exceeded the level recommended by the ICES many times. The Norwegian-Arctic cod stock is a case in point. To be fair, the regulatory council does not wield any influence over such international decisions. But for stocks like herring, where Norway has full control, TACs have time and again exceeded the recommendations by the ICES. In 1990, for instance, ICES recommended a ban on the herring fishery while the council proposed a quota of 60,000 tons. The ministry accepted the latter, probably to protect market share. That same year the council also endorsed a saithe (pollock) quota that was 70,000 tons higher than recommended by ICES. Furthermore, as of 1996, the council had not accepted ICES's proposal to expand mesh sizes from 135 to 155 millimetres to protect juvenile groundfish.

Such decisions have undermined the authority of the council. Fisheries management issues are increasingly in the public eye, and the present system of consultative management is criticized for giving too much influence to the industry.

The council's authority, and hence the legitimacy of regulatory policy, have been questioned by inshore fishermen in the north (Sagdahl, 1992) as well as by people outside the industry (Eriksen and Mikalsen, 1990). This raises a question, to which we will return, about the degree to which corporatist institutions are becoming more open and transparent, so that increased public awareness helps counter the problem of industry influence without formal responsibility. Another question is how this situation may eventually affect the selection of participants, the definition of management problems and objectives, and the choice of regulatory strategies.

Enforcement and Control

The importance of monitoring and enforcement has increased with the number and complexity of regulations and the use of individual vessel quotas. The introduction of individual vessel quotas (IVQs) to the inshore fleet in 1989 (they had been in the offshore trawler fleet since 1978) brought the issue onto the national political agenda as reports of increasing violations reached the media. At one point there was even talk of creating a special 'fraud squad' for the fisheries to combat cheating.

The debate on cheating and enforcement has focused on cod fishing, the single most important fishery.

According to one official, the mackerel fishery is probably the hardest to control. The problem goes back to 1988–9 when Japan became a lucrative market for big mackerel (over 0.6 kilograms) – with prices at about Nkr 5 per kilo. Smaller fish only fetched one-fifth of that. Given that the purse seine fleet had individual vessel quotas, there was a strong incentive to highgrade. The problem is exacerbated by the fact that the mackerel does not have a swim-bladder, so when discarded – it sinks. 'You dump it, and it's gone' – contrary to other species such as cod and saithe which will be floating as highly visible evidence of discarding.

Cheating is difficult to control because of the number of participants as well as because of the incentives and opportunities to cheat. In 1992, for example, about 8,000 vessels participated. Of these, 3,500 were on individual quotas (including 150 offshore vessels), and more than 4,000 smaller vessels were on a competitive quota. Cheating is probably a more serious problem in the latter group, because of the inaccurate reporting of catches.

Overall and formal responsibility resides with the Ministry of Fisheries, but the directorate plays a key role in implementation and coordination. Of particular importance is the National Agency for Enforcement and Control (Fiskeridirektoratets Kontrollverk), a regionally organized branch of the directorate employing some 160 people, whose main tasks are quality checks and quota control through dockside monitoring.

The agency is divided into five regional offices: Finnmark/Tromsø, Nordland, Møre and Romsdal, Stadt/Svenskegrensen, and Trøndelag, and it has about eighty inspectors working out of thirty-six communities along the coast. At three of the regional offices there are also laboratory facilities for scientific quality checks. The agency as such has a budget of approximately Nkr 52 million (1992).

Spot checks at processing plants are important, and the frequency of these has clearly increased in recent years because of the growth in landings from foreign vessels, especially Russian. The directorate cooperates with other fisheries nations such as Russia, Great Britain, and Denmark on issues of control and enforcement. There is exchange of information as well as observers, and mutual assistance in inspection of vessels and dockside monitoring. Inspections at sea are also part of the agency's activities, particularly outside the coast of north Norway. In this capacity it cooperates with the Coast Guard (established in 1980 as part of the Ministry of Defence). As for sanctions, the agency is authorized to confiscate the value of the catch that exceeds the quota. Major violations are a matter for the police and eventually the courts, where reactions may vary from fines to confiscation of gear and vessel.

The Coast Guard, a large player with over 700 employees, is responsible for a wide range of tasks within the Norwegian 200-mile zone, from the control of catches and quotas to the enforcement of gear restrictions. It has authority over Norwegian as well as foreign vessels.

The Coast Guard's northern division, which is the most important from a fisheries point of view, has at its disposal three large navy vessels with Lynx helicopters. In addition, twelve smaller vessels are used for monitoring the use of gear, and four more are leased for covering special areas during seasonal fisheries. The Coast Guard had a budget of Nkr 458 million in 1993, but as it has other tasks besides monitoring the fisheries, it is difficult to calculate the exact costs of its 'pure' fisheries operations (Hallenstvedt, 1993). To monitor the activities of foreign vessels in the Norwegian 200-mile zone, it uses a computer-based system set up by the Directorate of Fisheries. In case of serious violations, foreign vessels will be arrested and the owner or skipper prosecuted (Hallenstvedt, 1993).

The mandated sales organizations (MSOs) are the third important player in this arena. They monitor landings and 'measure' them against individual vessel quotas. They rely mainly on collecting and checking sales slips and dock-side monitoring (spot checks). Violations are reported to the directorate, and violators may eventually end up in the courts.

Despite the elaborate enforcement machinery, there is clearly need for improvement. However, the general perception, among the public and in Parliament, that the fisheries bureaucracy does little about cheating, is not tenable. Enforcement has been tightened since the mid-1980s, spurred, among other things, by an increase in Russian landings. The fact that several agencies participate raises an organizational problem, and a working group is currently looking into better ways of coordinating the activities of the institutions involved. The main problem, however, may be the complexity of the regulations themselves. Not only are these numerous, they are also ambiguous and difficult for the individual fisherman to understand. In this sense problems of enforcement and control will be roughly proportional to the complexity and scope of management schemes. To improve control the government's policy is to simplify the management regime and further expand its capacity to enforce regulations.

The Politics of Management

Fisheries management is, if anyone was ever in doubt, political: it is about the articulation and coordination of conflicting interests and demands and the exercise of 'cruel' and controversial choices. That said, the intensity of conflicts and the scope of controversies may, of course, vary. The initial coordination of interests within the industry itself is one point of controversy. A key actor at this stage is the Fishermen's Association, where most of those directly affected by government policy hold membership. Formed in 1926 as a federation of regional associations, the association's decision-making structure was, initially, based exclusively on territorial representation, by county. Increasing diversification and capitalization within the industry, led to the emergence of cross-cutting organizations such as the Norwegian Trawler Association, the Boat Owners Association, and the Association of Seiners. In the 1960s these were granted 'associate membership' in the Fishermen's Association. They represent the offshore and technologically more advanced part of the fleet, with home bases mainly along the southern part of Norway's coast. Accordingly, the association is a fragile coalition of conflicting interests pertaining to region, technology, and economics. This makes it difficult and time consuming to hammer out a consistent regulatory policy. Whatever stand the association takes on particular issues is thus either a carefully worked-out compromise or the product of a winning coalition.

The internal process in the Fishermen's Association is dominated by its executive committee. This is where the associations's stand or policies are decided, after consulting the regional associations as well as the functional groups. The regional associations of (mostly) inshore fishermen have a clear majority at this level, as the functional groups currently hold only five of eighteen representative places on the executive committee. However, the latter are able to compensate for their minority status on the executive in several ways (Jentoft and Mikalsen, 1987).

First, overlapping membership provides for alternative sources of influence and representation. More than half of the members of the Boat Owners Association, for example, also hold membership in regional organizations of the Fishermen's Association. Furthermore, some of them are also affiliated with the Association of Seiners. Consequently, owners of purse seiners may have three channels of interest representation within the association, whereas inshore cod fishers may have only one. Second, most of the functional groups have a well-equipped administration of their own and sufficient resources to engage themselves in a broad set of fisheries policy issues. The Trawler Association has a modest administration, but a homogeneous and clear-cut membership which may be essential for political efficiency. In comparison with most of the regional associations, the functional groups score higher on administrative capacity as well as political coherence. This may prove decisive for the exercise of political influence (Jentoft and Mikalsen, 1987: 227).

What these patterns mean for the aggregation and articulation of interests is still unclear, but a few qualified guesses can be made. The fact that the Fishermen's Association is a coalition of groups with conflicting interests clearly complicates the coordination of demands and bespeaks its fragility. Recently, a significant number of inshore fishermen broke with the association and formed their own organization. Dissatisfaction with the association's stand on management issues was given as a major reason for the split. On top of this, offshore groups have recently threatened to leave the association because of disagreements over quota allocations. Second, the fact that the association's representatives on the regulatory council must stick to the decisions of the national executive of the association is, in itself, a sign of complex processes and fragile compromises. They are not supposed to engage in a process of give and take in the council, and they are, it seems, the only participants with little or no room for manoeuvre.

TABLE 1
Recommended and agreed TACs for Norway and Russia

Year	1982	1983	1984	1985	1986	1987	1988
Recommended TAC	434	380	150	170	446	645	363
Agreed TAC	300	300	220	220	400	560	451

Weight in 1000 tonnes
Source: *Report of the Advisory Committee on Fishery Management* (Copenhagen: ICES), 1988.

Decisions pertaining to the size of the TAC and subsequent allocations, in particular, are not just a question of abiding by the scientific advice from ICES. Negotiations between Norway and Russia, and with the EU, are highly political. Obvious national interests are at stake, as illustrated by the annual 'tug of war' within the joint Norwegian–Russian fishery commission. The negotiation process is susceptible also to 'internal' pressures, that is, from interest groups like the Norwegian Fishermen's Association which is represented in the joint commission. The fact that the overall quota has often been different from the one recommended by ICES indicates that biological assessments are not the only factor at work. Table 1 illustrates this.

The proposals of the Directorate set the agenda for the domestic process, and they will invariably be influenced by considerations of the political, social, and economic effects of different measures and policies. The work of the directorate is, as already pointed out, part of a sounding-out process and, as such, susceptible to inputs and reactions from different sections of the industry (especially from the Fishermen's Association, which enjoys a privileged position at this stage).

The regulatory council is supposed to represent the expertise of fisheries management, but it is also an arena for the articulation and aggregation of political demands. Representatives of the relevant interest groups, be they fishermen or processors, do have a fair amount of professional or technical knowledge, but they also have political points to score.

The council must face questions of allocation which do not have a 'technical' solution, and its recommendations will not always be unanimous. The final decision by the Ministry of Fisheries may not be based exclusively on the council's recommendations. Alternatives and adjustments will be considered in view of administrative feasibility, distributional effects, and pressures and protests from discontented groups. The policies implemented should, then, be understood as a compromise between what can be defended biologically, legitimized

148 Section 2: Resource Regimes

politically, and accepted on social and economic grounds. The final decision on the allocation of the capelin quota for 1991 is a good illustration. The council recommended a 25/75 per cent distribution between north Norway and the rest of the coast, in line with proposals from the Fishermen's Association. The ministry, however, opted for a 30/70 solution after vociferous opposition from northern fishermen.

Management by Decentralized Consultation: Atlantic Canada

Centralized federal government control is often cast as the defining diagnostic feature of Canada's approach to fisheries management (Bannister, 1989: 70–6; Marchak, Guppy, and McMullan, 1987; Pross and McCorquodale, 1987; Sinclair, 1984). The federal government of Canada has taken a more directive role in attempting to preserve fisheries stocks, as well as in regulating harvesting, processing, and, to some degree, marketing activities than has been the case for other 'federal' nations such as the United States. In that sense Canada is more like Norway. Accordingly, we argue that fisheries management in Nova Scotia as elsewhere in Canada continues to be centralized in the sense of a continuing commitment to what Ostrom (1990: 52) calls 'externally organized collective action.'

Fisheries managers in Canada continue to believe that a combination of state- and firm-level organization can be utilized to manage the fisheries successfully. Nevertheless, there has been a major shift in Canadian thinking about the relative contribution of the public and private sectors to fisheries management. Specifically, we will point to a trend, reflecting developments in the larger economy, towards deregulation of the allocative tasks of management, with the state retaining its overall responsibility for conservation of resources. In the fisheries this is partly captured by discussions by social scientists (for example, Davis, 1984; Kearney, 1989) and fishing industry advocates about community-based management and co-management, but also by increased reliance on individual transferable quotas and other 'rights-based' management systems that allow market forces to play more directly.

The Role of the Federal Government

Jurisdictional disputes are more common in federal political systems like Canada's than in unitary states such as Norway. At present formal authority for managing the fisheries is divided between the federal government and its provincial counterparts, jurisdiction and resources being heavily concentrated at the federal level (Parsons, 1993b). The power and functions of the federal Minister of Fisheries originate in the British North America Act (the Constitution Act of 1867) and are reaffirmed in the subsequent legal framework of management of the fisheries. However, this has been the focus of jurisdictional conflicts and disagreements between federal and provincial governments since at least the late 1800s.

The Constitution Act of 1867 gave the federal government of Canada more or less exclusive authority for the management of Canadian fisheries, a prerogative confirmed in the first Dominion Fisheries Act of 1868 where the Cabinet was given the authority to manage both sea-coast and inland fisheries. Licensing and leasing of fishing rights were the 'tools of the trade' at that time, but they were mainly used for statistical purposes (Parsons, 1993b: 19). Even if the tasks and measures of management were limited, legislation put the *authority* to manage firmly in the hands of the federal government.

This authority and jurisdiction of the federal government has been repeatedly challenged in the courts. Put briefly (see Parsons, 1993b), court judgments have confirmed the authority of the federal government to manage the fisheries with reference to social and economic as well as conservation objectives.

It has been suggested that the level of ministerial discretion regarding the Canadian fisheries 'would surprise fishermen and managers from other nations' (Lamson and Hanson, 1984: 4). This by and large reflects the central position of the Department of Fisheries and Oceans (DFO) in matters of management policy. The management functions of DFO include all matters within the jurisdiction of the Parliament of Canada that have not by law been assigned to other departments, boards, or agencies. The minister's jurisdiction is furthermore derived from the Fisheries Act (a new Fisheries Act bill has been tabled in Parliament for the autumn of 1997) and includes matters relating to coastal and inland fisheries, hydrography and marine sciences, fishing and recreational harbours, and the coordination of policies and programs pertaining to oceans. The federal department is by far the most powerful public sector player in the fisheries. Provincial departments are, by comparison, policy 'takers,' not policy-makers (Pross and McCorquodale, 1987: 80).

Provincial Responsibilities

Regional issues in the management of Canada's fisheries reflect important differences in industrial structure, historical background, and culture (Apostle and Barrett, 1992; Marchak, Guppy, and McMullan, 1987; Neis, 1991). The Newfoundland, Scotia–Fundy, and Pacific regions account for a major proportion of Canada's financial and human resources devoted to fisheries management. Fisheries account for a minor part of Canada's GNP or total employment (around 1 per cent of both), but as in Norway their regional significance is far greater, especially on the east coast. The fishing industry is a major employer in both Newfoundland and the Scotia–Fundy region.

The provincial governments are mainly responsible for the type, number, and location of processing plants. Federal–provincial and interprovincial issues are dealt with by a special committee at the deputy minister level, and through the Atlantic Council of Fisheries Ministers (Kirby, 1982: 340). The provincial ministries, however, lack both the jurisdictional authority and the spending power necessary to make a decisive impact on policy. They do, however, hold a recognized position within the fisheries policy community and can, on occasions, challenge or augment federal policies (Pross and McCorquodale, 1987: 65–7).

The authority of the federal government has been challenged not only in the courts but also in the political sphere – as a question of constitutional reform. Provincial governments, particularly that of Newfoundland and Labrador, have repeatedly challenged federal powers over the fisheries at federal–provincial meetings and constitutional conferences. However, both processors and fishermen have come out firmly in support of the federal government as being paramount.

Provincial responsibilities are more limited than federal ones, concentrating on the land-based aspects of the fishery, development and regulation of aquaculture, and industrial financing. In the case of aquaculture details may differ from one province to another, but there are memoranda of understanding (MOUs) which clearly define what is to be done and who is to take the lead. Because one crosses into provincial jurisdiction when it is a question of attaching gear to the ocean floor, for example, there is more provincial involvement and responsibility in aquaculture. On the other hand, the issue of free passage on a waterway is a federal transport issue, and there are sometimes federal health implications (for example, paralytic shellfish poisoning) and federal environmental concerns.

Limits to Federal Management?

In Canada the issue of jurisdiction also pertains to scope: what can and should be the purpose of fisheries management? A court decision in 1984 stated that federal powers over marine fisheries were limited to the conservation of resources. This decision prompted government action to amend the Fisheries Act to include social and economic considerations as legitimate objectives of fisheries management. Other court decisions as well as government commissions and task forces convey the message that management is not just about protecting the fish, but also about maintaining the economic viability and livelihood of fishermen and others employed in the industry. The message is realized in the design of Canadian management institutions from the 1980s on, with their emphasis on consultations between government and user groups.

Formal Organization

As already pointed out, the duties, powers, and functions of the Minister of Fisheries are extensive. The minister heads the Department of Fisheries and Oceans (DFO), many of the functions of which are delegated to regional offices headed by regional director generals. Science, policy-making, and the implementation of management policy are carried out within this regionalized federal agency.

In Atlantic Canada the Fisheries Act is particularly significant as the legal framework for management. For instance, one section (7) on licensing, and another (43) on management gives the governor general in council and, by extension, the minister, the power to govern seasonal and stock variations. There are also associated regulations at the regional level. Sections 12 to 32 of the Atlantic Fisheries Regulations deal with licensing and registration of vessels, and section 33 deals with conditions of licences. The regulations generally give the regional director general authority to regulate variations in species and areas. In addition, the Coastal Fisheries Protection Act governs the activities of foreign vessels. A number of policies are enabled by these acts and regulations, including limited entry licensing for groundfish and lobster; quota management in the form of competitive or enterprise allocations (offshore vessels), vessel quotas (herring), or individual transferable quotas (inshore mobile gear less than sixty-five feet); closed areas; protection of gear; seasons (lobster); fish size (groundfish, lobster); stock plans; and aquaculture.[1]

In general DFO stands out as more directive and effective than its federal counterpart in the United States, the National Marine Fisheries Service. In contrast to Norway, where the Ministry of Municipal Affairs is responsible for regional development, and where the Ministry of Fisheries tends to take a sectoral rather than a territorial perspective, the Canadian fisheries administration pays closer attention to regional variation. For example, in 1992, in the context of the cod crisis, the Canadian fisheries minister was also responsible for the

---

1 This paragraph is based on personal notes from a talk given by John Angel, Assistant Regional Director, Department of Fisheries and Oceans Canada, at St Mary's University, Halifax, Fisheries Seminar Series, 22 February 1991.

major regional development agency in Atlantic Canada, the Atlantic Canada Opportunities Agency (ACOA). Department officials relate this to the extensive discretionary power given to the federal fisheries minister by the Canadian constitution.

The Newfoundland and Scotia–Fundy regions on the east coasts and the Pacific region of the west coast account for a substantial proportion of the financial and human resources devoted to fisheries management. Over 61 per cent of the DFO's person-years and 47 per cent of the financial resources are concentrated in these three regions. The two major Atlantic Canada regions, Newfoundland and Scotia–Fundy, together account for a plurality of the personnel in Canadian fisheries management and are second only to the headquarters, in Ottawa (the national capital) in financial expenditures. Furthermore, science, as a discrete program sector in DFO's structure, is widely dispersed throughout the regions, and accounts for 2,179 person-years and $194.100 million dollars of expenditures.

DFO is one of the most decentralized units in the federal government, with over three-quarters of the individuals employed by the department working outside Ottawa. Officials in Ottawa see their primary task as policy coordination rather than day-to-day management of the fishery. On this point there is a clear division of authority with the provinces. The mandate to manage the fishery is, as observed earlier, clearly federal, with processing being a provincial responsibility. There are, however, frequent formal consultations between the federal and provincial governments, both at the minister and deputy minister level. The Atlantic groups include the four Atlantic provinces (New Brunswick, Nova Scotia, Prince Edward Island, and Newfoundland), as well as the Northwest Territories. From a regional point of view, there is a fine balance between operations in the regions and matters requiring consultation with Ottawa. Regional managers see themselves as having considerable discretion in what they 'should send, must send, and can send' to Ottawa. One complexity is dealing with attempts by private interests to set regional against national levels in the department. Where there is disagreement, particularly on interregional stocks, like northern cod, or Gulf redfish, these issues tend to get pushed up in the system. They may go to federal–provincial committees, at either the deputy minister or minister levels or a council of Atlantic fisheries ministers. Again, views may be polarized, leaving decisions on contentious or controversial issues to the minister.

Enforcement and Control

In both Norway and Canada measures have been taken to increase enforcement capacity. DFO has greatly increased its enforcement powers since the 1980s. Its legal staff coordinates all prosecutions for the Atlantic region. Fisheries legislation provides for prosecution or administrative sanctions. In the prosecution area, penalties have been increased from Cdn $5,000 to $100,000 for summary offences and $500,000 for indictable offences. The Department of Justice is now permitted to charge corporate officers, employers, or licence holders, rather than just specific active fishermen.

The foreign fleet is covered by other legislation. A reciprocal agreement with the United States has been negotiated to provide much higher penalties (up to Cdn $100,000) for cross-boundary offenders, and the United States Coast Guard has been increasingly helpful in enforcing the 'Hague Line' (between the two countries).

Administrative sanctions are typically applied to individuals who fish enclosed areas or break other conditions of their licences. The fisheries minister has the power to decide whether to issue and to cancel or suspend licences, and the minister's administrative powers have been strengthened. Some one-year suspensions of licences for non-compliance with fisheries regulations have already been issued. There are few prospects for appealing the minister's decision, save for review in the Federal Court of Canada. Administrative sanctions are begun at the area level, go to regional headquarters, and are ultimately approved by the minister for decision. Those penalized have thirty days to inform the minister of reasons for not applying the sanctions.

DFO can proceed through the courts or by way of administrative action in dealing with offences, but it cannot do both. DFO prefers to avoid the courts, which are expensive, time consuming, and unpredictable. For instance, it might cost Cdn $10,000 or $20,000 in legal fees to obtain a Cdn $2,000 penalty. Further, it became clear during recent hearings (conducted by the Haché Task Force on the Scotia–Fundy groundfish fisheries) that court-imposed sanctions were merely regarded as the cost of doing business. In addition, the middle levels of the enforcement system are acutely aware of the impact that budget cut-backs have on their effectiveness. Area-level officers not infrequently find themselves being squeezed to become more active, both by industry and the department itself, at a time when financial shortages are increasing.

Quota or licence sanctions through administrative action may be necessary to make rules effective. For example, the IQ Group in the Scotia–Fundy region, established for the implementation of individual vessel quotas in the mid-size dragger fleet, now has the power to impose penalties ranging from 5 per cent to 100 per cent of quota or to suspend licences for two weeks to a year. Area officials interviewed are very supportive of increases in administrative sanctions. As one official put it, the criminal requirement of 'beyond a reasonable doubt' is 'ridiculous.' The new administrative criterion of 'balance of probabilities' is a better one, given the need for some flexibility in the application of rules to fisheries: 'There are 10 per cent who always break the law, and 10 per cent who never break it. The 80 per cent in the middle can go either way, and they have to focus efforts on them.'

Through amendments to the Fisheries Act, a new administrative sanctions and licence appeal system was planned for 1996. The system requires independent tribunals, one on each coast, to decide licence appeals and deal with violations of the Fisheries Act. The tribunals replace the criminal court system in enforcing sanctions. They operate at arm's length from the ministry, and their decisions are not appealable.

Once quotas are used to control fishing, monitoring becomes a necessary and costly requirement, which is only intensified with fleet or individual quotas. DFO has vessel observer, offshore air surveillance, and dockside monitoring programs. These services are mostly contracted out, and the strong trend is to require industry financing.

Conservation: The Role of Science

In the post–Second World War era Canada's management process has been dominated by a bioeconomic model of scientific analysis. Canada has concentrated on limited-entry licensing and quota systems of various sorts to restrict both effort and catches. For a number of reasons, to be discussed below, these strategies have not been particularly successful. Their lack of success has stimulated critical appraisals of fisheries science, which in the bioeconomic resource management model is based primarily on single-species population dynamics, as well as the application of that science to policy.

A major problem facing scientists in DFO is incomplete and inaccurate data. The information requirements of any management system based on output controls, or quotas on fish caught, are relatively high. By contrast, systems based on input controls, like trap limits in the lobster fishery, have modest information needs. Because Canada has tended to favour output controls, there is considerable pressure to misreport.

One provincial official observed that the herring seine fishery, which is run as an ITQ system, only has 50,000 metric tonnes being reported for landings when, in fact, the real figure is twice that. Since the system has been in place for a decade, and involves approximately sixty vessels landing at five ports, there is real concern about the viability of the system.

When systems involve other complexities, like a multiplicity of species, or numerous boats, licence holders, or ports, the information is even more suspect. In the recent past, DFO scientists in the Scotia–Fundy region have had to refrain from making groundfish estimates because the level of misrepresentation for particular areas and species has been too great. Misreporting also contributes to a decline in the credibility of scientific advice because fishers are well aware that some of them are providing bad data.

Relationships between fishermen and scientists affect the reliability and accuracy of landings data. One structural impediment is a timing problem. Groundfish assessments for the following year are conducted in May because some data are not available until March. Scientists must provide results for a July submission of preliminary plans in the advisory system. The pressure put on them to do the assessments in a short time substantially reduces the prospects for meaningful dialogue with industry.

Second is a communication problem. Scientists can have problems taking seriously the localized, anecdotal types of knowledge fishermen possess, and the differences in education levels sometimes create status barriers.

Third is tension among scientists about the extent to which they will interact with national and international peer groups (that is, the International Commission on the Exploration of the Seas) versus local fishermen. Scientific–industry cooperation appears to be greater for fisheries without output controls, such as lobsters and other invertebrate stocks, than for groundfish or pelagics, where output controls foster 'data fouling' (Copes, 1986). It may also be affected negatively by the reorganization of the fisheries agency to clearly separate science and management.

The need for assessment of fisheries stock to develop TACs and quotas puts considerable pressure on a modelling system, given its reliance on questionable data and its vulnerability to error in prediction because of its reliance on 'quick and dirty' procedures. One DFO scientist described the scientific error involved in the northern cod crisis of the late 1980s as 'relatively minor,' with scientists overestimating the growth rate by only 4 per cent. The problem was the result of compounding this error over four or five years. Some fisheries scientists are now interested in retreating from year-by-year predictions and in examining the prospect of greater emphasis on input controls.

Scientists interviewed in the Scotia–Fundy region acknowledge that their primary stock assessment techniques frequently employ inadequate data and simplistic models. As a consequence their methods often provide an imprecise method for setting quotas. However, they often find themselves under pressure to provide hard estimates upon which resource allocations can be made. The subjective perception is that they are being pressured to provide guarantees of certainty where their scientific procedures continually remind them of the substantial error ranges one must associate with projections. There is considerable unease about the fishing industry's inability to understand elementary probability concepts in their work.[2]

In each region scientific activity is embedded in a bureaucratic hierarchy, with ten or more layers of administration between data gatherers, scientists, and policy changes. Geographic and institutional fragmentation further complicates the matter. In the Scotia–Fundy region, the Lower Water Street establishment in Halifax is responsible for shellfish, the Bedford Institute of Oceanography for habitat and ecology, and St Andrew's, New Brunswick, for fin-fish.

Scientists also find themselves caught up in micro- and macro-political processes. Some scientists think they received an undue amount of criticism for their involvement in the construction of advice on cod stocks and, perhaps more importantly, believe they were not permitted to defend their activities adequately to the general public. Some of them also feel threatened by a new policy shift to environment, recreation, and Aboriginal issues. These new policy orientations may presage a reduction in the level of resources devoted to scientific effort.

Scientists in Atlantic Canada are prepared to acknowledge recent errors that have caused them to reconsider the extent to which they should be involved in the business of providing hard estimates. These and other problems with projections have stimulated some interest within the scientific community with new forms of science, including tentative moves towards modelling approaches that emphasize the non-linear dynamics of complex systems and build around the notion that equilibrium solutions and deterministic predictability may not be possible (Smith, 1990; Steele et al., 1992; Wilson and Kleban, 1992; Wilson et al.,1990). But for the most part, fisheries scientists continue to focus on refining and using existing models.[3]

Fishery Management Planning and Consultation

The major planning processes in Atlantic Canada centre on the creation of annual management plans for particular species. Until recently the process was initiated using scientific input from the Canadian Atlantic Fisheries Scientific Advisory Committee (CAFSAC), which provided the top-level adminis-

---

2 We are indebted to David Gray, Financial and Management Sciences, St Mary's University, for this point.
3 This paragraph is based on personal notes from a talk by Dr Michael Sinclair, Director, Biological Sciences Branch, Department of Fisheries and Oceans, Canada, given at St Mary's University, Halifax, Fisheries Seminar Series, 1 February 1991.

Figure 8. Management procedures for groundfish in Atlantic Canada.

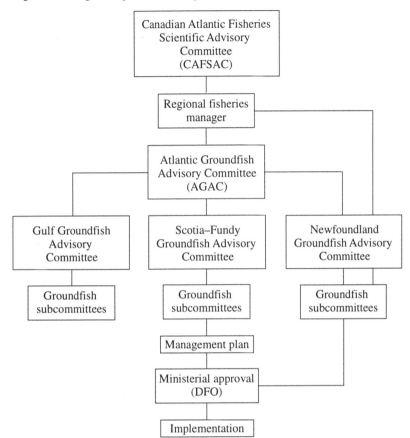

tration in DFO with preliminary estimates for total allowable catches (TACs) for the upcoming year (Figure 8). Final authority for resource assessments and recommending quotas and other conservation strategies to the minister was moved from CAFSAC and the science branch and vested in the Fisheries Resource Conservation Council (FRCC), comprised of scientists, academics, leaders of industry groups, and other experts outside DFO. The estimates are then filtered through the management process, particularly the Atlantic director general's committee (ADGC), to the various advisory committees in Atlantic Canada for comment. After consultation with various levels of advisory groups, the ADGC prepares a final plan for the Minister of Fisheries and

Oceans, who decides (Bannister, 1989: 72–3, 179–82; Hurley and Gray, 1988: 4/1 – 4/50).

Consultation is a key part of the process. Within the regions, advice is sought from industry on matters related to particular species, including the condition of the stock, allocation of the resource among fishermen, methods of harvesting, division of catch among processors, research needs and techniques, licensing policies, regulatory amendments, market opportunities, and management strategies. This advice proceeds to regional advisory committee members for deliberation and resolution. Subsequent meetings may be necessary at the area level to seek compromises and/or amendments. Regional fisheries managers must then develop the management plan for ministerial approval. Once received, the final plan is presented to the appropriate advisory committee members for advice and recommendations on implementation (Government of Canada, 1990: 3–4).

Recently there have been attempts by DFO to encourage greater industry involvement in the management of the fisheries in Atlantic Canada (ibid.). In the Scotia–Fundy region DFO has created a large number of consultative groups (36, with approximately 600 members, which meet about 90 times a year in total). To quote a major DFO official, 'The department is giving the industry more of a say in managing the fishery.' Accordingly, the official view of the activities of the advisory groups in the management process of the early 1990s was, '[S]cientific advice is assembled at the Canadian Atlantic Fisheries Scientific Advisory Committee (CAFSAC) for each major Atlantic stock and presented to regional fisheries managers, who in turn, present the advice to area advisory committee members for consultation on the development of the management plan' (Government of Canada, 1991: 3).'

Consultation has a wide range of meanings within DFO:

> It is seen as a continuum of interaction between government and the public it serves. It may range from simply listening at one end, to acting upon jointly agreed solutions at the other. In between, there is dialogue, debate and analysis. Consultation may refer to individual steps along the continuum or to the continuum as a whole. The *primary purpose* of consultation is to provide a forum where the department and the public it serves can exchange information and points of view on a broad array of issues and initiatives. (Government of Canada, 1990: 3)

The limitations of consultation are also recognized and emphasized. It 'does not necessarily lead to agreement on a particular course of action, though this may result. Nor will consultation remove the need for governments to decide, which is what they are elected to do' (Government of Canada, 1991: 1, 8).

The system is in no sense one of shared responsibility and decision-making power, or co-management (Pinkerton, 1989). It is quite clear to all parties involved that DFO maintains control of the type and amount of input the industry representatives are allowed to make and that final decisions remain with the Minister of Fisheries. Rather, the consultative process fits into a system that had been restructured to make clearer distinctions between conservation and allocation, reserving data gathering and planning about conservation to scientific experts and government officials, but relying more than before on industry for advice on allocation. Advocates of the new system see it as a solution for current difficulties in the existing regulations, which are regarded as too slow and cumbersome to provide adequate flexibility in decisions of allocation.

One concern with the advisory group system, a concern that also appears in the new FRCC, is that those involved may not adequately represent the industry, and that leadership roles may have been appropriated, in some instances, by individuals with special agendas. The nomination of representatives to advisory committees is largely done by polling individual fishermen. DFO provides a form that fishermen complete when renewing their licences. On this form they are to name the representative of their choice. The results are tallied by DFO staff, who also notify the prospective representative and ask him or her to confirm his or her willingness to serve on a committee. This process is repeated every two years.

Unions such as the Maritime Fishermen's Union can nominate their own representatives. Representatives from provincial governments are chosen by the provincial deputy minister (Government of Canada, 1990). However, it may be a matter of years before people recognize the possibility of inadequate representation and are able to replace their representatives.

Moreover, it can be difficult to reach agreement when there are representatives of very different sectors and communities. One method that has been devised for circumventing the self-interested nature of discussions in advisory groups is the 'working group.' Once an advisory group is able to identify a specific problem, or set of problems, DFO not infrequently moves to create a representative subgroup that will try to generate 'ideas, options, and alternatives' for the larger advisory group. For example, it became clear, at the June 1992 meeting of the Scotia–Fundy Groundfish Advisory Committee, that there was some common concern with the future of fisheries management under a condition of substantial cut-backs in allocations. John Angel obtained consent for the idea that it would be useful to have a subgroup chosen from the enterprise allocation (EA), individual quota (IQ), and fixed-gear sectors to examine 'the 1993 fishery in the light of the existing scientific advice.' In establishing the working group, Angel emphasized that people are not 'necessarily there to put your point of view forward.'

Critics of the consultative system are concerned that the proliferation of advisory groups will simply intensify the existing rigidities and will not, in fact, provide the kind of flexibility that the Scotia–Fundy region, in particular, requires. Some observers argue that the new consultative groups will not help to eliminate the complex scheme for opening and closing seasons, will not facilitate utilization of unexpected pulses in stock recruitment, and will not reward good fishermen for improved quality in their activities.[4]

Others suggest that consultative management is something that applies only to the midshore sector, particularly the midshore fleet of seiners, draggers, and crabbers which are connected to intermediate-sized companies. For example, the herring industry, which DFO regards as being quite cooperative, has a group of forty seiners that have been operating under an independent quota system for eight years. The offshore sector participates little in the advisory group, relying more on organizations such as the Fisheries Council of Canada to provide advice and influence policy. The inshore fishermen can participate, but their effectiveness may be compromised by greater difficulties organizing, as well as the possibility that consultative management intensifies divisions among fishermen because the advisory process is predicated on licensing and quota principles which undermine independent fishing activity and small-scale rural communities in Nova Scotia (MacInnes and Davis, 1990).

The Groundfish Management Plan

We have had the opportunity to attend a number of groundfish advisory group meetings over five years. What follows is an attempt to convey how these committees function as arenas of consultation, mostly focused on the annual groundfish management plan and its implementation.

As can be seen, the issues discussed are far-ranging, from the technical and scientific to regional competition among fleets. At a meeting in the fall of 1991, for example, most of the day was devoted to a detailed evaluation of the connections among current data on landings, CAFSAC documents, and the Atlantic Groundfish Management Plan. For example, in the discussion of cod within a particular area (4Vn), there was extended discussion of whether the planned reduction from 48,000 tonnes to 43,000 tonnes in the Scotia–Fundy region

---

4 These points are developed from talks given by Glenn Jefferson, Assistant Director of Resource Management, Department of Fisheries and Oceans Canada, and Art Longard, Director of Marine Resources, Nova Scotia Department of Fisheries, at St Mary's University, Halifax, Fisheries Seminar Series, 8 March 1991.

Managing the Fisheries: Procedures and Politics 161

should occur in 1992, what sharing formula by gear type should be employed, and how one should deal with the assertion by CAFSAC that some of the catch in the 4Vn area were fish migrating from another area (4T) during the January-to-April season.

Subsequent discussion proceeded through the relevant Northwest Atlantic Fisheries Organization (NAFO) areas by stock to consider the connections among landing data, CAFSAC input, and the management plan. Two issues recurred. One had to do with the acceptability of the scientific advice available, and the second had to do with questions of regional differences and allocations. For example, in a discussion of 4Vn cod, DFO scientists gave a slide presentation to argue that their catches indicate that some 4T cod were moving out of the Gulf of St Lawrence in front of the ice cover into 4Vn and 4Vs during the period January to April, and then moving back as the ice receded. However, they acknowledged that DFO had only two years of data to support this analysis, and they indicated that this was not the most desirable data base from which to make projections. Several individual representatives then challenged the 4Vn projection, saying that the evidence did not justify a decrease in quota. Rather, it was a 'political decision' to favour the gulf region. The representative of the Seafood Producers Association of Nova Scotia (SPANS) supported this statement, saying that '43,000 tonnes is not a scientific decision, but a political one.'

DFO scientists were often quite disarming in their recognition of the relatively modest amounts of good evidence underlying their statements. For example, it was argued that the scientific assessment of fish in one area (4Vs) was 'quite shaky scientifically, and that the DFO scientists do not know why.' Industry representatives speculated that this uncertainty may be related to a misrepresentation of catches. Further, when challenged by a representative of a major processing firm regarding a proposal to re-map certain stocks to create separate TACs for different regions, the biologists agreed that their acoustic surveys did not provide 'solid evidence' to establish such a policy. At this same meeting, the scientists, under challenge, admitted that there is little knowledge of flounder stocks or biology, and that it is very difficult to get information on separate species. However, they argued that the industry might have to accept indirect inferences based on price differentials in the marketplace, given the costs of scientific investigations.

An undercurrent of resentment towards fishermen in other regions was evident. For example, one Nova Scotia fisherman was quite vehement in his assertion that Newfoundland fishermen were destroying an offshore (3NO) cod fishery that Nova Scotians use as a hook-and-line fishery to obtain hake and halibut for salt-fish plants. In addition, there were concerns about American excursions into international waters (5Y). As one fisherman pointed out, 'Them 85-footers aren't sports fishermen.'

An important issue at the groundfish advisory group meetings was the new system for monitoring catch. The dockside monitoring program, which began in 1990, is designed to put enough equipment, infrastructure, and people in place to create a system adequate to police the IQ systems in the region. In general, the industry was strongly opposed to the DFO's cost recovery approach. They maintained that an elaborate system of independent- or industry-related weigh masters is not necessary, as one can have spot checkers and an adequate 'paper trail' from fishermen's logs and fish processors' books. DFO responded that its accountants, well versed in doing audits of fishery firms, were concerned about the poor quality of current industry record-keeping, and that it was very difficult to trace catches back to particular fishermen. After extensive discussion, DFO agreed to an open advisory subgroup to evaluate alternatives on the question.

Since proposals for institutional reform were under way, which resulted in the FRCC (discussed below), there were no Scotia–Fundy advisory committee meetings between January 1992 and December 1993. The 9 December 1993 meeting considered whether to resuscitate the committee to fill the void created by the delay of the reform process. Industry representatives were generally of the opinion that they would like to continue management in this forum to counterbalance the growing influence of the new FRCC, which had industry representatives but only very few, making it impossible to represent all regions. Technically there is no conflict between the mandates of these groups. Nevertheless, there was a widespread perception at the Scotia–Fundy meeting that the FRCC was beginning to encroach on management concerns. Further, committee members believed that continuation of their group would be helpful in dispelling a considerable amount of misinformation that was circulating among the various sectors. Finally, there were a number of intersectoral issues, like spawning and nursery closures, gear limits, mesh size, and hook size, which they thought would benefit from broader discussion.

Much of this meeting addressed itself to differences in advice coming from the fixed gear and IQ Subcommittees, with some efforts being made to resolve contradictions in advice to the minister regarding allocations. The technical discussions had a new tone of cooperation that stood in marked contrast to the acrimony that had characterized earlier meetings. For example, when the fixed- and mobile-gear representatives got into a dispute about the relative merits of their types of gear, the exchange began with the usual charges about the destructive capacity of mobile gear (that is, otter trawls), as well as gillnets and small hooks. However, the meeting then turned to the question of creating an observer program in which unemployed mobile- and fixed-gear fishermen would serve as observers for fixed-gear and mobile vessels, respectively. In this way, as the proponent tartly pointed out, they could be sure to have 'good, biased' observers. Further, another industrial representative remarked that the Scotia–Fundy group should try to stay away from 'little conflicts' between gear sectors.

Also there was a new sense of realism in these deliberations. When one of the industrial representatives argued that they seemed to be creating a new set of rules that could not be enforced ('making regulations to regulate regulations'), a DFO representative agreed, saying that is 'exactly right. To any great degree, we are not going to be able to enforce all that.' Nevertheless, it is 'still worthwhile to establish a consensus and put it on the licence.' However, there were limits to this new mood of cooperation. When the group argued that it would be useful, for practical purposes, to diminish the overall size of the group to facilitate its work, no agreement could be reached on who should be in the reconstituted advisory group.

This meeting was also noteworthy because the Scotia–Fundy region had begun formal contacts with their American counterparts on the New England Management Council through a Gulf of Maine advisory council. While it is clear that the American and Canadian areas are very different in style and philosophy, they were ready to discuss joint conservation measures, particularly as they affect overlapping cod and haddock stocks. (Concurrently, American and Canadian scientists developed more cooperation in data collection and stock assessment.)

Lobster Fishery Advisory Groups

The consultative arenas vary not only over time, as we have seen, but also across fisheries. Advisory groups for the lobster fishery, although still within the consultative management framework, have a different dynamic, judging from meetings of the 'LFA 34' lobster committee. The committee represents the most lucrative single geographic area for Atlantic Canada's lobster fishery, an area which runs roughly from Baccaro to Digby in southwest Nova Scotia. In contrast to the groundfish meetings, the general mood was more cooperative and conciliatory, even when there were clear differences of opinion on controversial matters such as trap limits and Aboriginal fishing rights. This extended to science and scientists, who were rarely challenged.

The fourteen elected members who sit on this committee are clearly representative of specific geographic locales in the area. Their constituencies are well defined, and there is a sense of equality among the individuals who attend. By comparison with the general groundfish meetings, there is little grumbling about the basis for inclusion in the advisory committee. Nevertheless, there is some difficulty in ensuring regular attendance and continuity in committee affairs. At the two meetings we attended only half of the regular elected members were there, along with a few alternates. At the October 1992 meeting it was reported that adjacent lobster fishing area 33 had chosen to deal with the question of rotation by having two-year terms, with half of the people rotating out each year. Nominations would be made in the fall, when people get their trap tags, and voting would take place in the spring, when they pick up their spring tags. The April 1993 meeting decided to leave the issue of membership in abeyance until there was a better sense of the implications of the ongoing reform process for advisory groups.

Much of the early parts of the meetings are devoted to local reports from elected representatives. The issues covered included catch levels, pot limits, enforcement, the effects of weather on gear and catches, and safety conditions on opening day during the rush to set out traps. While satisfaction with the high catch levels and existing trap limits was reasonable, there were a number of concerns about enforcement. Representatives complained that some people cheated by leaving pots early on the designated starting date for the fall season, and there were numerous reports of draggers and gillnetters illegally catching lobster. The DFO area manager defended his operations by pointing out that officers, three vessels, and a helicopter had to cover a huge area, and that he sometimes felt as if they were 'going into the OK Corral with no bullets.'

Complex discussions took place about creating a more flexible definition of the fall 'dumping' date because of concerns about pressure to set out traps in unsafe weather and sea conditions. One fisherman said that five of the last ten years had been 'a nightmare for safety ... Safety is our major objective, and we are open to any suggestions.' He reported that one captain told him, 'My men were not scared, they were petrified.' However, DFO officials argued that they could not accept the legal responsibility for safety implied in having the department define unacceptable weather conditions.

The scientific presentations at these meetings were well received, with considerable interest in issues like shell disease and the effects of water temperature on abundance, as well as support for expanding research resources. Although the presentations sometimes showed high levels of scientific uncertainty, industry representatives did not appear as concerned as they were in groundfish meetings, perhaps because their lobster catches had been quite good.

Directed Participation?

The consultative system is one of directed participation, with DFO making final decisions in virtually all instances. As a DFO official stated when a sharp debate occurred over quota allocations at one meeting, the committee is 'only advisory.' To clarify how limited this form of democracy is, he said, 'We don't entertain votes here, [just] advice that will be passed along.' The meetings are chaired by a DFO official who typically sits at the head of the table, surrounded by the senior management advisers, as well as DFO biologists and economists. Industry representatives are arrayed around the table and contain a mix of fishermen and fish processors. The more important the advisory committee, the greater the likelihood that processors, especially larger ones, will be represented. Members are expected to represent the interests of their respective peers. However, it is clear that a number of the people who attend these meetings with voting privileges represent relatively few people, while others may have reasonably large constituencies. The fishermen's representatives can be divided by type of gear, size of vessel, and geographic location. Their different perspectives frequently lead to sharp exchanges among the fishermen representatives, as well as between fishermen and DFO.

The fish processors, a different and competing set of interests, are usually more guarded in their criticisms of one another and of DFO activities. In part, this difference in style represents differences in education, class, and experience that affect the ability to deal with the sophistication of DFO managers. However, there are also power differences, with the processors having more influence over the outcome of different issues. Sometimes the consultative structure can lead to considerable frustration among the fishermen. For example, at one meeting, one of the more vociferous fishermen's representatives began the afternoon session by walking in late to a full meeting, slamming his attache case against the wall, and declaring, 'Let's get this ——— meeting going. I don't have time to waste.' More commonly, the power differentials result in direct, but reasonably polite, challenges to DFO policy. For example, one critic of the new ITQ scheme for inshore draggers said that it 'is a curse to this area'; Digby Neck [his town] is going to be 'wiped off the map. We are desperate and I do not think you people understand.'

For their part, the DFO officials are adept at absorbing criticism, deflecting anger towards higher levels in the organizational structure, and making a few points on their own behalf. One DFO official wryly commented, when it became apparent that misrepresentation of catches was undermining estimations for a particular stock, that DFO might proceed with existing plans so long as fishermen 'lie consistently.'

A second example of directed participation as well as the issue of representation appeared in the selection of a management committee for the inshore mobile-gear fleet that was to be managed under individual quotas in the Scotia-

Fundy region. DFO sent out a questionnaire about representation. It provided three options: (1) representation by the amount of quota held; (2) individual selection in which people would have to get endorsement from at least 20 per cent of the IQ holders in their area to become representatives; and (3) any other solutions, including having DFO select the management committee. Approximately 60 per cent of those answering replied that they wanted DFO to select the committee. The fishermen's vote split relatively evenly among the various options, while the fish processors strongly favoured DFO selection. The balance was tipped by the relative uniformity of the processor vote.

Consultative Management: Problem or Panacea?

Both the Norwegian and Atlantic Canadian fishery management systems are marked by a heavy emphasis on representation from user groups to provide information and advice as well as legitimacy and support. They are, however, strikingly different when it comes to the organization and coordination of group interests and in the nature of the consultative process. In Norway interests and demands are coordinated through a dominant peak organization. Participation works mainly through formalized cooperation between central government and representatives of industry. In Atlantic Canada interest coordination is probably less efficient, and participation is more fluid and less well structured. It is channeled through a multitude of advisory committees and working groups at the provincial and regional levels.

The Norwegian regime is clearly corporatist. Relationships between industry and government are formalized and extensive, and representation is direct and selective. Management decisions are, first and foremost, the domain of a relatively coherent policy community that dominates the process by virtue of its specialized knowledge, vested interests, and administrative responsibilities. The core members of this community include scientists and managers from the Directorate of Fisheries in Bergen, officials in the Ministry of Fisheries, and representatives of the Norwegian Fishermen's Association. The legitimacy of this arrangement is largely sustained by history and tradition and by the fact that it does produce viable compromises. There are, however, several points worth pondering with regard to the 'democracy' as well as the 'efficiency' of the system, which apply as well to Canada and other nations.

One is the 'classical' issue of group demands versus public concerns. Corporatist arrangements are biased in the sense that representation is exclusive, granted only to 'affected interests,' even though decisions and policies may affect wider groups. Fisheries management has long been based on the presumption that policies are, or should be, the exclusive domain of the industry, even of a particular segment of it (for example, harvesting). However, there are other groups that could claim a stake in the management of fishery resources such as consumers, local and regional authorities, environmental groups, and ethnic communities. These groups constitute what Pross (1992: 121) has called the attentive public; those who are interested in, or somehow affected by, management policies. Some of these groups or interests have recently been awarded 'associate membership,' but they can, as yet, hardly be considered core members of the policy community.

Second is the 'efficiency' question versus power sharing, logrolling, and compromise. Corporatism is highly conducive to interest group politics; it represents a recognition of special interests and provides a structural framework for their articulation, but sometimes at the expense of public governance and long-term planning. Special interest groups, such as fishermen's associations, are recognized as partners by government, and sometimes they are allowed to operate as veto groups, leaving relatively little scope for the exercise of ministerial discretion and departmental power. In this context a commendable policy, almost by definition, is the measures to which those involved can agree, those that are not vetoed by any major player, or those that are favoured by a winning coalition.

Logrolling, consensus and legitimacy rather than efficiency and governance are the core values. That said, it is important to note that user groups have more influence over implementation, for example, the allocation of quotas among groups of fishermen, than over the more basic decisions of overall fishing effort. On this point scientific advice carries greater weight than 'anecdotal' information from user groups.

Third, there is always the question, even within systems conducive to collective action, of whose interests are being served or accommodated. The fact that the industry as such is well represented in policy-making does not necessarily imply that all groups and interests are given a fair hearing. The coordination of interests and demands within the Fishermen's Association is decisive in this respect. Without a doubt this aggregation is difficult and time consuming, mainly because of the heterogeneous nature of this organization. It is basically a fragile coalition of offshore and inshore groups of boat owners, crew members and independent operators – groups whose administrative capacity and political clout differ considerably (Hallenstvedt and Dynna, 1976; Jentoft and Mikalsen, 1987). Opposition to particular management policies and decisions is frequent, especially among inshore fishermen in the north, who contend that the association's policies are biased in favour of larger operators, namely, the offshore fleet. As already mentioned, a group of inshore fishermen has, in fact, broken with the association and formed their own organization.

Although the Canadian system generates similar issues of organizational fragility and representation, most striking in contrast are the dominance of ministerial and departmental power in fisheries management decision-making. The Department of Fisheries and Oceans retains control of the type and amount of input industry representatives as well as provincial governments are allowed to make. In this sense there is no real sharing of power with the industry. Openness, purpose, genuine consultation, and adequate information are said to be conspicuously absent from the management process, and there is clearly an element of mutual distrust in relations between government and industry.

The lack of power sharing and real consultation was noted by a government task force even back in 1982: 'The Department of Fisheries and Oceans is sometimes accused of being secretive and arbitrary and of taking decisions without adequate knowledge or advice from knowledgeable sources (that is processors and fishermen). Provincial governments complained that they are treated as merely one of many interest groups rather than as another level of government (Kirby, 1982: 339). This was acknowledged in a recent government report on fisheries management which states, 'The Minister of Fisheries and Oceans and departmental staff make the critical decisions about who gets to fish – the licensing decision – and how much – the allocation decision – behind closed doors. Lack of information about how and on what basis decisions are made sometimes make them seem arbitrary and unfair' (Government of Canada, 1993: 1).

Consequently, in Canada there is little that resembles the Norwegian version of a more formalized partnership between government and industry. In fact, relationships sometimes seem to be based more on mutual distrust than genuine cooperation.

Second, the system of advisory committees is complex and built around a highly fragmented organizational structure. With the exception of Newfoundland, where a fishermen's union plays a key advisory role, the level of organization among fishermen is low, and there is no peak association to coordinate demands. Particularly in Nova Scotia government initiatives to create representative associations have met with little enthusiasm. Traditions of independence and individualism are strong, reinforced by technological differences and economic circumstances, and they have proved a real barrier to collective action (Pross and McCorquodale, 1987). Because of the lack of truly representative associations, the selection of representatives to advisory committees seems rather arbitrary and ill-structured. There are few stringent rules governing appointments to these committees, and those appointed are seldom formally responsible to 'their' group. No formal criteria exist to ensure a balanced participation across the variety of groups and interests, and committee meetings tend to be fairly open, almost public events, where consensus is difficult to achieve. In the words of one of our respondents, 'It seems that as the resource has declined, the demand for a say has increased. Some groups you would never see ten years ago ... There may be 80–100 people at a round-the-table discussion now. It gives everyone an opportunity to have their say, but from the top manager's point of view it makes things more difficult, it's hard to get consensus.'

Third, there is the familiar problem of ensuring that the ultimate decisions of representatives correspond with the values, expectations, and demands of those represented and affected. Again, the lack of clear-cut rules and procedures of membership selection make the consultative process fairly open and indiscriminatory, often to the extent that it is not clear whether those turning up at meetings represent anyone but themselves. For instance, there is the problem of 'community notables': prominent persons turning up at advisory group meetings without a mandate from any particular group or association. There is also a problem of representation, as emphasized in a task force report on the Scotia–Fundy groundfish industry: 'Some members of Advisory Committees do not represent significant numbers of fishermen and often do not have any mandate or instructions to make commitments on behalf of the fishermen they do represent. Too often, Advisory Committee members express their own views and are quick to point out that they do not wish to be seen speaking for other fishermen' (Haché, 1989: 63).

Some Preliminary Comparisons

The issue of jurisdiction illustrates differences in legal foundations and institutional histories. In Canada the authority of central government to manage is often a matter of dispute. Federal legislation has frequently been challenged by the provinces, and disagreement over the sharing of functions and power between these two levels of government continues. What sets Norway apart from Canada in this particular respect is the general acceptance of fisheries management as the (exclusive) jurisdiction of central government, as may be appropriate and uncontroversial within the framework of a unitary state in contrast with a federal state.

There is a paradox, though. Looking more closely at the management institutions and the ways in which decisions are actually taken, the Canadian system appears more centralized than the Norwegian one. Despite challenges to federal authority, executive discretion of the Minister of Fisheries is considerable. In Norway, where there is little controversy about the constitutional right of central government to manage the fisheries, ministerial power is much more confined. As we shall see below, this may, in part, be because of the power and privileged position of user groups.

Both regimes are systems of consultative management. Decisions are made within administrative structures where the need for centralized control is balanced against demands for user-group participation. Assuming consensus that such participation ought to prevail, two problems arise. First is the question of who should be represented when policies are discussed and decided, that is, the problem of inclusion (Dahl, 1986: 191). Second, how do we know that those participating in fact speak on behalf of those they claim to represent? This is the problem of representability.

To begin with, inclusion is more often than not a question of organization, of being able to coordinate common interests and demands and act upon them. However, all affected interests may not be organized or may not have equal status or access. The Norwegian Fishermen's Association, for example, has long had privileged access to government. Other organizations such as the Norwegian Association of Fish Processors, not to mention the plant workers, have not had the same political 'clout,' and they have just recently been represented in management policy-making. One reason is that processors were late in joining forces for collective action (Hallenstvedt, 1982), thus lacking both the power and the skill to press for representation and put their stamp on the relevant agendas.

In Atlantic Canada fishermen are the ones who have largely failed to build organizational capability. With the exception of Newfoundland, there is more fragmentation than unity – in spite of efforts from both government and industrial 'entrepreneurs' to encourage the formation of more comprehensive and representative associations. There is not necessarily a lack of associations per se. The problem is rather that they are too many and too small – with few prospects of forming a 'federation,' or peak organization along the lines of the Norwegian Fishermen's Association. The Canadian fish processors have been more successful in organizing. The Fisheries Council of Canada, established in 1945 as a vehicle for the processors, has been an important link between the federal government and the industry. Moreover, the big processing companies have sufficient administrative capacity of their own to present their views directly to the federal government (Pross and McCorquodale, 1987: 68–9). Fish buyers and processors are well represented on advisory committees, usually by the same individuals acting as representatives on all relevant committees (MacInnes and Davis, 1990).

In Atlantic Canada, then, it is not at all certain that the voices heard represent a cross-section of industry opinion or that representatives have in fact been granted the authority to speak on behalf of the larger group. At the root of this problem, distinguishing Atlantic Canada from Norway, is poor organization. A task force of the late 1980s identified the problem and reported the unlikelihood that it would change in the forseeable future (Haché, 1989: 63). In the context of the 1990s resource crises, serious attempts are being made to improve the situation, including legislation that requires all licensed fishers to pay mandatory dues to an association of their choice.

In addition, and again compared with Norway, the sheer complexity of the consultative process in Canada, with its multitude of committees and subcommittees, makes it hard to know exactly how and why decisions are made, adding to the general frustration and alienation of ordinary fishermen. It is partly against this background that a recent initiative to change the institutional framework of management decision-making must be seen. Before turning to the issue of management reform, however, we will outline management objectives and policies and discuss how current institutional frameworks may affect the scope and content of regulatory measures.

5

From Procedures to Policies

Three issues loom large on the management agenda for the fisheries of both Norway and Atlantic Canada: resource conservation, economic efficiency and regional development and employment. Conservation of resources, how to avoid resource depletion, has been there since the state started to take an interest in the industry, even if it did not – with a few exceptions – become its major focus until the late 1960s. The second, how to make the fisheries economically viable, has spurred a wide range of state policies including the safeguarding of collective institutions in Norwegian fisheries through formal legislation (Hallenstvedt, 1982), modernization programs, and a fairly elaborate system of limited entry on both sides of the Atlantic (Mikalsen, 1987; Parsons, 1993b; Ørebech, 1982). The third issue pertains to the regional and employment repercussions of management programs and policies: how these may affect the geographic distribution of resources, employment opportunities, and harvesting rights. In Norway, ever since limited entry was introduced, there has been concern about the regional effects of regulatory decisions. Measures have been introduced to prevent concentration of licences and quotas, such as restrictions on the transferability of licences across regions and the possibility, grounded in the revised Salt Water Fishing Act of 1983, of retaining a part of the overall quota for fishermen in particular regions. The old schism between the south and the north has, however, taken on a new significance as a consequence of dwindling stocks and smaller quotas, raising equity questions that have attracted increasing attention in recent discussions about the (re)design of management policies and institutions.

In Atlantic Canada there is the perennial discussion of whether the fishery should be an economic or a social fishery, an economically viable industry or an employer of last resort. While the usefulness and relevance of this particular distinction is debatable, it has come to denote two different approaches to management – one germane to the industrialized, resource-centred, offshore sector, the other more firmly rooted in the problems of local communities and the adaptations of inshore fishermen. As pointed out in a recent task force report (Cashin, 1993), the fishery is in fact an employer of last resort in much of Atlantic Canada, where alternative means of employment are scarce or nonexistent. This is particularly true for Newfoundland's south and northeast coasts, but in parts of the Maritimes other sectors of the economy are heavily dependent on the fishery. There are also regional differences in the resource base. Newfoundland, for example, has been heavily dependent on groundfish, while the fisheries of Nova Scotia have long been more diversified.

Common to both countries is that political pressures, social considerations, and regional variations have served to mellow the basic approach of industrialization and economic efficiency – as can be seen from the management objectives adopted by the two regimes.

Management Objectives: Norway

Probably the best source on official management policies in Norway is a recent report from the Ministry of Fisheries to Parliament (Stortinget) (Government of Norway 1991–2). The document is one of analysis and recommendations, and it roughly corresponds to what the British would call a 'white paper.' It has been 'cleared' by the Cabinet, and is therefore a valid indication of how the present government conceives of fisheries management issues. This document contains a statement of policy objectives, an analysis of current problems in the management of fish stocks, and a set of proposals of how these could be dealt with.

With a few exceptions the current restrictions on effort and catch in Norwegian fisheries are fairly recent, the oldest dating to the early 1970s. Since then there has been a continuous increase in the number and scope of regulations, and today close to 100 per cent of the catch (in terms of economic value) comes from limited entry fisheries. This figure was 50 per cent ten years ago and only 15 per cent in 1970. Tighter restrictions and more complex rules notwithstanding, the main objectives of fisheries management have been fairly stable over time. In its report to Parliament, the Ministry of Fisheries in fact goes a long way to assure the continuing validity of general objectives stated more than ten years ago.

In its discussion of goals, the report draws on several other 'white papers.' All have particular emphasis on *resource conservation* as the primary objective of fisheries management. The main challenge of fisheries management is seen as the rebuilding or strengthening of the stocks. The 1991–2 report renews an emphasis on *economic efficiency*, both directly through assertions that the profitability of the industry leaves much to be desired, and indirectly through the fact that significant reductions and the eventual elimination of government subsidies have become a major concern.

The report's diagnosis of *fleet overcapacity* as the main problem ties in with the twin goals of conservation and efficiency. That being said, there is also an emphasis on the role and significance of fisheries management within the wider context of regional policy, and on the need to balance economic rationalization against social concerns. Preserving the dispersed settlement structure of coastal Norway and securing employment opportunities in 'marginal' regions have always been important concerns of Norwegian fisheries management. The main dilemma has long been how to improve economic efficiency without fundamentally altering the geographic distribution of vessels and harvesting rights. This dilemma is reiterated in the present report, again by quoting, and commending, a recent white paper on regional policy:

It is necessary to design the regulations in such a way that the basis for future harvesting is not destroyed in the most fishery-dependent regions along the coast. This requires a regional redistribution of harvesting rights and catch supplies ... The government will enforce a transformation of fisheries policy towards a stronger emphasis on resource management and on the regional significance of the fishing industry. (Norway 1991–2: 15)

The ministry thus takes great care not just to emphasize the consistency in its management goals over time, but also to show how the regulation of fisheries fits into a more coherent policy of regional development.

The potential conflict of goals, acknowledged in several places in the report, is perhaps most apparent when the requirements of economic efficiency are set against the goal of regional development and consolidation. While the latter implies intervention and control, that is, socially motivated redistribution and positive discrimination, the former points towards autonomy and competition, that is, towards the use of marketlike solutions in fisheries management. As pointed out in the report, fleet rationalization by reducing catching capacity may strengthen profitability, but it will also increase unemployment along the coast. That said, it is argued that such conflicts are not insurmountable, in as much as economic efficiency may be critical to the future viability of coastal communities, particularly if government subsidies are gradually ended.

Fisheries management is not only about preserving the resource, but also about protecting bona fide fishermen from, say, part-timers and outside investors. The professionalization of fishing is an objective alluded to in several white papers on management. It is based on the assumption that the conservation of resources as well as economic viability can be improved by giving preference to bona fide operators when rights to access are allocated. At present the scope and duration of previous participation is the main criterion for distinguishing 'full-timers' from 'part-timers.'

Although most government reports and white papers go a long way in acknowledging the obvious conflict of objectives, at least in the short run, the basic tone is one of economic efficiency. Resource conservation is clearly the overriding, long-term objective, but it is considered less as an end in itself than as a step towards improving the overall profitability and efficiency of the industry. Missing, however, is an explicit attempt to establish criteria for weighing goals and priorities within the framework of 'hard' management decisions. This is not surprising, given the controversial and overtly political nature of the task. Parliament has contributed little to clarify the issue, other than reiterating that there is a trade-off to be made between economic objectives and social considerations.

Summing up, the goals of management outlined in the ministry's 1991–2 report are strikingly familiar: resource conservation, economic efficiency, and regional consolidation and development. The first is largely a question of controlling effort, through licences and quotas and of strengthening research efforts; the second addresses the issue of fleet rationalization more directly, and, in the present situation of overcapacity, seems to call for a significant reduction in the number of vessels in most fisheries. The third is a political 'must,' given the dispersed settlement structure along the coast and the long-standing commitment to regional development in Norwegian politics.

From Objectives to Policy: The Case for Fleet Rationalization

The basic approach of modern Norwegian fisheries management springs from the premise that there are, and have been for about thirty years now, far too many (and expensive) vessels chasing too few fish. The result, as depicted in various government reports, is that expenditures (especially the costs of capital investments) tend to exceed earnings, accelerating the need for government subsidies that, in turn, stimulate further investment. Figures cited in the report demonstrate the problem, although it is recognized that the actual effort at any given time may be less than technical capacity. The main strategy advocated is thus one of 'efficiency through fleet rationalization,' of adjusting harvesting capacity to the quotas set.

On this point general characteristics of fisheries policy as well as deficiencies of the management system are offered as explanations for the present malaise. Government subsidies have generated incentives that have slowed down the 'natural' reduction of harvesting capacity. There has been scant motivation to withdraw from a fishery when income support has made it possible to balance earnings and costs, especially in a situation where alternative employment was unavailable. Moreover, licensing schemes have largely been based on single fisheries, and there have been continuous pressures from the industry to expand the number of licences. The harvesting capacity thus generated has not been affected by downturns in overall quotas, mainly because of the safety net provided by government subsidies.

In addition, management of the fisheries has been a question of making and enforcing a set of intricate rules through fairly complex administrative procedures. Although not mentioned in the 1991–2 report, one gets the impression that rules and regulations have been open to interpretation and dispute, making management decisions an exercise in administrative discretion under pressure. Finally, the fact that most licences have been restricted to a particular fishery has made licensing schemes extremely inflexible. The need to tailor management programs to the diversities of fishermen's adaptations has largely been neglected, thus making it virtually impossible to change patterns of harvesting in response to fluctuations in stocks and markets.

At the core of the analysis, the problem of excess capacity is largely ascribed to deficiencies in regulatory schemes. The basic assumption seems to be that the industry has reached a point where the market failures of the 1960s and 1970s have been replaced by the management failures of the 1980s. Hence, the challenge is framed as one of restructuring the incentives of individuals and groups by (re)introducing market forces. In other, and more bureaucratic, words (Government of Norway, 1991–2):

An important task in connection with a restructuring of the management system is to find those methods of regulation that, to a greater extent than is the case today, facilitate the use of market forces should these provide for a more efficient use of resources. (p. 111)

Individual transferable quotas [ITQs] is the model that with the greatest probability will secure a continuous adaptation of harvesting capacity to the resource base. This, in turn, ensures that the industry becomes profitable. (p. 120)

In the next chapter we will return to this topic, asking why, then, the ministry has not yet adopted ITQs in Norway. To some extent the answer is in this report, where the ministry adds that a marketlike approach must not be at odds with the public interest or societal goals ('samfunnsmessige mal'). It also concedes that ITQs, in their pure form, do not pay sufficient attention to the goals of regional development and consolidation. Nonetheless, the basic approach is clearly one of 'more market, less government,' and is neatly summarized in the

TABLE 2
Allocation of Norwegian quota between offshore and inshore sectors

Norwegian quota (000 tonnes)	Offshore (%)	Inshore (%)
100–150	25	75
150–200	28	72
200–300	31	69
300	35	65

report's characterization of the preferred system of management as one that is largely self-regulatory, secures greater freedom and autonomy for the individual fisherman to plan his activities, and reduces the need for government intervention in solving the structural problems of the industry.

The Proposals

As is true in many other countries, management policy in Norway requires distinguishing between inshore and offshore fleets – defined in terms of type of gear, scale of operation, and other factors that correlate with distance from shore. The need to maintain a diversified fishery is one of the reasons given for this: 'In a free competition between the inshore and offshore fleet, there is the danger that the inshore fisheries will lose because the fishing will take place far off the coast, and fewer fish will be available inshore. In order to maintain a diversified fleet, while at the same time granting individual fishermen greater autonomy and responsibility ..., the inshore fleet must be shielded' (p. 135). This argument supported continuing Norway's traditional 'quota ladder' arrangement that defines the respective shares of the two sectors under different TACs. This, in turn, underpins the continuation of a two-tier management system whereby regulatory measures are tailored to the particular problems of each sector, and, ironically given the emphasis on marketlike solutions and efficiency, a movement away from that for the inshore sector.

The comparison, in the report, between offshore and inshore, is based on the so-called quota ladder for the allocation of total allowable catch (TAC) between these two segments of the fleet. The ladder is really a rule or guideline – agreed to by the Norwegian Fishermen's Union – whereby the share of each sector varies by the size of the Norwegian TAC. The ladder is illustrated in Table 2.

The context is a significant change of circumstances for the inshore fleet since the late 1980s. Pressure from the former Soviet Union put an end to the privilege of exceeding the agreed-upon quota in 1989. This meant that fishing had to stop when the quota was taken, whereas in the past the inshore fleet had been allowed to overfish considerably. At the same time, growing suspicion among marine biologists that the Arctic-Norwegian cod stock was far weaker than originally anticipated led to a dramatic reduction of the TAC which, in turn, meant smaller quotas for the inshore fleet, the bulk of which is located in north Norway. Meanwhile, government subsidies have been greatly reduced, and there has been a growing interest among government officials, in market-like solutions to problems of management. To talk about a major shift in fisheries policy may be premature at this point since it is not at all clear whether recent adjustments of 'paradigms' and perspectives are part of a long-term strategy to 'modernize' the industry.

The inshore sector, which in the report is largely synonymous with the 'conventional' gear (such as longline, nets, and handline) cod fisheries, was open access until 1984. From then on the prevalent form of management has been a combination of overall quotas and periods of no fishing. Since 1990, however, these measures have been replaced by vessel quotas. The impetus for change is largely to be found in the alleged deficiencies of this system: lack of flexibility and a 'hidden' or 'black' market for quotas. Also, the legitimacy of vessel quotas within the industry is somewhat shaky, especially among those who benefited from the competitive race of the overall quota system.

By and large, the ministry seemed to favour a return to the policy of the 1980s. The plan, as outlined in the report, is to gradually phase out individual vessel quotas, replacing them with a system based on group quotas divided into 'segments' or parts that will have to be taken within designated time periods. There will be no licences in inshore fisheries. An annual fee for a listing in the vessel register is, however, proposed. This is to be graded according to vessel size. Further, the requirements for ownership will be tightened, securing better professional protection for bona fide fishermen.

According to the Ownership Act of 1951, a person must have been actively involved in fishing for at least three of the past ten years to own a fishing vessel that is fifteen metres or longer. The current proposal implies activity for three of the last five years, and extends this rule to all vessels regardless of size.

An overriding objective of management in the offshore sector is the reduction of harvesting capacity through enterprise quotas and a tightening of licence policies. Enterprise quotas ('rederikvoter') allow an enterprise with several quota shares to transfer one or more of these to a single vessel – on the condition that the other vessels are withdrawn from the fishery. It has also been proposed that there should be fewer restrictions on the transfer of licences within enterprises that own more than one vessel, in order to reduce the number of vessels and, hence, capacity. Most offshore fisheries, with the exception of offshore longlining, are based on limited entry through licensing. This will continue, but adjustments in current rules will be made to create greater flexibility and less specialization, in other words, a system that makes it possible to take part in several fisheries throughout the year.

Minor Adjustments or New Policies?

Recent statements like the 1991–2 white paper discussed above tend to favour incremental adjustments rather than comprehensive change. Great care is taken in emphasizing the continuity between current practices and future policies and in underscoring that changes will be gradual and long term. Considering that the industry had just been through a major crisis, there is surprisingly little drama, revolution, or even urgency. However, a report like this is intended as a basis for further discussion rather than as a final statement of policy. Phrases like 'further work is needed on ...,' 'the Ministry will have to consider ...,' 'other options to be taken into account are ...' are used throughout and indicate a fair amount of unfinished business. The final shaping of management policy does not take place until Parliament has given its views and reactions (of which more below).

Within the continuity and incrementalism of the report can be seen a greater debt than before to economics. The analysis of the industry's problems strongly emphasizes excess capacity and the lack of economic efficiency. Its overall strategy for change is rationalization of fleets through a controlled use of market processes. There is also, in the discussion of changes and adjustments, a clear emphasis on individual autonomy and choice and on the need to secure greater flexibility in fishing operations. On this point, however, the report has a cautious air, stressing the need to retain some form of government control. As such, it comes across as a compromise between the economically desirable and the politically feasible. This is explicitly acknowledged by the ministry through frequent references to legitimacy as a crucial prerequisite of effective management. It is, on the whole, interesting to note the frequent use of terms such as 'trust,' 'faith,' and 'confidence,' and the political realism it entails. There seems to be a clear understanding that no management policy will work unless it is conceived as reasonable and fair by those affected, a point that probably goes a long way in explaining the discrepancy between the ministry's overall approach, emphasizing fleet rationalization and efficiency, and the more specific and rather cautious policy proposals.

Management Objectives: Atlantic Canada

Canadian fisheries policies have rested on three fundamental assumptions

(House, 1986). One is that fish are comparable to any other national resource and should be exploited accordingly, in the most rational and efficient manner. The second holds that efficiency can best be achieved through a strategy of industrialization, and the third that state intervention is necessary to rationalize fishing effort, given the common property nature of the resource. The first two assumptions have been highly influential in legitimizing attempts to modernize, that is, rationalize, the fisheries, while the third has been the basic premise behind the introduction of limited entry – for reasons of economic efficiency as well as conservation of resources.

A recent institutional manifestation of the government's commitment to resource conservation is the newly created Fisheries Resource Conservation Council (FRCC, see below). In a report to the minister on conservation requirements for Atlantic groundfish, the new council identifies two primary conservation objectives: 'Rebuilding stocks to their "optimum" levels and thereafter maintaining them at or near these levels, subject to natural fluctuations, and with "sufficient" spawning biomass to allow a continuing strong production of young fish; and managing the pattern of fishing over the sizes and ages present in fish stocks and catching fish of optimal size' (Government of Canada, 1993: 1).

While these may be the ultimate goals, and there are signs that they have become paramount during the 1980s and 1990s, they are not the only focus of strategies of management. Three other objectives that are reiterated in various government sources (Cashin, 1993; Haché, 1989; Kirby, 1982) seem to be at the core of the Atlantic Canada fisheries management system. One goal is national control over the fisheries, and it implies the 'Canadianization' of the fishing industry. Resources within the Canadian 200-mile zone should be harvested and processed by firms located in Canada and owned by Canadians.

Second is the goal of economic efficiency, of creating an economically viable industry. As in the Norwegian case, this is now viewed largely in terms of controlling harvesting capacity and cutting costs, thereby reducing dependency on government subsidies (Kirby, 1982: 186).

Much the same point was made by a government task force when it stated that 'control over the composition and size of the fleet should be an important feature of fisheries management' (Haché, 1989: 10), and by another in asserting that the first step towards solving the 1990s resource crisis should be 'to bring harvesting and processing capacity into balance with the sustainable limits of the resource' (Cashin, 1993: 36).

The third goal pertains to the social implications of a purely efficiency approach, and it is usually stated as an ambition to secure employment in the fishing industry. Economic efficiency will have to be balanced against social stability and the well-being of coastal communities. In the words of the Atlantic Fisheries Task Force this goal 'emphasizes the need for the fishery to employ as many people as possible, given that it is located in an economically disadvantaged region of Canada and that in large parts of that region the fishing industry is the only source of employment' (Kirby, 1982: 187). This point is reiterated by the recent Task Force on Incomes and Adjustment in the Atlantic Fishery, with the new caveat that this must be done within the limits of the resource and with a view to the profitability of enterprises and the quality of work generated (Cashin, 1993: 35–36).

Conflicts among these goals are acknowledged by the task forces referred to. Strengthening the economic efficiency of the industry will not always be compatible with a goal of maximizing employment opportunities, although the choice may not necessarily be between an 'economic' and a 'social' fishery. The point is, rather, that trade-offs will have to be made, the nature of which will be shaped by power relationships and institutional structures, and reflected in policies of management.

Policies: The Quest for Efficiency

Canadian management of groundfish, under the bioeconomic resource management model, is based on quotas, or TACs, to some extent an inheritance of the international regime that preceded extended jurisdiction. The present regime for managing Atlantic fisheries has developed from measures proposed by the International Commission for the Northwest Atlantic Fisheries (ICNAF) – the most important being the introduction of TACs for Atlantic groundfish in the early 1970s (Parsons, 1993b).

In tandem with quotas are efforts to limit entry. The economic diagnosis of too many people for too few fish has played a major role in Canada, as it has in Norway. In 1968 limited entry was introduced in the lobster and salmon fisheries. By 1980 some form of entry limitation had been instituted in most Atlantic fisheries, although the inshore cod fisheries remained essentially open access into the early 1980s, paralleling the situation in Norway.

A government review of fisheries management in 1974–5 focused on overcapacity, and it resulted in a new policy framework where more government intervention was advocated. This was seen as essential given the problems facing the industry: overcapacity and the need to control effort. There were, of course, also controversies surrounding the allocation of resources among competing offshore and inshore interests. The new approach laid down three strategies for dealing with the problems of overcapacity and allocation: allocation of access on the basis of a trade-off between economic efficiency and fleet dependency, extension of limited entry to all commercial fisheries, and the reduction of excess capacity (Hurley and Gray, 1988). In practice, the emphasis on reduction of effort proved more symbolic than real, as few steps were taken to actually reduce the number of vessels in the groundfish fleet. In fact, the number of vessels between 35 and 100 feet increased in the latter part of the 1970s, to some extent fuelled by expectations created by extended jurisdiction in 1977 (Parsons, 1993b: 176) and by a government eager to support the replacement of foreign harvesting by Canadian effort (Hurley and Gray, 1988). Since then overcapacity has been an acknowledged problem, and management policies have largely been designed to deal with it. Individual quotas have gradually become an integral part of the tool-kit for dealing with overcapacity since they were first introduced in the Bay of Fundy herring seine fishery in 1976 (Crowley and Palsson, 1992).

Taking on the Crisis: A Note on the Cashin Task Force

The most recent policy document for the Atlantic fisheries is the report from a task force on 'incomes and adjustment' chaired by Richard Cashin (Cashin, 1993). While not a report exclusively on management issues, and not yet a part of official government policy, it represents an attempt at defining the basic (management) problems of the industry and at discussing how these might be alleviated. The background is the 1990s 'collapse of the resource base' in the Atlantic groundfish fisheries. Strong words perhaps, but they are certainly justified in view of the drastic decline in groundfish catches, which the task force estimated as 90 per cent between 1988 and 1994 (Cashin, 1993: 19). The general tone of the report and its more specific proposals are heavily influenced by the magnitude of the crisis:

We are dealing here with a famine of biblical scale – a great destruction. The social and economic consequences of this great destruction are a challenge to be met and a burden to be borne by the nation, not just those who are its victims ... Conventional approaches to adjustment, including conventional ways of delivering programs to individuals, will be far from adequate. The task is too great to be undertaken as things have been done in the past. (ibid.: vii)

Throughout there is an emphasis on the special responsibility of federal and provincial governments to help individuals and communities adjust, and the report contains some forty-two recommendations, of which almost half pertain directly or indirectly to management. These range from a recommendation that policies and criteria for a reduction in harvesting capacity should be established, to the proposal that all gear types should be put under limited entry licences.

Overcapacity, in harvesting as well as processing, is clearly recognized as the major problem, and the need for reduction of capacity is emphasized again and again in the report. The main deficiencies of the current regime pertain to elements such as open entry (as anyone over sixteen years of age can register as a commercial fisherman), little if any differentiation between bona fide fishermen and marginal actors, and a system of limited entry that has failed to curb effort. What is needed, among other things, are stronger professionalization of fishermen, an extension of limited entry licensing, scientific and technical reviews of harvesting technologies, and greater emphasis on licence retirement and buy-outs.

It is further maintained that the responsibility of DFO goes beyond conservation and includes measures geared at preserving the commercial sustainability of the industry. On this point the task force calls for 'fundamental change' – based on long-term planning rather than ad hoc adjustments, on consultation and partnership rather than ministerial directives.

The need for cooperation between government and industry, and for industrial participation in decision-making, is emphasized throughout. However, a distinction is made between decision and implementation, and it is suggested that the latter is best left to an independent body. The report thus proposes the creation of fishing industry renewal boards composed of knowledgeable but independent people at arm's length from government and industry. In this respect, the report ties neatly in with recent DFO attempts at reforming the overall institutional structure of management decision-making. The proposal for independent boards has subsequently been acted upon by the federal government with the creation of four harvesting adjustment boards. The chief task of these boards will be to develop detailed plans to reduce capacity in the harvesting sector. The boards will review existing harvesting capacity and advise the federal government on how to implement a voluntary licence retirement program.

The Cashin report proposed that the boards should be a joint effort between federal and provincial governments – with shared responsibility for appointments. According to DFO, consultations had shown that not all provinces were ready to participate, hence the federally appointed harvesting adjustment boards. There will be one board for each of the eastern regional fishing areas managed by DFO, that is, one for Newfoundland, one for Scotia–Fundy, one for the Gulf of St Lawrence region, and one for Quebec. Each board will have a chair, two members, and a small support staff. They are supposed to complete their work of planning for reduction of harvesting capacity within six to twelve months.

Fisheries Management as Crisis Management?

In assessing the objectives and policy processes in the fisheries of Atlantic Canada and Norway several points deserve notice. First, looking at objectives, one must conclude that management of the fisheries is a multipurpose activity and a balancing act. Not only are objectives fairly general, they are also – to some extent – mutually incompatible. Economic efficiency, for example, is not necessarily compatible with the goal of regional dispersion of harvesting rights; resource conservation may collide with the need to maintain employment opportunities and preserve local communities. As a result, decision-makers in both Atlantic Canada and Norway have been reluctant to adopt clear, well-ordered, and hierarchical sets of management objectives.

Second is the gap between promise and performance. Policies aimed at conservation of resources leave much to be desired – in spite of the apparent success of Norwegian management of the Barents Sea cod. Uncertainty about stock assessments, political pressures, and social concerns make it difficult to accept measures that would adjust harvesting capacity to stock size, much less to a 'precautionary' or conservative level of effort. 'Downsizing' is seldom accepted as legitimate unless there is a crisis in a fishery, that is, a situation where the resource is clearly being depleted. Even then, the possible dislocation of those excluded, and the political problem of deciding who should go, may limit the options available. In face of possible social disruptions, the commitment to economic viability has also been tenuous. From the industry's point of view, resource management has not necessarily been a question of improving the overall efficiency of the fleet. Fishermen, for example, tend to see entry restrictions as a way of protecting their earnings and livelihoods – and there is certainly a case to be made for that. Restrictions on effort and catches are seldom 'neutral.' They constitute property relations in as much as they specify, protect, and reproduce conditions of access. Affecting traditional rights and vested interests, quotas and licences can rarely be enforced without the consent of the industry. If this applies, protection rather than efficiency may come to dominate management programs – which probably goes a long way in accounting for the aforementioned gap between promise and performance.

Third, government task forces are frequently used for purposes of analysis and review. This is not only a strategy to collect information and clarify options, it is probably as much a method of consultation. Representatives of industry are often appointed as members, and when they are not (as was the case with the latest Norwegian white paper), interest groups, are given the opportunity to respond. In this sense, the appointment of a task force is but the first step in a more comprehensive 'sounding out' process. In the reports that inevitably follow, the tone tends to be pragmatic and responsible, with a penchant for incremental adjustments rather than grand schemes. This frequently translates into efforts to create solutions which share the wealth, or lack thereof, among existing interests. When confronted with scarcity, as policy-makers and fisheries managers frequently are, there is a perception in both Atlantic Canada and Norway that decisions must be regarded as 'fair' by the various economic sectors, interest groups, and lower level agencies involved. As one Canadian official put it, in discussing a crisis, 'You can't have a solution on the back of one section of industry.'

Fourth, in both countries, overcapacity, whether in harvesting or processing, is viewed as the major problem confronting the fisheries. On this point management policies clearly reflect the influence of economic theories of open access as well as reliance on scientific advice and assessments. There is a long-standing commitment to fleet rationalization through limited entry policies. While managers and policy-makers do not view economics and biology as integrated disciplines in the day-to-day operation of the bureaucracy, there is considerable affinity between procedures of stock assessment and orthodox economic conceptions of what constitutes an efficient fishery. On this point there seem to be standard solutions to the mismatch between harvesting capacity and stock size, and little that sets our two countries apart. On the other hand, management policy-making is clearly embedded in a broader framework of regional development where the goal of economic rationalization is pitted against the survival of coastal communities. In both countries debates on policy reveal an inherent conflict between economic criteria and social concerns, and management decisions are often based on a series of trade-offs between efficiency and equity, between the wish to 'modernize' and the need to preserve employment opportunities.

All is not rosy. One observer we interviewed argued that structured scientific advice is primarily biological in content, and commented, 'Biologists and economists are not together in trying to quantify the impacts of alternative scenarios.' A Canadian regional natural scientist observed that economists are not direct participants in their activities. While economists may study fleet viability, and costs and earnings, they come closest in modelling the effects of population changes.

Fifth, and perhaps most interesting, is the recognition that supply and marketing uncertainties mean that 'fisheries management is crisis management.' Significant problems are often treated as normal events, as the 'way things are in this business.' However, the collapse of the Barents Sea cod stock in the late 1980s and the collapse of the northern cod of Newfoundland and Labrador in the early 1990s came as something of a shock even to the most cynical of observers.

It is an old truism that crisis breeds reform, and it may be true for resource management too. C.S. Holling has suggested that it is only at points of deep crisis in both the ecosystem and the social system that fundamental conceptual and structural change is possible (Holling, 1986). In the next chapter we examine one of the directions taken in making such changes, the use of individual transferable quotas (ITQs). It should be noted, however, that ITQs do not represent fundamental structural change as much as the logical end-point of the long-term policy diagnoses of the problems of open access and incomplete property rights and of policy prescriptions for more market-based solutions.

6

Institutional Structures and Management Policies: The Case of Individual Quotas

As we have seen, individual vessel quotas have become an integral part of management policy in both Norway and Atlantic Canada. In Atlantic Canada this sort of 'rights-based' fishing dates to the late 1970s and early 1980s, with the introduction of IQs in the Scotia–Fundy herring fishery (Parsons, 1993b) and enterprise allocations (EAs) in the offshore groundfish fishery. These quota allocations have since been made fully transferable and were extended to parts of the inshore dragger groundfish fleet. In Norway group quotas and individual vessel quotas have been in place in offshore fisheries since the late 1970s. They were introduced to parts of the inshore fisheries in 1983 and extended to most of the inshore fleet in 1990.

A system of enterprise allocations for the offshore fleet has recently been put in place. Proposals for a more fully fledged ITQ system have, however, been turned down, and there is currently little enthusiasm for a more marketlike approach to fisheries management in the Norwegian context. There is thus little doubt that the question of transferability has been handled differently in the fisheries of Norway and Atlantic Canada – at least on a formal level. In other words, there is a difference of policy here that needs to be explored and accounted for.

In the Canadian context we focus mainly on the transferable quota system for inshore draggers in the Scotia–Fundy region. Almost all inshore mobile-gear vessels operating in southwest Nova Scotia and southeast New Brunswick are now under this system, and those in eastern Nova Scotia participate on a voluntary basis. By the end of 1993, of the 455 draggers, 366 had declared status under the scheme. In the Norwegian context we take a closer look at a recent attempt to introduce a modified ITQ system for the entire inshore groundfish fleet, giving special attention to the question of why this failed.

The idea of ITQs has been taken further in Atlantic Canada than in Norway fisheries. If transferability is a key issue, and we contend that it is, there is currently a major difference between our two regimes in their basic approach to problems of management. Whereas Canadian policy-makers and industrial representatives see market dynamics as a workable tool (for some fisheries), the idea that quotas should be transferable at the 'owner's' discretion has found little favour with their Norwegian counterparts. In our attempt to account for this, we emphasize variations in process, institutional structures, and political context, and ask how institutional factors may facilitate or obstruct innovations in public policy. An ITQ system may not be an innovation in any strict sense of the term, but it does represent a new trend in fisheries management. There is a specification of property rights to the resource and the application of market principles (by making these rights transferable) to problems of conservation and efficiency.

Implementing the ITQ System: Atlantic Canada

The Canadian system we describe originated in a task force report on groundfish in the Scotia–Fundy region released by the Department of Fisheries and Oceans (DFO) in January 1990 (Haché, 1989). The Haché task force saw declining stocks and overcapacity as the two major problems. With reduced TACs, vessels would rush out in January to try and catch as much as possible before the overall quota was reached and the fishery was shut down. In most years the fishery remained open, though, because of pressure to reallocate it from other sectors. However, in 1989 the fleet caught its entire quota by mid-June and was forced to shut down for the rest of the year, helping precipitate change in the system. The regime of trip and catch limits that preceded IQs was also known as a 'bureaucratic nightmare.' In response to these problems and recommendations of the task force, on 2 January 1990, the federal Minister of Fisheries announced that he would introduce 'boat quotas.'

Although the decision was apparently made by the exercise of ministerial discretion and with little or no consultation with the industry, definition and implementation of this 'boat quota' was done more democratically. More specifically, the task force report states, 'More effective long-term measures to keep harvesting capacity and effort in balance with the amount of fish available for harvest must be put in place and strictly enforced. High capacity/investment operators must be persuaded to address their own capacity problems in a manner that does not have a negative effect on the balance of the industry' (Haché, 1989: 3). According to both DFO and a key industry representative, there were about four times too many boats for the stocks available (120,000 tonnes capacity). The overcapacity has generally been associated with two major changes.

Extended jurisdiction after 1977 encouraged fleet expansion, and high prices for cod, haddock, and pollock in the years 1978–9 facilitated another round of growth.

At first the boat quota system was conducted by a special group (implementation committee) within DFO headed by the director of the resource allocation and development branch for the Scotia–Fundy region. It eventually became consultative, known as the IQ working group (later simply the IQ group), which involved representatives of relevant fisheries in both eastern and western Nova Scotia. The IQ group, which at the outset had representatives from both mobile- and fixed-gear sectors, was to decide which stocks were to be included in the IQ system, establish the fleet sectors to be involved, decide on initial allocations, outline the appeal process, and take part in the establishment of a system monitoring catch.

Two organizational meetings were held with the groundfish advisory committees in late January 1990. The chairman described the task of the new group as addressing the restructuring of the inshore fleet, identifying inactive licences and establishing individual vessel quotas. He further indicated that it had a target date of 1 May 1990 for completion of its work.[1] The two area advisory committees nominated members to serve on working groups which would deal with issues affecting mobile-gear, fixed-gear, and both gear sectors jointly. Two workshops were held during February and March to develop options, alternatives, and recommend approaches. Among the key questions addressed were limits on transferability and the accumulation of quota as well as bases for initial allocations, appeal procedures, and monitoring and enforcement.

Three more meetings took place before the summer, one in April to 'finalize the latest options for the CI fleet boat quota system' and consider historical catch information which might be applied to these options; the others in May to look at final sharing formulas and presentation of 1990 individual vessel quota calculations. In a detailed twenty-page document, the chairman recorded support for the idea of temporary, as opposed to no or permanent transfers. The working group did look at the possibility of permanent transfers, but decided that it would like to operate with a temporary transfer system. No consensus was apparent on the question of a ceiling on the amount of quota anyone might hold. In part this uncertainty was related to the fact that temporary transfers would bring quotas back to the original holders at year's end.

The working group agreed that the boat quota system should be restricted to cod, haddock, and pollock and that the initial program should run for five years.

---

1 We are grateful to the Director, Mr John Angel, for sharing his 1990 chronology of individual quotas with us.

A special committee was also recommended to adjudicate individual appeals, with the committee to be comprised of 'industry representation with some DFO participation.' It was acknowledged that the federal Minister's statement of 2 January restricted implementation of the boat quota scheme to southwestern Nova Scotia. As a consequence, fishermen in eastern Nova Scotia would have to be 'canvassed' about their desire to participate in a quota system or continue with a competitive fishery.

Meetings with groups of fifteen to twenty holders of licences were held around the region to ensure that fishermen had 'the fullest opportunity to understand the system' and express their views. Twenty-eight meetings were held between 18 June and 11 July, with a majority of them in southwest Nova Scotia ports. On the basis of these meetings, the implementation committee recorded 'general acceptance, in principle, of the boat quota system concept.' However, it recognized that 'fishermen were nearly unanimous in saying that without more fish or money for a buy-back program or both, the IQ system was unacceptable and would be rejected.' An important contextual feature of this attempt at institutional change is that it took place at a time of severe cut-backs in allowable catches, which meant that individual quota allocations based on previous history of performance were likely to be cut back sharply, with cut-backs in overall TACS. The committee also gave special emphasis to frequent reports of dumping and discarding juvenile and other fish, which they viewed as an argument favouring an IQ system.

The fishermen involved in the series of consultations were sensitive to the various forms of influence that DFO employed. One attendee described the process as being one in which they started by telling the government, 'Go to hell. If you want to get rid of us, buy us out.' At the next meeting they said, 'We don't want an IQ system. If you want to get rid of us, buy us out.' Then, after a total of three months, 'It finally began to dawn on us that they just weren't going to change their minds.' At that point, they started thinking seriously about an IQ system where the allocation would be based on historical catches.

At the end of July the committee began an appeal process. Fishermen were given the opportunity to appeal their recorded catch histories (based on purchase slips and/or fishing logs and records collected by DFO) or make cases for exceptional circumstances. The appeal board, made up of the three area managers within the Scotia–Fundy region, dealt with 143 appeals in nine meetings held between 15 and 25 October. Observers from the inshore mobile groundfish fleet were permitted to attend these hearings, but chose to go to only five. While the observers were required to sign an oath regarding the confidentiality of the hearings, there was some indication that appellants were obtaining information 'that could only have come from the hearings.' This breach was regarded as justification for holding some 'in camera' sessions which excluded observers in order to 'consider sensitive issues' and 'findings' in a 'candid fashion.'

On 30 November the implementation committee sent ballots out that outlined a series of choices that the fleet had to make with regard to the introduction of an individual quota system. Fishermen were given about two weeks to return the ballots through which they would be 'deciding on their participation in the IQ program and the features of the program itself.'

As a result of these choices, the mobile-gear sector had 336 draggers between 45 and 65 feet operating under the IQ system as of the summer of 1992, of which over half were at or near the upper length limit established by law. An additional 60 midshore vessels decided to concentrate exclusively on fixed-gear fisheries and took off all trawls. Fifty-five vessels thirty feet in length or less chose to be classified as 'generalists' which fished mainly for flounder, with cod or haddock as by-catch. Area quotas for cod, haddock, and pollock were divided among these groups.

The Issue of Transferability

Major issues during the first two years of the IQ system included permanent transfers. On 12 June 1992 the Scotia–Fundy region of DFO announced the introduction of permanent transfers of IQs. DFO announced a 2 per cent limit of the total fleet quota for any one licence holder to limit the possibility of over-concentration of IQs. Industry representatives, however, were concerned that individuals might be able to accumulate more than 2 per cent of quota by acquiring additional licences, or they would try to transfer quota for security reasons to 'Mickey Mouse' companies. There was agreement that one should not be able to bank quota under a permanent transfer system, but no consensus could be reached on the desirability of licence accumulation. The IQ Group also agreed that the 2 per cent limit should be permitted to increase in its tonnage equivalent as more stocks are added to the IQ program. One industry representative argued that a tonnage limit would be a disincentive for good conservation practices.

As for the consequences of the program for ownership and control, there are few solid facts available. Fisheries managers do, however, report the beginnings of considerable concentration in the fleet. One observer said that as many as one-third of the licences may already be affected, many of them by under-the-table dealings.

A DFO official interviewed acknowledged that lawyers can find ways to circumvent most rules. There are suggestions that a considerable number of vessel owners have private agreements that bind them to processors. One processor involved with the dragger fleet, commenting on vessel ownership, acknowledged, 'There are two numbers – the real number and the DFO number.' They have trust arrangements with fishermen, and they believe could not otherwise survive. 'Government is forcing us to shrink, but they won't give us the legal mechanisms to do it. We're making the accountants and lawyers rich, really rich.' Another official suggested, probably accurately, that the IQ fleet would decline by 20 per cent in the next few years.

It is theoretically possible to talk about the fleet declining to as few as fifty vessels, based on current TACs, but it is expected that several hundred vessels will remain after permanent transfers have taken place. The network of contracts and connections which bind licence holders to local fish processors and communities are factors that will probably inhibit most plans to intensify concentration. At a more economic level, other observers point out that the highly competitive and opportunistic Scotia–Fundy fishery is not one that encourages monolithic fish companies. Fishing is more of a business in which 'whoever is on the water has an ownership position.' However, the fact that 85 per cent of the landings for the IQ fleet take place in only fifteen communities does suggest that the IQ system may help crystallize fishing activity in a relatively small number of more 'urbanized' communities outside the regional centres. Tendencies towards concentration already existed. One observer estimated that 45 to 55 per cent of the IQ fleet was formally or informally owned by processors prior to the introduction of the IQ system. Some were 'grandfathered' into the IQ system even though they were owned by processors, and there were a number of informal connections that date to the price surges of the late 1980s.

As of 1996 there were 200 to 250 licences with IQ quota, of which 60 to 70 per cent were processor owned or controlled. Most of the shrinkage took place among marginal quota holders who were getting out. The optimistic view of this process was that one may be observing an 'industry buy back' which permits actors who are less viable than some others to leave with some capital and their dignity intact. The more pessimistic interpretation suggests that a new round of 'capital stuffing' had begun, where the investment object became fish rather than technology or the ability to catch fish.

There are also concerns that the first-generation windfalls that come about from initial allocations of individual quotas will make it difficult for future generations to make a living in the fishery. For example, a 65-foot groundfish dragger worth 100,000 CAD may have 500,000 or 600,000 pounds of quota, worth around 1.00 CAD per pound as a permanent transfer. This means that the value of the enterprise has increased by a factor of five or six. A young person trying to get into the fishery would be lucky to have the capital to purchase the vessel itself much less the quota. However, federal officials remain committed

Institutional Structures and Management Policies 193

to creating an economically viable fishery and are prepared to see at least some individuals move out of the industry.

Consultative Management or the Management of Consultation?

The ITQ experiment offers an opportunity to examine the Canadian consultative management system and to address the question of whether it gives fishers, and others, a voice or whether it serves merely to 'educate' the industry and legitimize decisions already taken (Arnstein, 1969). Seen from the perspective of DFO, the ITQ experiment is first and foremost an attempt to deal with some fundamental problems in the inshore sector, notably the growing mismatch between harvesting capacity and size of stock. To quote John Angel, then assistant regional director, Department of Fisheries and Oceans Canada: 'There are not enough fish for everyone [in this sector]. We all know that.' The ultimate aim is rationalization of fleets as well as reducing the administrative and political burdens that surround questions of allocation. It is, however, also evident that DFO views the advisory group set up for the design of IQs in the inshore mobile-gear fleet as successful.

The IQ group may indeed improve on the type of consultative management present in the system in Atlantic Canada. Many of the general groundfish advisory meetings have been acrimonious events with little prospect of building consensus. The IQ group is homogeneous enough in composition to make consultations worthwhile and efficient. One participant contrasted the IQ group meetings with some of the old groundfish advisory meetings, where the representatives of the offshore, mobile-gear, and fixed-gear sectors were all present in the same room, 'Instead of dealing with the issues, we sat in the room and said, "I hate you, you hate me, and we all hate each other."' The IQ group, by contrast, is quieter and more productive. 'Most people there seem to be genuinely concerned. I've seen some politicking, but not a lot ... They've come a long way in three years.'

In contrast with many participatory situations, where people are asked for input but the agency appears not to listen, in the IQ group DFO seems to have genuinely solicited input. This regime has also produced a greater sense of equity in allocative processes. The formulaic allocation rules, as well as the various appeal processes, have created a sense that the system operates more fairly than it used to, even though interregional tensions remain.

The flow of information has also improved. Although there is some misreporting, it is widely regarded to be on the decline, perhaps in a major way. Nevertheless, serious problems with highgrading and transshipping remain. Finally, while the legitimacy of the system among its members has clearly increased, growing resource shortages have intensified conflicts with the offshore and the fixed-gear sectors. Several members of the IQ group report being the recipients of death threats for publicly representing their interests. The internal legitimacy of the system may thus owe something to its hostile environment and external disapproval.

Of course, there are also critical voices. Some 'insiders' point out that this ITQ system did not arise from a particularly democratic process of consultation as it did not include the entire range of participants within the fisheries involved. For example, soon after the system was put in effect, a representative of the Maritime Fishermen's Association claimed that the inshore dragger fleet was not asked *whether* a boat quota system would be instituted but only *how*. He also reported on legal disputes and threats of violence in fishing communities.

The regime of consultative management is designed to delimit individual participation within parameters set by DFO. In addition to the obvious fact of ministerial power in management policy-making, DFO has considerable influence in designing the advisory structure, creating timetables and agendas, and shaping the direction of any debate. For example, the alternatives for the ballot that led to adoption of the IQ system assumed that the default position was a move to inclusion in an IQ system. Voters were all instructed: 'If you do not make a selection or do not return the ballot you will automatically be placed in the individual quota system.' Further, fishermen were told that if they chose to fish fixed gear only, they would retain their mobile-gear licence, and they might enter the IQ system at a later point by acquiring quota through transfers. However, 'Once you enter the IQ system, your fishing operation is limited by the IQ and you are no longer able to fish competitively.' In addition, those mobile-gear fishermen in western Nova Scotia who had the option of fishing as 'generalists' under a competitive quota, and those in eastern Nova Scotia who could elect whether to enter an IQ system were told that selection of the IQ alternative would mean that they could not return to a competitive fishery.

Finally, there is the question of democracy for whom? Regardless of the kind of advisory group involved, whether it be the IQ group, or any of the many others in the region, the presence of 'local notables' at the table is striking. A problem of representation occurs here, as there are few stringent rules governing appointments to advisory committees, and those appointed may have no formal responsibilities *vis-à-vis* wider groups. The problem seems to be one of ensuring that the voices heard are in fact a representative cross-section of industry interests and opinions. It is also possible to argue that the introduction of quota systems is increasing the level of democracy within the particular advisory group, while at the same time reducing the number of individuals who are capable of enjoying this democracy (McCay et al., 1995).

ITQs in Norway: Preparing the Ground

In Norway proposals for a more marketlike approach to fisheries management, and a modified version of a 'pure' ITQ system, were floated by the Ministry of Fisheries in 1991, only to be turned down by the government as opposition mounted. The proposal was the outcome of the white paper process described in the previous chapter. It is standard procedure that a draft of a white paper is 'heard' by groups and interests affected by its proposals. The final report as submitted to Parliament is supposed to reflect the views and concerns of affected groups or at least contain a summary of them. The process thus has elements of consultation. This particular case is of interest as an attempt to consult with social scientists as well. Unusual is the organized consultation of presumably 'knowledgeable individuals,' such as academics and researchers, during the process of drafting a report. But this is exactly what the ministry did as part of preparing its final version.

A small group of civil servants in the ministry, headed by an economist, produced a working paper that was distributed to several persons outside the ministry, mainly academics. The ministry then tried to sound out this group of 'experts,' which included biologists, economists, sociologists, and political scientists, by arranging several seminars or workshops around the main topics of the report. There was time set aside for general discussions, and smaller groups were organized around more specific issues. The discussions were frank and open, and the ministry had to take strong criticism on several points, particularly on its idea of introducing ITQs. The objective was to test this particular proposal on a panel of presumably well-informed and interested people.

Whether such consultations will become part of standard procedures is still unclear. The question of how to interpret the ministry's initiative, however, is interesting. At one extreme it could be seen as an attempt to provide legitimacy and intellectual backing for controversial policies. If that was the case, expectations were hardly fulfilled. At the other end it could be interpreted as a genuine fact-finding and consultative exercise and a recognition of the relevance and potential of (particularly) the social sciences for management policy-making.

A Market for Quotas?

The issue of ITQs, as distinct from non-transferable vessel quotas, was first raised in 1989 in a report from a working group on the structure of the harvesting sector. The group comprised representatives from the ministry, the Directorate of Fisheries, and the Fishermen's Association. Its report was widely debated, especially the proposal to introduce enterprise allocations (EAs;

'rederikvoter') in the offshore fleet. They were largely conceived as tantamount to transferable quotas and a step towards conferring property rights to the resources on a limited number of people and companies. There was vociferous opposition from parts of the industry. The Fishermen's Association, representing both inshore and offshore groups, initially came out in support of the EA proposal, but retracted as the reactions from its grass roots became clear. The political establishment was largely sceptical, mainly because of the grass-root reactions, and the proposal was, for the time being, duly shelved.

Within the ministry there were doubts about how to follow up the report. One alternative was to dress it up as a white paper to Parliament. The minister supported the idea, but insisted that further work be initiated on a broader range of management policies. Four officials were assigned the task of drafting the paper. There was no direct participation from industry, but a broad and international attempt was made to get advice from academics and researchers in order to deliver a 'professionally sound' product (according to one of its members). Academics and scientists from research institutions in Oslo, Bergen, and Tromsö were consulted individually and through special seminars or workshops. The group also visited Australia, Britain, Canada, Denmark, Iceland, and New Zealand.

According to one of the participants, the group worked from the premise that a more marketlike and flexible approach to management was desirable. However, there was no general consensus within the ministry on this. Some officials, especially the lawyers, found the issue of transferability a waste of time and argued that the desired increase in flexibility could be achieved within the existing system, which, it should be noted, was largely the product of judicial rather than economic expertise and thinking. The outcome of the group's work, however, proved that objections like these had little weight. Its preliminary report (Høringsnotat) of June 1991 was mainly an account of various forms of transferable quotas. The alternative recommended in the report, and initially supported by the political leadership in the ministry, was ITQs with restrictions on transferability across regions. A free market for quotas was never a serious option, and there was a strong emphasis on preserving some basic steering mechanisms.

As seen by the ministry, there were three 'pure' principles or models of management: setting an overall quota (and leaving the rest to individual choice), non-transferable IQs, and ITQs. The first model implies unlimited competition within the framework of an annual quota. When this has been taken, fishing is stopped. The model is simple in that it entails few, if any, barriers to entry, and it is easy to administer and control. On the other hand, it generates untrammeled competition that invariably leads to higher costs, including political pressure.

The second model requires that all participants, identified through a national register of fishermen, are given equal shares of the overall annual quota. Equity and employment are the major concerns, and no weight is given to previous participation or regional ties. The main problem, according to the report, is one of administration and control. Another, of course, is the classical one of too many fishermen chasing too few fish.

The third model entails the creation of a market for quotas. The ministry, and other proponents of ITQs, have tended to emphasize economic efficiency as the cardinal argument for such a scheme, while at the same time observing that there is a risk of excessive concentration of ownership and control.

The report does, however, add some refinement to the general argument for ITQs by specifying various forms of transferability (or, rather, different ways of trading one's share of the overall quota: (1) ITQs as traditionally defined, that is, the transfer of an individual quota from A to B for a price. (2) The transfer of a fishing vessel with its quota, that is, the individual quota – originally granted to A – is purchased by B along with the vessel and included in the price; this form of transferability is familiar in Norwegian fisheries. (3) Enterprise allocations that permit a company (for a year at a time) to transfer its quotas from one or more vessels to another, that is, to concentrate its harvesting rights to one or a limited number of vessels. (4) 'Renting' quotas on an annual basis, that is, a temporary transfer of A's share to B – for an agreed-upon price. (5) Co-fishing ('samfiske') whereby two or more boat-owners agree to take their quotas with one boat. This form of cooperation (and transfer) is prohibited, mainly because it places the crew members on the boats taken out of fishing at a disadvantage.

The ministry's original preferred version pertained to vessels longer than eight metres and ran as follows:

- The overall quotas are allocated to the different groups of vessels and regions as defined under the current regime, their respective shares decided by historical catch.
- Individual shares are allocated through a system by which each boat owner leases (for a fee) a quota from government, for a period of, say, five years. The maximum size of the starting lease will be dependent on historical catch.
- Individual quotas thus allocated are really portions or shares of the overall quota of a particular fishery, regardless of the size of the vessel, and they can be traded freely *within* the particular group or region during the five-year leasing period. Prices are decided by the market and will thus be permitted to fluctuate with supply and demand.
- To be able to buy a vessel from another region, the buyer must lease the necessary quota shares within his own region. Transfers across vessel groups and regions require the permission of the ministry.

The model restricts the transferability of quotas, and thus to a certain extent vessels, *across* regions. In principle then, it tends to the problem of concentration and centralization. On the other hand, the model provides for some flexibility and autonomy in that it facilitates the trading of quotas among fishermen *within* a region. Both points were, in principle, widely applauded within the fishing industry. However, the proposal found scant support because, it seems, of doubts about the permanence and effectiveness of the built-in restrictions.

Controversy and Retreat

As part of the standard administrative procedures, a draft of the report was 'heard' by a select group of interest organizations, other ministries, and regional authorities, twenty-eight in all. In addition, more than twenty interest groups, from regional associations of the Fishermen's Association to local branches of the Labour party, filed written comments. With the exception of the feedback from other ministries and the political parties, most comments were summarized near the end of the final report.

There was scant support for an ITQ system, even with built-in restrictions on transferability. The hard core opposition included groups and institutions such as the Fishermen's Association, the newly formed Association of Inshore Fishermen, the mandated sales organizations (MSOs), and all county councils along the coast. The only organization willing to contemplate such a system (or rather a modified version of it) was the Confederation of Norwegian Industry, but only as a long-term goal. The overwhelming majority of those consulted were strongly against ITQs, even in the modified version suggested in the draft:

The majority of those heard is very sceptical about transferable quotas even with built-in restrictions on marketability. The most important reasons given are that such a system will lead to a concentration of quotas and to a centralization of the industry that will make it difficult to achieve objectives related to employment and settlement. Several of those heard fear that the system poses the danger that Norwegian harvesting rights will be bought by European companies. (Government of Norway, 1991–2: 126–7)

Few, if any, were convinced that restrictions on transferability would be effective in preventing concentration and outsider control. Rather, it was felt that once such a system went into operation there would be strong pressures to modify restrictions (as did in fact happen in the Nova Scotia case, McCay et al., 1995).

In spite of the widespread scepticism about ITQs, there was some support for management policies that provided for greater individual autonomy, discretion, and flexibility. Fewer restrictions on temporary exchanges and transfers of quotas and enterprise allocations were favoured options. Marketlike solutions were not considered anathema; rather, the dispute was about how much, and in what form.

The question of transferability became an issue in the local elections of 1991 and an object of more general political concern as the possible implications of ITQs caught the attention of campaigning politicians. Heated debate took place in the regional press, especially in north Norway, where supporters of the system were few and far between. The outcome was that the process was essentially taken over by a task force within the Labour party (then as now in office). The party held a special conference on fisheries policy where the issue of ITQs loomed large. The main conclusion from this event was that transferability was not acceptable. The national executive of the party also intervened with a statement to the effect that ITQs were not on the agenda. Even the prime minister let it be known that she was sceptical about ITQs (Moldenæs, 1993). Thus, there were strong signals from 'above' to the effect that the ministry should downplay the issue. In short, the political leadership demanded other conclusions and proposals, and the final paper was clearly more 'moderate' in tone than the preliminary report had been.

In Parliament the standing committee on fisheries issues did a little 'research' of its own, travelling to Iceland and Greenland to take a closer look at actual ITQ systems. The committee also consulted fishermen's associations, fish processors, fish-plant workers, spokesmen for local and regional government, and the Norwegian Trade Union Congress, among others. In its report to the full house, a majority of the committee's members rejected the ITQ option. The minority, comprised by its conservative and 'liberalistic' wing, asked the government to continue its work on a new management system, with the aim of introducing a program for ITQs, a position they continue to hold (Government of Norway 1992–3: 17).

The general debate in Parliament touched on most current issues of management and harvesting, but the issue of ITQs loomed large throughout. As one would expect from the report of the standing committee, there was little support for a more marketlike approach to fisheries management. A somewhat ideological, yet fairly typical, comment from a representative of the Socialist party sums up some of the basic arguments against ITQs:

The Conservative party and the Progress party will vote for a system of transferable quotas. That will make it possible for the forces of capital to buy up the resources in the ocean and along the coast and squeeze out those groups that until today have lived off and preserved those resources for thousands of years ... There is one important aspect

200 Section 2: Resource Regimes

and that is to view this in an EF [European Community] perspective; that it will lead to a system that sets the table for rich foreigners so that they can buy quotas in the Norwegian zone. (Government of Norway, 1992: 1986)

With hindsight, it is quite clear that not enough attention was paid to the political feasibility of the ITQ option. In this sense the officials in charge were almost too professional in their approach, embarrassing their minister in the process. The fact that there was no direct participation from the industry in the drafting may have contributed to this. The group did make some efforts to 'market' their proposals, for instance, by attending the annual meetings of fishermen's associations, but to little effect. That the ITQ proposal became a salient issue in the local elections, particularly in north Norway, took the ministry completely by surprise, and forced the Labour party leadership to arrange for a swift retreat.

Policy Decisions as Institutionally Constrained?

In Canada there seems to be an openness to new ideas and a willingness to 'experiment,' that is notably absent in Norway. Explaining this involves underscoring the fact that we are dealing with an innovation developed for a relatively small regional fishery and introduced as a limited experiment (Atlantic Canada), and a comprehensive mandatory scheme for an entire national industry (Norway).

That said, there was also resistance in the Canadian case, and the reform was initiated against a background of vociferous protest from the groups affected. That the scheme still went ahead may indicate that the conditions for mobilization of successful opposition were lacking (Immergut, 1992). Thus, the failure, so far, of ITQ proposals in Norway may be the result of a combination of coordinated opposition and privileged access. Virtually all interest associations within the fisheries, and most political parties, have been strongly against any step towards 'privatization.' The industry has thus been fairly unanimous in its rejection of ITQs. Even more it has been able to coordinate its opposition through a coherent network of associations, whereas in Canada the fishing industry has fragmented and incomplete organization, making it difficult to develop consensus and mobilize opposition. The Norwegian government's retreat on the issue may well have been because of strong opposition from an articulate coalition of northern politicians (in Parliament and local governments alike) and fishermen's associations. Moreover, the sheer amount of time spent consulting and debating the merits of the proposed policy made it possible for the opposition to organize, spread their message, mobilize support, and 'force' the issue of ITQs onto the agenda in the 1991 local elections campaign. It attracted maximum publicity almost from day one. The electoral campaign became a possible 'veto point,' an opportunity for examining and opposing the proposed policy.

In Norway the industry generally has excellent access to government, through corporatist arrangements as well as parliamentary channels. Participation is, as already pointed out, highly institutionalized through councils and committees. Interest groups will often hold the key to new policies in the sense that decisions are seldom made and implemented without their consent. The consultative framework thus represents a limit to executive power and a structure of veto opportunities in the sense that proposals can be questioned, opposed, and sometimes overturned at the preparatory stages. An interesting question, then, in light of the strong opposition that emerged, is why the issue was allowed to develop as far as it did. The reason may have been linked to the fact that there was no direct participation from user groups proper in the drafting of the white paper. This was left to a select group of officials, who were strongly oriented towards efficiency and profitability and encouraged by the minister, who liked the idea of ITQs. One could perhaps say that the institutional safeguards provided by standard operating procedures were bypassed. Therefore, it took some time before the Labour party leadership realized the highly political and controversial nature of the issue and decided to intervene.

In Canada, as we have already pointed out, there is more fragmentation among interest groups and a government more decisive and perhaps less inclined to listen during the policy-making process. Ministerial power leaves user groups with little influence on decisions of whether to adopt new policies. In the case of ITQs, for instance, the industry was simply told that an IQ system would be installed. Strong protests did not stop the process. Once the basic structure was in place, however, a strategy of cooptation was adopted. User groups were actively involved in the design and implementation of the new system, which clearly strengthened the legitimacy of government policy and smoothed the path towards a full-fledged ITQ system. Besides, there was a clear element of choice as individual fishermen were given the opportunity to 'declare' themselves in relation to the scheme.

The Canadian process, then, seems to have been characterized by a 'carrot and stick approach,' the stick being wielded initially through ministerial power and discretion ('this is what you will have'). No consultation on basic principles took place. The carrot was allowing for meaningful participation from user groups in planning and implementing the system (the IQ group). In short, the institutional structure, characterized by centralization and relative autonomy of management policy-making and an organizational fragmentation of the industry, provided few points at which government policy could be effectively challenged, let alone vetoed or overturned. Despite an avowedly consultative system in Atlantic Canada, user-group participation is more a matter of influencing the fine print of implementation than in making more fundamental decisions on basic principles and tools of fisheries management. However, the inclusion of user groups, even at later stages of the policy process, clearly adds to the legitimacy of management regimes and decisions. As an instrument of participation, the IQ committee seems to have improved on the current advisory system and facilitated the building of consensus on potentially controversial issues such as initial allocations, transferability, provisions of monitoring and licence accumulation.

Other forces may be at work. One is the status of fish stocks as seen within particular institutional contexts. The Canadian system originated as part of efforts to deal with declining stocks, in the context of an existing system that featured competitive quotas for inshore and nearshore boats and EAs for offshore boats. Drastic change was seen as necessary, and a familiar model for change was at hand and being promoted by economists. In Norway non-transferable vessel quotas were in place for most offshore fishers for some time already and were extended to the inshore fishery in response to the Barents Sea cod crisis of 1988–9. IQs had become an integral if not entirely uncontested part of management policies by the 1990s. However, the question of transferability surfaced when the immediate crisis had passed. By 1990–1 stocks of cod in the Barents Sea were apparently increasing, as were quotas. In contrast, in Canada stocks continued to decline.

One explanation is that ecological 'disaster' and economic downturns will 'force' a rethinking of old solutions, almost irrespective of institutional set-ups and industrial opposition, and thereby facilitate the adoption of new ideas and policies. This is the 'crisis as mother of all innovations' thesis. Applied to our two cases it would explain the differences of policy by pointing out that the crisis ran (and runs) deeper in Canada than in Norway and that this in itself accounts for the fundamental reorientation of management policy. Certainly, in both countries, concerns about collapsing fish stocks and sharply slashed quotas made it possible to foreground issues of economic efficiency and business survival as against those of employment and survival of coastal communities. Drastic change was not needed. The allocative tools of individual vessel quotas and enterprise allocations were already established in some fisheries and thus at hand when crisis struck, leading Norway to extend IVQs to the coastal fleet and leading Canada to do the same to the inshore dragger fleet at about the same time, 1990–1.

Institutional Structures and Management Policies 203

The difference that we have emphasized is the acceptance of transferability (ITQs) in Canada, but not in Norway. It is possible that the shorter duration of the resource crisis in Norway resulted in a lowering of pressures on the government to address the 'downsizing' issue in the inshore fisheries, whereas the continuation, and deepening, of the groundfish crises in Atlantic Canada into the mid-1990s gave the Department of Fisheries a mandate to push for harder choices. However, too much can be made of this argument. The differential-crisis thesis cannot account for the important fact, underplayed in this account thus far, that in Atlantic Canada the vast majority of inshore fishers remained under competitive sectoral quotas, even though in many areas, especially Newfoundland and eastern Nova Scotia, the quotas amounted to nothing, or next to nothing, as stock collapses rippled through the region from 1992 onwards. Yet, as of 1997, the question of individual vessel quotas – never mind transferability – remains hotly contested in Atlantic Canada, whereas it has become a (begrudged) fact of life in coastal Norway.

An alternative explanatory thesis, which refers to the freedom and power of government to make difficult decisions at times of distress, has problems too. Emphasized throughout has been the extraordinary 'ministerial discretion' found in Canada's federal fisheries ministry. It is true that the relative decision-making autonomy of the ministry enabled the creation of IQs for the inshore dragger fleet in 1990, whereas a more open and consultative process might have delayed that decision for many years, as happened with ITQs in the United States (McCay et al., 1995). On the other hand, the ministry appears reluctant to push IQs, IVQs, or ITQs upon other inshore sectors of the fishery, seemingly aware of the political difficulties and the conflicting goals entailed. In this regard, it is hardly different from the Norwegian ministry, even though in Norway the stumbling block is now transferability, not individual quotas.

Separating the question of transferability from the issue of creating individual (or enterprise or vessel) quotas reflects the concerns in the two countries, but it is somewhat artificial in that one is likely to lead to the other, particularly when stocks and hence quotas are declining, and it is difficult to obtain enough quota to maintain one vessel or enterprise.

The above account of the IQ regime in Nova Scotia for inshore draggers is also one where IQs initially met with little enthusiasm among inshore draggermen, to say the least, and permanent transfers were precluded, in part to protect the interests of small-scale fishers and coastal communities. However, once IQs were adopted, the advantages of transferability to holders of quotas became transparent, and eventually IQs became ITQs. In Norway there was relatively little opposition to extending IQs to inshore fisheries, at least during the crisis, but, as we have seen, vociferous resistance to making quota shares transferable. However, there too the 'T' may have entered through the back door, as there are signs of an unofficial and informal traffic in quotas. If this is indeed the case, the difference between Atlantic Canada and Norway as to management policy may be mainly one of rhetoric and principle rather than practice.

7

Management Reform: The Search for Appropriate Institutions

Fisheries management in both Norway and Atlantic Canada has long been based on a process of consultation between government and industry – in Norway chiefly through formal consultative arrangements at the national level, in Atlantic Canada through a complicated, and more loosely connected, network of advisory committees at the regional and local levels. Pressures for change exist within both systems: in Canada for a 'downsizing' of traditional consultative practices and a larger role for professional and independent bodies, or 'management by independent commission'; in Norway for broader representation within corporatist arrangements and more decentralization to lower level administrative agencies, or more 'participatory management.'

Three problems should be noted when discussing the pressures for change in institutions charged with management of the fisheries. First is a continuous increase in state intervention in the industry, perhaps to a point where it is appropriate to speak of an 'overloaded' government. Second is the problem of knowledge, information, and feedback, or the sensitivity of centralized institutions to local conditions. Third is the problem of compliance and legitimacy, expressed in the fact that government institutions and policies have come under increasing attack both from fishermen and their organizations and from environmental groups.

Norway: A Reluctant Reformer?

Although debates on management in Norway tend to be more about policies than organizational matters, in the 1991–2 white paper (Government of Norway, 1991–2), the Ministry of Fisheries discussed options for changing current procedures. The major options were: *decentralization* of certain decisions to regional authorities, *delegating* authority to the industry itself, and *deregulation* through market incentives. Only the first was given much credence.

The question of greater regionalization of fisheries management, and hence decentralization, has been on the agenda for some time. One of the strongest initiatives in this direction has come from the Regional Committee for North Norway and Namdalen, a body composed of politicians from the three northernmost counties, with its own administration located in Bodø. In its 1990 fisheries program this committee suggested more delegation of management authority to the regional level, which was linked to its proposal that a greater portion of the TAC should be reserved for fishermen in north Norway. This is one of the three forms of 'regionalization' considered in the ministry's white paper: regional discrimination through quotas and licences, for example, by favouring fishermen and vessels from particular regions when harvesting rights are allocated. Elements of such a policy are already in operation in Norway, as they are in Canada, but it is not decentralization of decision-making.

The second model is one where central government (that is, the ministry) allocates an overall quota (TAC) to each region, but how that quota is allocated is left to regional authorities. This implies a major delegation of decision-making authority to regional political bodies, such as the county council, fishermen's sales organizations (MSOs), or new 'regional regulatory councils.'

The third form considered would give each region property rights to 'their' resources (citing the U.S. and Australia state fisheries regimes as examples), leaving the authority to manage exclusively in the hands of regional bodies. This would represent a fundamental change as the functions and power of central government would be greatly reduced.

Not surprisingly, the two last alternatives have largely been written off as irrelevant and unsuitable, because most stocks are found in more than one region and because decentralization seems unnecessary for such a small nation (Government of Norway, 1991–2: 105). The official view is clearly that the organization of the system is not at fault. That said, the white paper does contain statements to the effect that the role of regional bodies in management will be considered within the framework of a general evaluation of the regional and local fisheries administration. In this sense, decentralization rather than delegation seems to be the favoured strategy for future change.

Towards a New Agenda?

Although there are few signs of major reorganization, there are pressures to broaden participation beyond the fishing industry. The system of 'self-management,' in the sense of fisheries policy being the exclusive domain of the ministry and the fisheries industry, is no longer accepted as a matter of course. The view that the management of fish stocks is too important to be left to a relatively closed policy community is gaining ground among regional authorities as well as among non-producer groups. An interesting question, then, is how a new emphasis on the fish stocks, and the larger marine community and ecosystem, as the property of the nation will affect the debate on management institutions.

Fisheries Management: A Case of Government Overload?

A striking feature of fisheries policy during the past ten to fifteen years is the increasing number of ever more detailed regulations. This trend can be illustrated by a few figures from Norway. In the period 1930 to 1940, twenty-two new laws concerning the fisheries were enacted. This number increased to fifty-five in the period 1970 to 1979. Looking at administrative regulations the trend is even more dramatic. Between 1930 and 1939 forty-five regulations were issued; between 1970 and 1979 this number had increased to about 1,200 (Hersoug, 1983: 65). This trend showed no sign of subsiding in the 1980s (Ørebech, 1986). Considering that most of these regulations are enforced by central government, and that fisheries remains the smallest of all government departments, the administrative capacity of the ministry to handle the increased workload may well be questioned.

Another aspect of the overload problem is the financing of government programs. In Norway annual subsidies through the 1964 Main Agreement for the Fisheries (Jentoft and Mikalsen, 1987) have shown a more or less continuous increase since the 1960s. In the period 1964 to 1986 the subsidies equalled, on average, about 18 per cent of the dockside value of the total catch. In 1980, this figure was 40 per cent, the highest ever (Holm, 1991). Subsidies were substantially reduced between 1987 and 1989, but they increased again sharply in 1990 in the context of crisis. Thus, there is no doubt that the financial capacity of government has been stretched to a point where further increases in transfer payments are almost inconceivable.

Financial capacity has possible connections to management policy. The size of TACs is obviously related to the extent and level of demand for transfer payments. Accordingly, governments share some incentives with the industry to push for higher quotas, particularly in negotiations with other nations (Hoel, Jentoft, and Mikalsen, 1991). Success on this point not only reduces the demand for subsidies, it also makes the national distribution of quotas less controversial and demanding (Sagdahl, 1992).

Expertise, Information, and Feedback

The management of fisheries in Norway, as in Canada, is largely a question of centralized consultation and departmental execution. To some extent this is rational and necessary. Restricting participation in the fisheries through licences and quotas is, by any standard, a significant step, given the long history of free enterprise and open access. Restricting participation pertains to values of justice and equity; it is also a question of rules and laws. The argument goes: If the distribution of quotas and licences is a question of enforcing or applying a set of given rules, there is a case to be made for centralized decision-making. The rule of law is probably best achieved by vesting authority in the legally trained staff of government bureaucracy.

The problem is that centralized systems of management tend to neglect issues and problems of regional variations and local peculiarities (Strand, 1978), which are not necessarily reflected in the laws and administrative procedures. Although there is no systematic evidence that this is the case in Norway, the government is aware of complaints that central government neither understands nor pays attention to local and regional dynamics. In its white paper the ministry conceded that this topic deserves further scrutiny. A regional and local fisheries administration may alleviate this, but it will hardly solve it so long as the ministry can choose how and when to pay attention to information and advice from below.

Effective policy requires timely and accurate evaluation or feedback on the effects of current policies. The situation now is one where the ability to learn from past experience is limited, the collection of information is erratic and ad hoc, and changes of policy are driven mainly by crises and other conspicuous events. When limited entry was first introduced in the early 1970s, Parliament explicitly assumed that it would be informed about the implementation and effects of the system through annual reports (Mikalsen, 1987: 106–7). However, these reports never materialized, and Parliament apparently did not pursue the matter. The ministry has thus lacked an important incentive to establish routines for the collection and analysis of data on the effects of limited entry, namely, the pressure and attention from Parliament. Information on the effects of policy is collected more indirectly and ad hoc, through communications from individuals and organizations, and via accounts in the news media. If this is correct, then feedback becomes what Thompson (1967) has called 'social tests': You do not study the effects directly but you try to establish how important actors see them. Consequently, views and complaints from resourceful individuals and conspicuous interest groups are substituted for an independent assessment of effects.

The Search for Appropriate Institutions 209

The Case for Broadening Representation and Participation

The Norwegian fisheries ministry has conceded in the 1991–2 white paper, and elsewhere, that problems of legitimacy and control remain in management of the fisheries, and that these problems may have something to do with the organization of regulatory decision-making. What should or could be changed, however, is far from clear. Some talk of giving the industry greater influence and responsibility, but with few specifications as to how this should be done. Others talk of the need for looking into the role of the existing regional fisheries administration, presumably with a decentralization of management functions in mind. Neither of these 'strategies' addresses how to broaden representation and participation without seriously undermining the efficiency of management decision-making.

This is the classical dilemma between internal democracy and external effectiveness in formal organizations. It reflects a steady 'crowding' of the policy environment, to borrow a phrase from Harrop (1992: 270), in the sense that non-producer groups and regional agencies are increasingly demanding representation and recognition. The successful pressures from environmental groups for representation on the regulatory council is but one example of this trend.

To the extent that management of the fish stocks is an issue of public interest, there should be some arrangement for balancing group interests against common concerns. Weighing the legitimate demands of user groups against the public interest is, of course, a highly political task that requires institutions where disputes can be settled authoritatively. The regional committee for North Norway and Namdalen, with the support of the elected assemblies of the three northernmost counties, has proposed decentralization of regulatory decision-making to regional administrative agencies and (limited) participation of genuine political institutions at this level of government.

Elected bodies are not necessarily appropriate arenas for deciding on licences and quotas. These (at least partly) technical decisions should be taken in administrative agencies. Setting objectives and determining the criteria of allocation are, however, questions about political priorities and as such are matters for political discourse. The proposal from the regional committee suggests a stronger role in this regard for the institutions of local government, such as the elected assemblies and boards of the counties ('fylkesting' and 'fylkesutvalg'). Their argument is that the participation of *political* institutions would give the public interest a fair hearing when goals are set, and, in turn, make management decisions more sensitive to the particular needs and problems of the region. Further, a broadening of representation along these lines would be tantamount to recognition that we all have a stake in the management of fish stocks.

In a regionalized framework, the distribution of licences and quotas could be handled by the regional fisheries board (Fiskeristyret) or by a regional regulatory council with representation extended to non-producer groups. The latter would have to be created and designed for the purpose. Both options would help sustain what is now accepted as a central principle, namely, that the authority to issue licences and individual quotas should be separated from the power to set TACs (a matter for bilateral negotiations and central government). Protecting this separation reduces the tendency to solve controversial problems of allocation by increasing the overall quota.

Towards More Inclusive Institutions?

Proposals to decentralize certain aspects of management and to broaden participation so as to include elected assemblies and non-producer groups may sound far-fetched to some ears, but they address the growing concerns of public interest groups about important problems of the current management regime. This is reflected in the growing perception that citizens, consumers, and ethnic groups are 'stakeholders' in the fishing industry and maritime resources are the 'common heritage of mankind,' or at least the nation. Problems of the regime include the overload of central institutions, perceived mismatches between authority and expertise and between administrative decisions and local problems, increasing alienation of fishermen from the institutions and decisions of government, and the growing concern of non-producer groups and representatives of the general public that an important part of the national heritage is being squandered. In short, environmental concerns and regional mobilization are changing the perception of who is affected by regulatory decisions and, therefore, of who should be represented in management institutions.

Atlantic Canada: From Advisory Committees to Independent Boards?

Atlantic Canada shares most of Norway's challenges and dilemmas concerning management of its fisheries, which can be summed up as the need for more broadly representative, responsive, and effective institutions. A recent direction in institutional reform is to develop independent decision-making bodies to provide advice to the fisheries minister.

The power given the Canadian Minister of Fisheries and Oceans under the Fisheries Act has long been considered an anomaly as a regulatory system. Compared with other federally regulated sectors of the economy, the fisheries are relatively unique in being entirely subject to ministerial discretion. Sectors such as communications, energy, and transportation are generally managed by administrative agencies at arm's length from government and through a process where the right of affected groups to be heard is codified (Kernaghan and Siegel, 1991). The fisheries, however, have been managed through a highly centralized and fairly closed system, with little scope for systematic and representative industry input except for the problematic advisory committee system. Such a closed structure is subject to accusations of political patronage and cronyism.

Further, a series of controversial decisions by ministers have raised questions of due process and equal treatment. For example, there was recently a case in the Scotia–Fundy region where the Minister had awarded a lucrative offshore licence and then – under pressure – he withdrew it. The company in question sued and won over half a million dollars in damages for reasonable expectations and investments made. The issue of reform, then, is partly about relieving the minister of some of his tasks, particularly those of allocation, and partly about strengthening industry input into the management process. As we shall see, however, a proposal for independent boards (Government of Canada, 1993) is somewhat ambiguous on the latter point as it emphasizes expertise and independence rather than economic interest and direct representation.

The Background: Problems with the Current Regime

The idea of independent boards actually dates to the early 1980s, when two major studies of Canadian fisheries (Kirby, 1982; Pearse, 1982) recommended that issues of allocation and licencing should be 'depoliticized through new independent agencies or boards, at arm's length from the federal Minister for Fisheries' (Parsons, 1993b: 482). Instead, the government chose to strengthen its advisory group system. The present proposal seems to have originated in the ministry of DFO. In a highly general and tentative form, it became the subject of broader consultations at the end of 1991. The minister claimed than under this new system, decisions taken every year on licencing and allocating the catch would be more consistent, more predictable and more open to the public.

The ministry formed two working groups (one for each coast) of 'knowledgeable people' from industry who, along with representatives from provincial governments, were consulted on the details of the new system. The result was reported in a DFO paper (Government of Canada, 1993), the basis of our sketch of the proposed system.

In the first proposal for management reform, the problems were said to be 'confusion about rules, lack of openness to the public, and a very high degree of discretionary authority on the part of the Minister of Fisheries and Oceans' (Government of Canada, 1993). The 1993 report discussed the problem of the lack of openness, noting that decisions that fundamentally affected the industry were often taken behind closed doors in the ministry. Thus, there was no guarantee that those affected would be properly heard before decisions were made. Participants from industry have also seen decision-making as too centralized and not sufficiently sensitive to regional and local needs.

Second, the existing consultative process, based on a complex structure of advisory committees, was criticized as being too cumbersome, powerless 'window dressing,' and 'an exercise in frustration.' Third, the system for enforcing the rules of management, and applying sanctions to those who break them, has proved time-consuming and expensive. The criminal courts, responsible for sanctions, tended to underestimate the seriousness of violations of the fisheries, and there has been no role for industry in enforcement. The reforms, then, are largely about replacing a closed and cumbersome system based on considerable ministerial discretion and a complex structure of advisory committees, with a simpler and more open structure that provides for more of a partnership between government and industry within the realm of fisheries management.

At the core of the reform strategy is a distinction between *conservation* and *allocation*: between the setting of TACs for particular stocks and other conservation measures on the one hand and the allocation of quotas and licences among individual fishermen on the other. This distinction has already been sharpened by separating the science branch from management within the DFO regional offices. Conservation will continue to be the responsibility of the minister, but he or she will be aided in this capacity by a new management institution, the Fisheries Resource Conservation Council (FRCC). This part of the reform is already under way; the FRCC began its work in May 1993. The second part of the reform strategy is that allocation would be taken over by two fisheries boards, one for each coast, made up of individuals 'knowledgeable about and with experience related to the industry,' but with no direct financial interest in it. These boards would take over some of the responsibilities and powers currently exercised by the minister and DFO such as licencing and allocation of quotas, as well as the role of criminal courts in enforcing sanctions for violations.

Conservation: The Role of the FRCC

The new council is generally described as 'a partnership of government, the scientific community, and industry' (Government of Canada, 1993: 1) that was created, in part, to achieve greater involvement of the fishing industry and a more open decision-making process than had existed before, when the Canadian Atlantic Fisheries Scientific Advisory Committee (CAFSAC), made up largely of scientists, was the major advisory body to the minister. The primary purpose of the new institution is to make recommendations to the Minister of Fisheries and Oceans regarding TACs and other conservation measures for the Atlantic coast fisheries. Currently, the Council's work pertains exclusively to groundfish, but there are plans for extending its responsibility to pelagic and shellfish stocks. In addition to written recommendations to the minister, the mandate of the council includes such activities as advising on research priorities and reviewing stock assessment information and conservation proposals and DFO data and methodologies.

The council comprises fourteen members, five of whom are academics. Of the other nine members, five have a background in processing, four in harvesting. These do not, however, represent any particular group or organization but have, on the face of it, been appointed for their experience and expertise. In addition, there are nine ex officio members, three from the Department of Fisheries and Oceans and six representing the provincial governments of Quebec, Nova Scotia, Newfoundland and Labrador, New Brunswick, Prince Edward Island, and the Northwest Territories.

Also attached to the council are six subcommittees: for stock assessment, historical analysis, environment and ecology, communications, management, and gear technology. These committees have written mandates within their respective domains, and membership includes both the full and ex officio members of the council.

The council mainly works by way of consultations and public hearings and, as such, is supposed to provide a forum where all stakeholders in the Atlantic fisheries can participate.

The 1993 process is illustrative, although it may not be typical because it took place a year after the drastic decline in groundfish stocks was recognized through measures such as the moratorium on northern cod fishing. The process started with a report from DFO to FRCC on the status of Atlantic groundfish stocks. The FRCC was asked to advise the minister on the stocks thought to be in a critical state (Government of Canada, 1993: 2). As a first step, the FRCC held a number of hearings (seven in all) with stakeholders to convey information about the crisis and elicit their views on what should be done. A mid-1993 report was presented to the minister. Shortly thereafter, the council held thirteen public hearings in various places along the Atlantic coast. These were well attended by various groups of stakeholders, one by about a hundred people, including representatives from the news media and DFO. In addition, the council held four meetings with representatives of all industry organizations. On the basis of these meetings, plus an updated report on the status of the stocks and yet another stockholder consultation on redfish, the council made its recommendations to the minister.

A possible criticism, and one that in fact has been launched, is the council's close linkages to DFO. The perception in some quarters is that its ability to be critical about government policies has been compromised.

One of the criticisms of the FRCC from the start has been its linkage to the DFO. What most fishermen had hoped was the council would perform the same function as that of the auditor general's department, an independent body capable of criticizing DFO and its policies. Appointments like that of Vermette [an Ottawa 'mandarin'], however, assured its compatibility with the aims of Ottawa and DFO. Even the symbolic gesture of locating the FRCC's office in a centre other than Ottawa might have shown some independence, but such was not to be the case.' (*Atlantic Fisherman*, October 1994, p. 7)

Having attended several FRCC meetings in 1993, 1994, and 1996, we have the impression that the council hearings work well as fora for the exchange of information and views among government, science, and industry. They are mostly well attended by representatives of groups in the fishing industry, who have been quick to seize this opportunity to air their concerns and to question, criticize, and instruct the scientists, both orally and in writing. In turn the scientists and industry representatives on the FRCC appear to make genuine efforts to solicit and listen to the input of these groups.

The effects of industry input on the council's recommendations are not easily discerned, and the impact of the new process on conservation policies remains to be assessed. At this point the FRCC certainly looks like a potentially powerful player on the management scene. In its 1993 report, for example, the FRCC recommended drastic measures in response to the groundfish crisis, most of which were eventually adopted by the minister.

Allocation: The Proposed Role of the Fisheries Boards

The Fisheries Resource Conservation Council (FRCC) is supposed to limit itself to conservation questions such as TACS and whether to close or reopen fisheries. The 1993 reform proposal included the idea of two fisheries boards to handle allocation issues. The core of these boards, its executive, so to speak, would consist of seven members (Atlantic) and five members (Pacific), appointed by the Cabinet on recommendations from DFO, with additional members appointed to panels of the board to give more flexibility. The panels would hold hearings with interested parties, and produce advice and recommen-

Figure 9. Proposed management structure in Atlantic Canada.

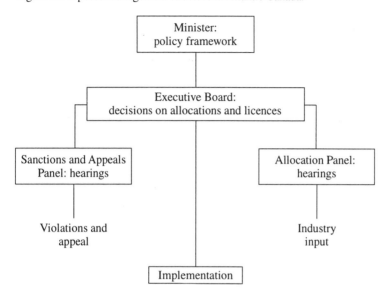

dations on allocations and licences to the executive board. The panels would replace the current advisory committees as arenas for the annual review of issues of allocation. The basic structure of the proposed system is shown in Figure 9.

As seen by the ministry, a reorganization along these lines offers advantages to both industry and governments. First, there should be greater openness through public hearings and the publication of the decisions and the arguments behind them (Figure 9). Key documents on policy and decisions would be available at DFO offices. Second, fairer and less arbitrary decisions may result from quasi-judicial procedures and the prospects of close public scrutiny. Third, greater consistency should result if the boards make decisions on the basis of a policy framework set by the ministry. The fact that the board's decisions, as well as the policy framework, will be public documents provides for checks on consistency over time. Finally, there will be more efficient enforcement, as violations will be heard and penalties set by the boards rather than the courts.

The executive board would have the authority to make decisions within three designated areas. First, it would have the power to issue licences in existing limited entry fisheries, to set and amend licencing rules, and to suspend or cancel licences where appropriate. Second, the board would be responsible for the allocation of individual quota shares within TACs as set by the minister. The issuing of licences as well as the allocation of individual quotas would be based on advice or input from the industry, mainly through panel hearings which would be open and whose motives for decisions would be available to the public. Third, the board would handle cases of illegal fishing and breaches of licencing conditions and apply sanctions where appropriate. As such, the board would replace the criminal courts as the primary source of enforcement.

As a vehicle for institutional change, the 1993 proposal implies a substantial delegation of decision-making power from the ministry to the boards, and it could well be conceived as a strategy, on the ministry's part, to shed some burdensome responsibilities. It is quite clear that the reform process has been driven by DFO with little, if any, industry involvement.

So far there have been few reactions from industry, but those that have come show little enthusiasm for the notion of independent boards. The crucial question, from the point of view of legitimacy, is really whether the boards, if they are created, will manage to establish themselves as genuinely independent and impartial agencies or, rather, be viewed as such by those affected by its decisions. On this point, there may well be problems and pitfalls ahead.

Why Independent Boards?

Apart from coping with ministerial 'overload' and simplifying a complex consultative system, the proposed changes in institutional structure have clearly been designed to bring the management of fisheries more in line with regulatory practices in other industries. On this point DFO, in its 1993 proposal, maintains that independent, quasi-judicial boards offer a tested and accepted model, suggesting a tradition of independence and impartiality in government regulation. To the extent that the boards will approach the standard format of regulatory agencies in Canadian public administration, their prospective contribution to the legitimacy of fisheries management should be assessed with a view to the role and aims of this kind of agency. For our purposes, a regulatory agency can be defined as a statutory body created to regulate an economic activity using specific rules of procedure. Regulatory decisions are supposed to be in the public interest and firmly within the framework of government policy. A regulatory agency is subject to direction by the legislature and a responsible minister with regard to matters of policy. It is, however, autonomous when it comes to making specific decisions within the policy guidelines of government (Kernaghan and Siegel, 1991: 225–6).

According to Kernaghan and Siegel (1991), regulatory agencies in the Canadian government are required to conduct their business in a judicial-like manner, while their decisions are to be based on policy guidelines and

The Search for Appropriate Institutions 217

considerations rather than on precise legislative standards. Furthermore, regulatory agencies are independent and autonomous when considering particular cases in that they are shielded from direct political intervention. The minister is fully responsible for the outcomes of the agencies' activities, but he or she must refrain from interfering in the making of individual decisions. Political control and direction are mainly exercised by the setting down of policies to which individual decisions must adhere. The reasons for creating autonomous bodies, at arm's length from government and the minister, can be manifold. In our case, three points deserve particular attention: First, a regulatory agency may be created to remove an issue, or issue area, from politics, that is, from processes that rely 'too much on partisanship, compromise, and expediency, and not enough on fairness and hard economic facts' (Kernaghan and Siegel, 1991: 229). In this sense, creating an agency may place an issue beyond the influence of special interest groups or powerful individuals with interests at stake.

Second, agencies may be set up to provide more openness within a particular area of public policy, by organizing impartial, judicial–like hearings when considering individual cases. Hearings will provide opportunities for those involved to state their case, and others to comment on and oppose it. To the extent that participation is inclusive, regulatory agencies may, in fact, contribute to fairness and democracy in public policy. Third, agencies may facilitate the use of specialized knowledge. An agency may be composed of people who are experts on particular industries, and it will, over time, accumulate specialized and detailed knowledge within its particular field of operation. In this sense regulation through independent agencies may contribute to more intelligent and more informed policy decisions.

How would the creation of fisheries boards affect the legitimacy of management decisions and policies? To what extent can the proposed reforms reduce the scope of 'politicking' in fisheries management, provide more transparent decision-making, and facilitate the utilization of specialized knowledge? All that can be said at this point is that the proposed new institutional set-up redefines the very nature of allocative issues. Decoupling conservation from allocation would probably facilitate a more 'technical,' 'professional,' and judicial approach to decision-making and play down the overtly political and adversarial nature of decisions pertaining to quotas and licences. In this sense the 1993 reform package may well be seen as a deliberate attempt to depoliticize the fisheries management process.

The Pros and Cons of Independence

The independent fishery board model bears further scrutiny, both for the case at hand and for what it offers to the broader question of ways to institutionalize partnerships between industry and government and co-management in fisheries. Official discussions of the model suggest that allocative issues can be handled mainly as matters of adequate information and impartial judgment. However, the boards would likely operate in an economic and ecological context of scarcity and in a political culture defined by a history in which political clout and ministerial discretion rather than predesigned policies, formal rules, and technical information have been at the core of management decisions.

The fisheries boards would also have to confront cases where the livelihood of individuals or even whole groups may be at stake, and where decisions are unlikely to be accepted as legitimate solely on the grounds that they have been reached through impartial and informed judgment. In as much as the legitimacy of management decisions hinges on their content as well as on the way in which they have been reached, the creation of independent boards may do little to improve compliance and support. The question of representation will remain. Appointments may prove decisive as the perception of independence and impartiality will be influenced by the personal and professional qualities of the board members.

The goal, as stated by the minister, is to create a decision-making system that is more open to public view and gives a more direct voice to those involved in the fishing industry. 'Openness' and 'shared responsibility' are, in other words, key principles behind the reform. Public hearings under explicit rules of procedure – an integral part of the activities of regulatory agencies – are the vehicle through which these principles will be fulfilled. This is clearly an alternative to the current system of consultations, where selected groups and individuals are permitted to give inputs to a complex, often confusing, and usually closed process.

A more open and transparent process may in itself strengthen its legitimacy, through its effects on public confidence in the fairness of the system, provided the boards are able to give due attention to the various, and conflicting, demands made upon them. Our observations of the regional fishery management councils of the United States, organized on a similar basis and even more transparent and open to public review, suggest that the linkages between these features and both legitimacy and effectiveness are not self-evident.

Whether the boards would facilitate a bigger role for industry in management, as envisaged by the minister, is even more doubtful. The policy framework would be created by the ministry – which presumably is not up for bargaining. The boards would operate according to rules of procedure that should give little room for logrolling, eliminate the significance of coalition building across groups and interests, and reduce the scope for the articulation of special interests. The boards would be arenas for the 'sounding out' of affected groups and individuals, rather than arrangements for the efficient representation of selected interests. This should make them more inclusive than, say, corporatist structures and consultative committees, and increase the legitimacy of management processes and policies among groups poorly served by existing arrangements.

Knowledge and experience rather than narrowly defined interest seem to be the primary criteria for appointment to the boards. Independence from political control as well as industrial influence is a core value, as is appropriate for a body modelled on the standard form of regulatory agencies, although it is difficult to measure. There is much to be said in favour of strengthening the role of knowledge and expertise in the management of fisheries. There is, however, always the question of who the experts are. What kinds of expertise and knowledge are particularly relevant to the problems of allocation in fisheries management, and how is this best acquired? Are there other sources of knowledge about the fishing industry besides the industry itself? These questions have particular resonance in the wake of the collapse of northern cod and post mortem critiques of the scientific establishment.

The legitimacy question is whether truly independent boards are possible. In the words of one industrial representative, 'Who's neutral and independent? You're either from a mentality where you're pro-business or from a background that says that we've got to preserve small communities and maximize employment and, maybe, minimize income.' In a polarized setting, such as that of the contemporary fisheries of Atlantic Canada, one would expect such attitudes to be fairly widespread. In that case the proof of the pudding would be in the eating; the board would have to 'prove' its independence and fairness through its work and decisions.

The composition of the boards is, of course, dependent on how members are appointed. The power of appointment normally resides with the minister and ultimately with the Cabinet (the Governor in Council), a procedure that, according to the 1993 proposal, will also be adopted in the present case. Control over appointments as well as policies may be crucial, given that the minister will remain politically accountable for the boards' decisions. On the other hand, the 'arm's length' principle requires that the boards be independent in deciding individual cases. A difficult balance must be struck between political accountability and regulatory autonomy, which will affect the reputation of the boards with both industry and the general public. Excessive political control, for example, may cause the boards to be perceived as just another branch of federal government, deflate the morale of their members, and undermine their independence. Little ministerial interest in appointments and vague policies, on the other hand, will make the boards an easy prey for powerful interest groups. Both situations would certainly affect the standing of the boards and the legitimacy of their decisions.

Is There a Case for Reform?

Participation by user groups has long been the standard response to problems of legitimacy and compliance in management of the fisheries. Also there is the basic principle, rooted in democratic theory, that those directly affected should have the opportunity to be heard before decisions are made on their behalf. Because of this there has been a proliferation of arrangements for greater industry involvement in fisheries management in both Canada and Norway: committees, councils, and boards geared to the provision of knowledge and advice as well as to the 'production' of support and compliance, that is, legitimacy.

There is a growing perception, however, particularly in Norway, that consultative management is tantamount to letting the fox into the henhouse; it helps the fox help himself to the goods without having to think of the long-term consequences. In other words, fisheries management may be too important, and have too many effects beyond the industry itself, to be left exclusively to a policy community dominated by government officials and industry representatives. Related is the fact of increase in demands for participation by other groups, particularly environmentalists, which then raises questions about the effects of new and more sources of interests, information, and conflict on efficient and effective policy-making. In addition, the extent to which industrial participation is representative may be questioned: in Atlantic Canada because of the fragmented organizational structure of the fisheries, in Norway because of politics within the Fishermen's Association. Accordingly, there is a risk of unforeseen resistance to agreed upon measures.

The 'independent commission' alternative to corporatist and other forms of consultative management that has arisen in Canada – the use of knowledgeable but disinterested and hence 'independent' people to make decisions, at least about issues of allocation – has risks too. Such bodies may be led to reduce inherently political questions to technical problems to be solved through the application of formal rules and disinterested expertise. Reducing the scope for political compromise, expediency, and lobbying may be a valid reason to introduce such structures. However, wherever there are interests at stake and conflicts to be tackled, political judgments are necessarily involved. Hence, the 'independent commissions,' like other regulatory agencies, boards, or commissions, must, of course, exercise political discretion, and do this partly on the basis of inputs from those affected by their decisions. Accordingly, open hearings conducted by special panels 'authorized' by the independent Commission's executive core are part of the Canadian proposal.

Another issue is how 'independent' decision-making bodies can be reconciled with the need for fair and thorough representation of the interests at stake. On this point it is important to note that the Canadian fishing industry has never been 'in' in quite the same way as is the case for Norway. Apart from a few very big players – the vertically integrated companies – access to government has been far more irregular and unequal and less corporatist and 'democratic' than in Norway. With this in mind, it might be suggested that the proposed fisheries boards, in combining expertise and consultation (through hearings) will be better equipped than are current arrangements to incorporate concerns of various sectors of the fishing industry. However, representation will at best be indirect. It is then far from obvious how the boards, in the words of the minister, would 'give a more direct voice to those who work in the industry.'

The use of independent commissions in fisheries management is not widespread, although the U.S. regional council system has elements of it in that voting members include appointees who may have very little if any relationship to the fishing industries. The 1993 Canadian proposal raises the question, not just of why the boards have been proposed, but also whether management by independent boards is more adequate, and easier to achieve, in some settings than in others. One might, for instance, argue that in Norway the case for delegation of ministerial responsibilities to an independent group is weaker than in Canada because there is less of a need for this kind of institutional change. The corporatist arrangements provide for a more workable trade-off between ministerial discretion and the influence of user groups than do the less systematic and more fragmented arrangements of Canada's proposed advisory system. Partly because of that, management of the Canadian fisheries has suffered from ministerial 'overload,' which was a major impetus for structural changes such as the FRCC, for conservation, and the proposed independent boards on allocation issues.

Another factor is how industry input has traditionally been obtained. In Canada the current system of advisory committees is cumbersome and clearly inadequate for securing representative input from industry. As there are no ways of knowing whether those consulted speak for a larger group, one might as well consult less. In Norway the consultative process is altogether more integrated and streamlined, and problems of representativity are less pressing. A third factor that may account for the choice of management institutions is what models are 'at hand.' In Canada one finds a long-standing tradition of public regulation through independent boards or agencies that work within a policy framework set by government but are autonomous in deciding particular cases. Fisheries management is anomalous in not being run this way. In contrast, public regulation in Norway is largely the domain of 'ordinary' government agencies such as ministerial departments and directorates. There are, of course, commissions and boards with regulatory tasks, but those tend to be representative (that is, corporatist) rather than independent.

Conclusions

In this part of the book we have examined differences and similarities between the regimes of fisheries management in Atlantic Canada and Norway. We have considered some of their distinctive features and how they compare with respect to institutional frameworks, management policies, and strategies for reform.

One of our respondents observed that 'fisheries management is crisis management.' If then, as the saying goes, crisis breeds reform, one should expect systems of fisheries management to be in the process of reform very often. Indeed, there has been some serious rethinking of institutions as well as policies in both systems, in response to the challenges posed by extended national jurisdiction and the recent crises in markets and supplier of natural resources. The extent to which ideas and proposals have materialized in the form of genuinely new structures and policies is perhaps more open to debate.

In Atlantic Canada institutional reform seemed to have gained momentum in the 1990s, apparently in response to decline in both fish stocks and the availability of government finances. New ideas about the appropriate roles of government and industry have materialized in legislative initiatives, some of which have led to concrete changes in management structures. The Fisheries Resource Conservation Council (FRCC), and proposals for independent management boards to take over allocation, are the most notable examples. An overall strategy of the ministry is 'partnership,' where management is conceived almost as a 'joint venture' between government and users of resources. This is a key feature of a new Fisheries Act which has been tabled to the autumn of 1997; the partnership provisions of the act are a major and contentious point of discussion in coastal Canada.

One of the concerns voiced is that through partnership agreements, the government will shirk its responsibilities to its constituents and the natural resources on which they depend. The concerns may be responding to a larger process, whereby the commitments of the federal government to management of the fisheries are being reduced, via devolution through partnerships and independent commissions, as noted, but also by increased reliance on market mechanisms for allocative efficiency. The introduction of enterprise allocations and individual quotas may thus be part of the same process.

In Norway similar economic and ecological pressures have generated a search for policies that give more than flexibility to participants in the fishing industry and for management procedures that are more attentive to local considerations. Rhetoric emphasizes the need for self-reliance and greater attention to market criteria, similar to the private enterprise rhetoric supporting ITQs in Canada. In Norway this has translated into significant reductions of government subsidies, on the grounds that they encourage market inefficiencies, as well as thus-far unsuccessful attempts to eliminate restrictions on the free transferability of individual fishing rights.

There has been a limited devolution of authority to the regional officers of the fisheries directorate. Regional directors are responsible for local developments in aquaculture. They have been delegated authority to allocate 'recruitment quotas' for new entrants to fishing, presumably because such allocations require careful attention to local considerations. Further, research indicating that there may be 'local' cod stocks, specific to particular areas or fjords, has led to demands for more local-level involvement in management (Maurstad and Sundet, in press).

Overall, the conjunction of institutional patterns, ecological imperatives, and trends within the international political economy interact to generate pressure towards more decentralized and privatized systems of management. In Norway privatization of fishing rights has been slowed down by organized resistance to proposals for individual transferable quotas. Norwegian fears may be Canadian realities, both in process and outcome. In the ITQ system we depict, concentration of quotas and centralization of fishing operations is increasingly apparent, as is the tendency for restrictions on full transferability to be gradually eliminated.

Our perspective has been to see management systems as a trade-off between government control and user-group influence, between executive discretion and democratic participation. How different are the two cases? We raised a question in the introduction to this part about whether management structures and policies reflect the imperatives of a particular issue area or the political and institutional characteristics of the nations in which they are developed. If the basic problem of fisheries management is limiting access or use so as to ensure the long-term economic viability of the resource, then it is not surprising that in both Norway and Atlantic Canada fisheries management has been about limited entry and fleet rationalization and downsizing, as well as the question of how to balance goals of resource conservation and economic efficiency against considerations of social and regional equity. Conceptions of the basic goals and tasks of management of the fisheries are remarkably alike, even though they have been developed and adjusted through distinct political processes and contexts.

This is partly because of the broad similarity of problems; it may also be the result of frequent and important exchanges of experts and expertise among the fisheries agencies of the North Atlantic nations, as seen in the consultations done for Norway's 1991–2 white paper.

Furthermore, both regimes are characterized by a combination of centralized control and user-group involvement, with scientific advice at the core of proceedings on management. A high level of central government control in both countries partly reflects their dependence on fish stocks that straddle the lines of extended jurisdiction and thus call for international negotiation. Centralized control is also important in dealing with issues raised by the mobility of fish and fishers across provincial or other subnational boundaries, particularly given that in both countries fish are considered by law to be resources of the nation. The emphasis on participation and consultation in both systems thus represents tradeoffs between hierarchy and participation and between science and politics.

Looking more closely at trade-offs between 'consultation' and 'command,' there are fairly conspicuous variations between the two regimes. In Norway consultation is done for almost everything and is ongoing. Relationships are firmly institutionalized, and bargaining tends to be consensual rather than confrontational. It is a corporatist system in the sense that user groups are accorded a central position in the policy-making process. They are treated as 'insiders,' and they do have some power over management decisions and policies, sometimes acting as veto groups. Industry is represented by a peak organization which expects to be heard before decisions are taken. Accordingly, the minister has little freedom to decide whether or not to consult. But there is clearly room for independent government action. The minister may 'overrule' industrial demands or play one group off against the other. Where industrial opinion is divided, which is often the case, the minister will risk little in pursuing a course of his or her own. The same applies, of course, in cases when international obligations need to be observed. The basic characteristic, nevertheless, is one of power sharing and partnership at the top, that is, a system where privileged access and ministerial discretion are traded for legitimacy and responsible interest group behaviour.

The balance between user-group influence and executive discretion may be shifting, however, moving the Norwegian system towards what might be called 'centralized consultation' as a result of change in the internal relationships of the management policy community. The most conspicuous of these is the growing tension within the Fishermen's Association between the offshore and inshore sectors. This is, as already pointed out, an old and lasting conflict, but one that has taken on a new significance as regulations have become tighter and more severe. The annual debate on quota allocations has become more passionate and public, and the controversy over the 'quota ladder' has intensified to a point where it may eventually split the association.

Another factor that could prove significant is the heightening of public awareness of and interest in how the fisheries are managed. Environmental groups are not yet full-fledged members of the management policy community, but they are certainly bringing public interest arguments to bear on the management process. Both these factors may work to strengthen the role of the minister in management policy-making and could move the Norwegian system closer to the Canadian one as far as executive discretion is concerned. Conflicts over allocations and demands from non-industrial groups for more 'responsible' and long-term policies could make the minister more of an independent arbiter and enhance the power of central institutions to act without consulting. Given the long-standing tradition of participation and bargaining, user groups will still be important and influential players. However, their ability to veto particular policies and decisions could be seriously weakened.

The Canadian system is more complex, yet fairly 'simple' as far as relations between interest groups and the state go. Its basic characteristic is a high degree of centralized control, where executive discretion is rooted in the constitution and underpinned by other legislation. The minister can act unilaterally, and he or she frequently does so. The question of when and who to consult, and whether to act on the advice given, is largely for him or her to decide. Nevertheless, the system comprises an elaborate network of consultative committees where user groups are given the opportunity to discuss management measures and plans. However, the sheer complexity of the system, and the lack of coordinated action on the part of industry (there is no peak organization as in Norway), do not facilitate user-group influence. With the possible exception of Newfoundland, where a union plays a major role, there is clearly more fragmentation than unity.

Industrial associations in Canada do not speak with the same authority as, for example, the Norwegian Fishermen's Association, and they act as pressure groups rather than as partners and co-managers. There is certainly consultation, but consultative practices have not been developed into full-fledged corporatist arrangements. This makes the fisheries regime in Atlantic Canada more pluralist and confrontational than corporatist and consensual. Adding to this complexity is the decentralized structure of DFO with much of the day-to-day management of the fisheries conducted by regional units or offices. This does not, however, reduce the centralized nature of the system, as the responsibility for policy and coordination rests firmly with the department in Ottawa.

Reforms in allocation may, however, move the system towards what might be called 'decentralized consultation.' The new Fisheries Act bill, tabled to the fall of 1997, reinforces a trend to separate allocation issues from other aspects of management, and in particular to separate management tasks from Cabinet responsibilities, as is done in other spheres of government regulation. Changes along these lines would relieve the minister of a burdensome task and vest considerable power in the experts replacing ministers and user groups as key actors. This is management by independent commission, where the role of government is to lay down guidelines and goals, and where user groups are 'heard' rather than 'consulted.' A reform along these lines seems highly improbable in Norway, as there are few, if any, traditions for separating regulatory tasks from Cabinet responsibilities. The former are generally located in ministries with a broad set of tasks.

Disregarding these plans for reform, the comparison of our two regimes presents a paradox in that the capacity for centralized control in management of the fisheries seems stronger in the federal state of Canada than in the unitary system of Norway. In the former executive power and ministerial discretion is anchored in the constitution and has survived repeated challenges from the provinces; in the latter the capacity for centralized control is under-cut, not by delegation of authority to lower administrative levels, but by a powerful industrial 'lobby' working closely with government officials within the framework of corporatist arrangements. In Canada federal power over management may also have to do with the organizational fragmentation of the industry, with the lack of a credible 'partner' for and an efficient 'opposition' to government. The fact that organizational strength and cohesion has probably been the single most important source of user-group influence in management of the Norwegian fisheries adds credibility to this point of view.

SECTION 3
COMMUNITIES AND ENTREPRENEURSHIP – RE-EMBEDDING
COASTAL COMMUNITIES

Introduction: Parallel Crises

In this section we will address two pertinent questions: First, what is rational behaviour for an industry like the fishery, in face of the two major challenges to its coastal communities in the 1990s (and beyond) – crisis in supply and globalization of markets? Second, how can governments enhance the fishery industry's own capacity to live with chronically unstable conditions? These questions have no easy answers. But certainly they can no longer be addressed with no reference to the people within the industry who have to cope with the situation every day.

Much can be learned from the ways households, fishers, and fish processors, individually and collectively, are handling their day-to-day-activities and their long-term strategies. We advocate an empirical approach to what amounts to an adequate response to crisis. On the basis of data gathered through a series of interviews with fish processors in Nova Scotia and north Norway, we explore the ways in which they have adapted. We delineate the logic of their business strategies and the choices they have been making. A comparative study is a useful method for doing this, particularly, as the Norwegian experience seems to have been a harbinger of the more recent, but similar crisis in Atlantic Canada.

Unemployment, marginality, and despair characterize the fishing communities these days. North Norway, contrary to all predictions and expectations, recovered quickly from the resource crisis that struck the industry in 1990 (see Jentoft, 1993). In 1993 Nova Scotia experienced parallel cuts in groundfish quotas. While by 1993 there was growing optimism in north Norway, in Nova Scotia pessimism still prevailed. In Shelburne, on Nova Scotia's south shore, one fish processor said: 'The fishery is a one-way street, and we are now almost at the dead end.' This is precisely the way people in coastal communities in north Norway saw it when the 1989 fisheries crisis hit them with drastic quota cuts. When the cod quotas were announced in 1989 a nightmare became real. The general feeling was that this meant the final blow. Indeed, problems skyrocketed. 'Finnmark falls overboard,' a newspaper editor wrote in October 1989. No other county in Norway is so dependent on cod as Finnmark is, and nowhere did the people suffer more (ibid.).

The cod fishery in the Barents Sea of northern Norway, was a disaster for a short period compared with the northern cod fishery off the east coast of Canada. In 1992 the government of Canada declared a two-year moratorium on all fishing in Canadian waters off Newfoundland and parts of eastern Nova Scotia for this Atlantic cod (*Gadus morhua*) stock. This necessitated a major relief program, the Atlantic Groundfish Strategy (TAGS), for about 30,000 fishermen and fish-plant workers. This 'black tragedy,' as a local newspaper described it at the time, is difficult to explain. Alleged causes range from scientific failure, overfishing by the industry, political malfeasance, union corruption, and failures of the management of the international fisheries, to profound ecological changes signalled by decline in water temperature, decline of relatively underfished species, and florescence of harp seals (Steele et al., 1992; Coady, 1993; Storey, 1993; Finlayson, 1994). The only certain fact is that no one knows for sure what are the causes of the decline in the fish stocks.

Rural communities in Atlantic Canada are still shell-shocked by the resource crisis, as their north Norwegian counterparts were a few years before. Fishers and plant workers, fish-plant owners, and ancillary community businesses are facing unprecedented uncertainty about the future. Although TAGS helped alleviate the worst short-term effects of the moratorium and cuts in quotas, a profound malaise settled over coastal communities that has deeply scarred both individual and collective psyches. No longer can parents assume that their son or daughter will live and work in their home community when he or she finishes school. No longer can people expect to live and work in their communities long enough to retire there. The skills and experience that for generations were highly valued in the community suddenly have become redundant and of little or no use in finding alternative prospects. Choices made ten or twenty years ago are being questioned and regretted. Frustration is affecting family life and long-standing relations with neighbours and friends. Small differences in income, property, or lifestyle have become magnified and an irritating source of tension and conflict that affect even the way schoolchildren relate to each other in the playground. Individuals and communities that for generations had adapted to periodic crises in the fisheries, through self-reliance and rugged individualism, are ill-equipped to handle structural crisis. Denial is a common response. 'It has happened before, it will turn around eventually if we just wait it out' or, 'It's not as bad as they say, there are lots of fish out there.'

Introduction: Parallel Crises 231

Crisis is a challenge. It may present opportunities that stir to action and innovation rather than leading to passivity and acquiescence (Bailey, 1995). The central government was not indifferent to the situation, neither in Norway nor Canada. Extra financial assistance brought some relief. In Norway, for instance, the county of Finnmark was designated a 'free commune experiment area,' which gave the county authorities special means to combat the effects of the crisis. (This experiment in decentralization of government initiative is explored in Greenwood (1991).)

One private initiative was particularly important in Finnmark. In the fishing community of Båtsfjord, a fish-processing firm showed a way that many would follow. In 1990 it started importing Russian cod. Two years later more than 50 per cent of total landings in Finnmark were Russian. Norway and the Soviet Union had exchanged quotas before this, but individual firms importing cod directly was something entirely new. Basically such importing had been forbidden without special permission from the government. Now the Lov om Norges Fiskerigrense (Sea Border Act) was changed: Landings of fish brought in by foreign vessels became legal unless the government decided otherwise. Most processors in Finnmark seized on this new opportunity to compensate the loss of Norwegian stocks with Russian fish.

North Norway, and Finnmark in particular, were fortunate. The fisheries crisis coincided with the reforms towards a market economy in Russia. Fish that the Russians traditionally would have consumed themselves now had become an export commodity of great economic value, and with its closest neighbour Finnmark desperate for fish and Russia itself desperate for hard currency, transactions could easily and quickly be arranged.

Another, no less sensational, change of circumstance occurred. Gradually and unexpectedly the cod stock improved, and so did quotas. From 113,000 tonnes in 1991, quotas increased by 15,000 tonnes in 1991 and by another 16,000 tonnes the year after. And from then on they really escalated. The 1993 total allowable catch (TAC) was 248,000 tonnes. The 1994 cod quota of 336,000 tonnes had not been exceeded since 1982.

The fishing industry in Finnmark, as in other parts of north Norway, was soon back on its feet. Fish plants were running full-time and modernizing their production lines. Unemployment in coastal Finnmark, which used to be the highest in Norway, essentially became non-existent. Those who wanted work could get it. No wonder optimism came back in the industry. But this time the people of Finnmark were cautious. They asked themselves how long this new prosperity would last. They knew from experience that another crisis could hit any time. Russian supplies might peter out or there might be another collapse of the cod stock.

When crisis hit again, as it did in the winter 1996, it was the market prices that plummeted. In 1996 unemployment sky-rocketed again, as fish plants went bankrupt or closed down temporarily. In many communities, some of them visited by us during the boom years, the situation again became critical. Some of the firms we describe in the following chapters are already out of business.

If there is a lesson here it is that nothing in the fishing industry is stable. This is the main challenge that the industry must face. The questions are: How can coastal communities become more prepared to face downturns? How can fishing communities become more ecologically, economically, and socially sustainable in an environment that is chronically unstable?

The resouce crisis that stuck Finnmark so hard in 1989 changed the fishing industry. So too did the government policies aimed at alleviating it. People within the industry learned some lessons, and for many of them business now is not as usual, despite the fact that the cod came back.

Canadian fishing communities are now deep into a fisheries resource crisis that appears to be of much longer endurance than the one for Norway. The federal government has announced that it cannot continue to allocate billions of dollars in subsidies to coastal communities. The Fisheries Council of Canada in fact launched a policy initiative designed to end what they call the 'social fishery' (Fisheries Council of Canada, 1994). But, in their struggle for survival, coastal communities in Canada also demonstrate active ingenuity, economic and political entrepreneurship, social solidarity, and responsibility. When crisis hits, communities discover that they have human and cultural resources to draw on that may be converted into an effective and forceful battle against economic and social disruption.

In the chapters that follow we describe what these resources are and how they are effectively put into practical use. Particular focus will be on the perspectives and rationales that guide small business management and the socioeconomic context within which it is embedded. To understand how small business managers operate, whether on land or at sea, we must depict how their businesses are connected to their markets and how they relate to their competitors. We must also consider the role they play in the local fisheries system, the ways in which they interact with the local labour force and with their fleets. Also we must understand the commitments and social responsibilities that fish processors face in their respective communities. In short, this is what we call the embeddedness perspective, and in this study this perspective will be our main theoretical thrust.

A comparative analysis of small business in north Norway and Nova Scotia is instructive for detecting the range of alternatives in crisis management and how different institutional frameworks enable or constrain the fish processors

Introduction: Parallel Crises 233

within the two systems. Thus, a comparative approach is particularly instrumental in identifying policy implications and measures that governments can take to alleviate the effects of fisheries crises in the short as well as the long run.

The data presented here, primarily in qualitative form, was collected through various field trips in north Norway and Atlantic Canada throughout the summer of 1993 and the summer and fall of 1994 and winter of 1995. In north Norway we conducted a series of interviews with fish processors, fishers, representatives of fishing industry organizations, and local government in Troms county. We also visited eleven fishing communities along the coast of Finnmark where we conducted similar interviews. In Nova Scotia we visited fourteen communities and had interviews with fish processors on the east, south, and southwest coasts. In all of these areas we were allowed to see the plants from the inside and were told in detail how the production and machinery worked. Three in-depth community studies were carried out, one in Finnmark, one in Nova Scotia, and one in Newfoundland. Together they give a comprehensive picture of how these coastal communities are structured and how they adapted to crisis.

The outline of this section is as follows: in Chapter 8 we explore conceptual issues in small business management and community organization, particularly as they relate to economies of scale and social change. The literature on social embeddedness, flexible specialization, and community development is reviewed in relation to the situation that prevails in crisis-ridden coastal fishing communities. In Chapter 9 we give a historical overview of how our two fisheries systems were formed in the postwar era, that is, how, with the strong support of public policy, they came to adopt the Fordist model of industrial organization and how the industry at the levels of both the firm and community was affected by the crisis. Chapter 10 portrays the traditionally embedded fishing community and the patterns of adaptation that have allowed coastal communities to survive times of crises. In embedded communities small firms have typically drawn heavily upon social capital in the form of social solidarity and interpersonal trust that stems from the social relations within which these firms and their managers are involved. We focus on various forms of small firms to help identify the types of people that are able to cope with the current crisis by taking advantage of the opportunities presented by globalization.

In Chapter 11 we describe how large-scale Fordist firms have been affected by crisis and globalization. We describe a restructuring process that has affected production and technology, organizational forms, and management practices. Despite government efforts to counteract some of the worst side-effects of Fordism, we show how these processes lead to further disembedding of the industry from the local community. Chapter 12 describes an alternative scenario, one characterized by modernization of small firms and global marketing. In this case interfirm networking and close ties to community-based sources of social capital make systems more robust. We argue that such post-Fordist firms are instrumental in the re-embedding of coastal communities.

The next three chapters present community-based case studies of post-Fordist rural renewal from north Norway, Newfoundland, and Nova Scotia. Each captures a different dimension of the re-embedding process: an ethnically based study of bounded solidarity, community development based on cooperativism, and a community transforming individual self-reliance into collective action.

Chapter 16 explores a number of complexities represented by globalization and the role of the state. Neither process is predetermined or monolithic in terms of its impact on rural communities. This presents social researchers with a complex problem area, but offers communities some interesting options if they can organize themselves and demand more community autonomy and control over their resources.

8

Community Sustainability, Small Firms, and Embeddedness

This chapter presents the theoretical framework for our study of small business management and coastal fishing communities in Atlantic Canada and north Norway. It lays the foundation for our central thesis: coastal communities need to be re-embedded in order to be flexible and, hence, sustainable. The relevant literature on industrial organization and community development is summarized and our key analytical concepts are introduced: embeddedness and flexible specialization.

For liberal economies the relationship between business and community is at best an instrumental question and at worst a non-issue. Why in making business decisions would managers take the needs of their communities or workers into account? Only for instrumental–rational reasons – is it best for business? This logic is no less true for individuals in the labour force. Why choose one job and place to live over another? The choice is made in the individual's own best interest. The rational individual is the ideal standard against which choice in various realms like the labour market or the firm are judged. The individual rational maximizer is a universal ideal type. To the extent that sociological factors such as values and morality or power and exploitation are seen to affect the choices that people make in their lives, liberal economies treat these as externalities or irrational influences.

What are the consequences of dismissing the factors associated with daily life in communities? In so doing we risk missing important variables that not only explain how things work, but offer solutions to the kinds of problems people are experiencing in coastal communities. But in following rational choice theory, are we inadvertently buying into the 'modernist' world-view that underlies so much policy designed to restructure economic and social life along rigid efficiency criteria (Holton, 1992)?

Embeddedness

In a seminal paper published in 1985, Mark Granovetter resurrected a term that had been coined by Karl Polanyi nearly thirty years earlier to account for the social and cultural constraints on economic action in premarket societies. The term was 'embeddedness.' In the chapters that follow, we adopt a hermeneutic perspective on embeddedness. To us the term refers to two distinct levels: it refers to the 'embedded' nature of individuals belonging in social groups such as family, kin group, community, church, or occupational group. It also refers to the 'embedded' nature of institutional interconnectivity, to the ways in which rules and procedures and normative standards of conduct in various institutional realms such as economic, cultural, and social life impinge on and shape each other.

In other words, we distinguish an interpretive and a normative side of embeddedness. The first involves the realm of choice, interaction, and collective action. The second concerns the overarching parameters within which choices are made, shaped, and have emergent meaning for the actors themselves. Conceptually, therefore, the choices that rational *homo economicus* makes cannot be adequately understood except in terms of the manner in which they are 'embedded' within a wider social and cultural matrix. This matrix manifests itself in terms of a macro-social and -cultural environment, and in terms of a micro-environment associated with individual attachment to groups and communities. Further, social groups, acting as coherent collectivities also make choices, mould and shape their social and economic environment, and, in turn are constrained by it. Agency therefore has both an individual and a collective manifestation. Collective agency cannot exist without individual members being conscious of their situation and making purposive decisions as part of the group. Individual agency, however, can exist independently of any collective manifestation.

Given that our focus is rural coastal fishing communities, we are primarily concerned with two forms of embeddedness: community and ecological. Concern with community has a long and distinguished lineage in the classical tradition of sociology, dating back to the work of Ferdinand Tönnies (Tönnies, 1963 [1887]). For us, community embeddedness refers to an emergent sense of identity and belonging that is the product of interactive density and role homogeneity in a social setting with definable boundaries. Identity springs from a depth of knowing – 'memoryscape,' attachment to people and place, and a normative structure that gives the physical boundaries of place social closure (Cohen, 1985).

Community in the sociological sense can be found in a variety of settings, urban or rural. Ecological embeddedness, for the most part, is distinctive to rural communities and emphasized in the human ecology tradition in anthropology (Moran, 1982). At issue is the centrality of natural resources to the rural economy and way of life. Culture and religion, economy and technology, kinship, marriage, and residence are often all closely conditioned by the local environment and resources. Economic decisions, for example, are often framed within a wider cultural context that in itself is framed by an overarching interest in maintaining an adaptative relationship to the environment in the long term. Local knowledge and indigenous systems of management of the natural resource are well-documented illustrations of such embedded systems around the world (Acheson and McCay, 1987; Dyer and McGoodwin, 1994; McCay and Jentoft, 1996).

Community embeddedness and ecological embeddedness are usually intimately tied in rural communities through egalitarianism. As Cohen (1985) pointed out, it is the appearance of egalitarianism that matters. It is how people act towards one another that really counts. Abiding by community norms of conduct such as neighbourliness, the everyday etiquette of interaction, generosity, and so on, reaffirms a common bond of identity, and supercedes, at least for the purpose of daily interaction, factors that divide people such as class or kinship. Nicknames are usually the symbolic expression of such community-based identities.

The norm of sharing and reciprocity is a way of redistributing scarce resources, while at the same time it functions as a strategy for risk management for individuals and families facing a life of uncertainty. The norm preserves relative equality among individuals in the community, discouraging individual greed and avarice, and encouraging cooperation and altruism. This serves as an effective check on overharvesting of scarce resources for individual profit and at the expense of others in the community.

Conflict avoidance, another oft-mentioned norm of community embeddedness, has a similar dual effect in reinforcing community and ecology. As with egalitarianism, appearance is more important than the substance. Cultures vary greatly in terms of what kinds of issues can be openly discussed and argued over, who has the right to do so, and under what circumstances. Flora et al. (1992) pointed to the importance of 'sacred values' in defining some of these dimensions in small-town North America. Challenges to certain values will invoke a communitywide moral response. The norms associated with community embeddedness – sharing, generosity, honesty, industriousness, and conflict avoidance itself – are a good place to begin in defining these sacred norms. Important cultural rituals and styles of interaction often emerge as a safety valve in communities and provide a culturally sanctioned way for individuals and groups to informally check one another's behaviour. Gossip, witchcraft, role-reversal, and masquerade festivals all usually serve this purpose.

Avoiding open conflict over certain 'sacred' issues preserves community cohesion and strengthens social solidarity. Such group cohesion is a vital feature of economic cooperation in many resource endeavours based on community work groups and collective economic security. Avoidance of conflict not only preserves group cohesion but also reinforces the normative controls that act as a check on individual opportunism or community factionalization.

Disembeddedness

Karl Polanyi concluded that modern industrial society spelled the end to traditional patterns of embeddedness – a process he described as 'differentiation.' More recently, Giddens (1990) referred to the process of atomization under modernity as one of disembeddedness: a process through which the economic activities, rationalities, and social relations are 'lifted out' from local contexts of interaction.

We focus on two useful ways of looking at disembeddedness in the global context: the institutional level and the level of individuals and social groups. Disembeddedness refers to a process of institutional disarticulation. Institutional disarticulation in this context means that a particular institution, for example, economic life, gets ripped out of an embedded situation in two ways. Linkages are severed and transformed into leakages: the spin-off potential is leaked and another location benefits from it. The spin-offs go to someone else; value-added production benefits someone else. In this context a variety of things become externalized. Rather than having an internal focus, for local markets or local needs, economic activity is pursued to meet external needs and an export bias. For example, resource activity becomes refocused for foreign markets.

Institutional disarticulation also refers to loss of control. In the disarticulated system, decision-making is no longer locally oriented and controlled. It becomes oriented to some universal criteria such as the general laws of supply and demand as they affect prices, wages, or interest rates, profit rates, and shareholders interests. Outside managers and owners influence how decisions are made, who is hired, what products are produced, where they are sold, at what price, and who you buy from.

Our central argument therefore is that in rural communities modern capitalist society severs the most essential connections between economic, social, and ecological life. Modern society creates schisms disembedding the community from the environment, undermining traditional systems of resource management, and eroding the social and cultural bonds on which communities are based.

Our research focuses on the relationship between disembeddedness and Fordism in coastal fishing communities. In particular we consider the extent to which unsustainable systems can be said to have emerged through the operation of Fordist systems and whether the transformation in the fishing industry following the crises of the late 1980s changed matters for the worse or the better.

Re-embeddedness

Granovetter (1985) argued against the notion that embeddedness disappeared altogether in a heartless, self-interested modern world. He submitted that value-oriented social action can be seen in all levels of modern economic life and that a new social economy is needed to account for the nature of individual action. Granovetter's basic argument was that emergent normative standards within interaction networks facilitate trust and obligations in situations that would otherwise be wreaked by opportunism, mistrust, and disorder. Coleman (1988) and Portes and Sensenbrenner (1993) have since amplified this theme by examining the connection between trust, obligations, and 'closed' – or 'bounded' – collectivities or interaction networks. The emergent nature of 'social capital' facilitates various kinds of social and economic advantages that the parties to the interaction can draw upon. To their credit, Portes and Sensenbrenner (1993) also comment on the down side of embeddedness with 'closed' communities, notably, clientism and 'free riders.' To this list we would add patriarchy, 'captive' super-exploitation, and xenophobia.

Gherardi and Masiero (1990) attempted to explore the issues of trust and solidarity without the implicit utilitarian bias of earlier treatments. For example, Coleman (1988) argued that social capital is a public good that is subject to certain inherent 'inefficiencies,' since it is more the unintentional by-product of interaction than conscious agency. He submitted that individuals may be prone to 'underinvest' in social capital, since they cannot be assured of reaping the rewards of their investment, or, worse, someone else will benefit as much or more from their efforts. Gherardi and Masiero (1990) would no doubt suggest that Coleman's utilitarian bias forced him to adopt a fairly restrictive position. Altruism frees the individual from such self-centred considerations and stimulates the growth of social capital. Gherardi and Masiero argued that a strong element of altruism frames the nature of interaction in certain kinds of moral, or ideologically shrouded, networks. Such identity generates a second form of social capital, or a second aspect of re-embeddedness, called 'solidarity.'

Solidarity creates an environment of tolerance that is far more expansive and flexible in terms of individual behaviour than under traditional embeddedness. Trust, according to Gherardi and Masiero, offers organizations the opportunity of reducing certain kinds of internal control and supervisory costs, while solidarity between organizations cushions entire communities against the vagaries of the market. Each is a source of social capital that emerges through the spontaneous and unintended outcomes of interaction, communication, and sociability. The continuity of such patterns of interaction creates a climate of familiarity, predictability, and confidence. As in the case of traditionally embedded communities, strong social boundaries provide social closure and back up emergent norms of conduct with the power of group sanctions.

With the incorporation of rural communities into the global economy traditional forms did not vanish altogether. We argue that if one is to develop an adequate understanding of modern forms of re-embeddedness, particularly in business organizations, then the 'individualist' approach to social action needs to be tempered with more traditional notions of collectivism. We are interested in the interface between community re-embeddedness and industrial organization. In particular, we are interested in what in the literature has been termed 'flexible specialization,' originally introduced by Piore and Sabel (1984).

Gherardi and Masiero (1990) suggested that the impact of interorganizational solidarity among certain Italian cooperatives in Emilia-Romagna has been to create a certain degree of homogeneity. Networking based on solidaristic exchanges therefore achieved a two-fold result: it defended cooperatives against the effects of competition and, in general, against the principle of market exchange, but it also tended to homogenize their behaviour. Solidaristic exchanges multiplied and became institutionalized; they transmitted information, interpretations, and constructions of reality that reflected the aims of the cooperative movement. Now entry and exit to the process of exchange become much more 'constrained,' which amounts to saying that the system of solidaristic relations was more structured and that it foreshadowed a fully developed organizational network (ibid.: 568). Such homogeneity tends not to characterize small communities that are diversifying their economies through the private sector – flexible specialization – but does that mean that 'embeddedness' and, in particular, community-based solidarity, does not characterize this development? We argue that there is a close connection between 'embeddedness' and certain kinds of flexibility.

Re-embeddedness and flexible specialization are intrinsically related concepts particularly in considerations of rural restructuring and decentralization – the 'new' countryside.

Community re-embeddedness provides a number of opportunities for small businesses and cooperatives alike. This works at various levels and, over time, generates new levels of solidarity, which in turn facilitate emergent types of flexibility within firms, industrial networks, and communities.

There are at least five types of flexibility that characterize small businesses in general and small fish plants in particular: production flexibility, input-sourcing flexibility, and marketing flexibility, plus two kinds stemming from relationships external to the firm per se, networking and embeddedness (or attachment of the firm to the community). While there may be considerable overlap when it comes to the types of firms and their utilization of these strategies, differences of ownership and scale separate them. Overall, however, it is the social and moral economy of this structure that generates a sustainable system, not only the structural traits of firms and communities, for instance, pertaining to scale, technology, and form of organization.

From the perspective of the firm, trust and commitment within the organization, between management and employees, reduce supervision costs, free up managers' time for other things, and form the basis of 'functional flexibility' strategies on the shopfloor.

Generally, flexibility in production operations can be obtained in two ways, in what is often referred to as 'numerical' versus 'functional' flexibility (Nåtti, 1993). The former is ensured through a fluctuating workforce. Rather than keeping workers employed on a permanent basis, people work on short-term contracts. In fish processing, local or transient workers are called when landings are high and released when they are low. Often, as is typically the case for women, they work only part-time. Female workers whose spouses are fishermen carry the burden at the home front alone during the fishing season. The instability of fish-plant work often provides them with the flexibility they need for their household responsibilities.

Functional flexibility is the ability to do multiple work tasks and to diversify production according to need. Individuals can increase their functional flexibility by learning more, through formal education or by widening their practical experience. If flexibility is crucial for fish processors in times of crisis, so also are managerial and workforce qualifications. Consequently, investing in functional flexibility, that is, in upgrading the competence of the management and the workers, becomes an important component of crisis management.

Organization of production is another key realm for flexibility. The normal rule of thumb in the fish-processing industry is that the larger the scale of production, the more specialized the division of labour, and the more reliant the firms on 'numerical flexibility' in adjusting their labour force and work schedules to fluctuations in the supply of fish. 'Functional flexibility,' by contrast, seems to characterize smaller scale operations, where the vagaries of the supply of fish are coupled with diverse niche marketing and production decisions. This is seen to require a workforce whose absolute numbers are not so important as is their versatility in being deployed to different, unspecialized jobs.

For their part, workers gain satisfaction from improved work conditions, stimulation, and challenge from a situation where they can exercise greater initiative and experience improved social worth. Such conditions in turn reduce turnover, training costs, and absenteeism for the employer, and they will begin to generate tertiary benefits in the form of worker innovation and synergism. The kinds of solidaristic relations that exist in many situations between company and trade unions – 'bargained flexibility' (Perulli, 1990) – are a formalization of these relations at the interpersonal level within firms.

Trust and solidarity between organizations – firms in the same field, vertically linked firms, and firms and municipal authorities – create a second dynamic of social capital that has its own logic and emergent stages. Shared purchasing and subcontracting allow small-scale firms to reap the benefits of economies of scale without losing managerial control and autonomy. Over time, trust between players leads to more permanent forms of cooperation that are mutually beneficial: sponsorship or support for third-party ventures that will be of either mutual benefit or benefit to the community. Thereafter it is only a short step to form direct bilateral ventures that involve formal commitments such as joint marketing and production ventures, worker exchange, and a second form of synergism that stems from sharing information and technology. Trust and solidarity strengthen at each stage, generating new layers of social capital that the participants can 'accumulate' and build on. Improved levels of trust reduce uncertainty, and, hence, transaction costs. The economic stability that emerges with this kind of networking in turn forms the basis for a diversified system of employment, which from a worker viewpoint is often called 'pluriactive,' but from the point of view of the firm is termed 'numerical flexibility.' That is, the whole employment system maintains a type of homeostasis, while facilitating employment flexibility at the level of the individual firm in response to market conditions.

Being highly localized and obliquely constituted, the system is both robust and vulnerable at the same time. Its robustness emerges over time, much as Rostow (1960) argued would happen in developing economies as systems distanced themselves from overdependence on one or two activities and generated a self-sustaining dynamic.

Small firms are a vital beginning point and an ongoing basis for stability in the system. First, and most importantly, because they offer local people an opportunity to work in the community. This in turn facilitates interaction, time for socializing, face-to-face familiarity, and trust within the community. Second, small firms provide a local economic base that has the effect of drawing money, information, and expertise into the community rather than the other way around. The ripple effects of this can be seen in a host of respects: in cultural realms such as entrepreneurialism, confidence, and self-reliance; in the field of recreation through funding for facilities, teams, and festivals; and in municipal finances through an improved tax base for physical infrastructure, education, and health care services.

The system's vulnerability stems from a variety of sources that corrode trust and solidarity at both the interpersonal and the organizational levels. Foremost among these is competition. At an interpersonal level, envy, jealousy, and suspicion poison relations of trust and altruism. If firms are having to carry individuals who are essentially 'free riders' (Portes and Sensenbrenner, 1993), such a threshold will threaten survival if the problem becomes systemic rather than simply anomalous and individual. At an organizational level, solidaristic relations break down when firms are perceived to be breaking tacit agreements, seizing opportunities for themselves at the expense of others, or violating understood codes of conduct.

Sustainable Development and Re-embeddedness

Following Lelé (1991), we can approach the question of sustainable community development in terms of three issues: ecological, economic, and social sustainability. Ecological criteria stress the preservation of habitat and indirect environmental processes needed for the sustenance of biological processes. In terms of direct action, ecological sustainability also refers to the rate of resource exploitation – a rate that has to be within the range necessary for regeneration – and to the form of exploitation – one that is subject to control and management. Sustainable exploitation also has to be within the realm of minimal subsidiary or indirect effects. As a rule of thumb, therefore, we argue that for re-embeddedness to also be sustainable, processes have to be put in place that are subject to broadly based conservationist criteria, they have to be diverse rather than specialized applications, and involve technologies that facilitate periodic shifts or switches in resource focus rather than those that are individualist in orientation, specialized, and inflexible in application.

Economic sustainability refers first and foremost to the criteria that govern decision-making in firms or otherwise. Are these long-term streams of income as opposed to short-term maximization and the discounting of future rewards? Each can be profit driven, but with quite different implications: (a) appropriate technological capacity that does not set in motion certain cost- or debt-driven 'technological imperatives' (avoidance of idle capacity costs), such as insatiable appetites for resources; (b) articulated systems that capture and generate downstream and upstream linkages for the local economic system rather than losing spin-off potential to systems elsewhere or externalizing costs. Relatedly, local ownership and control dynamics are seen to be superior to absentee decision-making power in responding to local needs and concerns. The combination of the preceding factors concomitantly maximizes local autonomy and the dynamics of autocentric growth, factors that spur indigenous entrepreneurship, innovation, and risk taking. Last, economic diversity and flexibility provide re-embedding communities with a range of complementary activities that in totality generate stability, rather than one speciality that generates little more than a social psychology of dependence and an addictive malaise to subsidies and other forms of external inputs and demands (Freudenburg, 1992).

Social sustainability refers to livelihoods that are sustainable in the long term rather than quick and dirty bubbles that burst after some short economic boom. In distributional terms, sustainable social processes stress the importance of equity. This has economic spillovers in terms of maintaining effective demand, preserving tax bases and social infrastructure, and maintaining an achievement ethic in the workforce.

Social sustainability in re-embedding rural communities also refers to quality of life and work. Labour processes that stimulate problem solving provide a sense of social worth, and involvement of workers in consultation and decision-making have been shown to reduce the worst forms of alienation and anomie that accompany the organization of modern industrial work. A cultural dimension is also vital to social sustainability. This is found in a coherent expression of collective identity, personal security, and self-image.

To what extent can people continue to possess a sense of belonging and maintain a way of life that they perceive is their own and in line with their heritage? It depends on the degree to which they will identify with norms and values that characterized the community, and whether or not they are able to maintain these standards from one generation to the next. Central to the notion of cultural sustainability is the argument that the re-embedded community cannot only be understood in structural and geographic terms; it also exists in symbolic terms in the minds of its inhabitants (Cohen, 1985).

Cultural values are nurtured through rituals, celebrations, and festivities that create opportunities and arenas for members to come together in expressing social cohesiveness and collective identity (Aronoff, 1993). Cultural norms and values also provide meaning and emotional stability, which under unfavourable conditions such as those that are created during crises, may be eroded. To be sustainable, cultural values must not resist change. Neither must they become encapsulated from external influences such as those communicated though media. To be sustainable, cultural values and norms must adapt within frames controlled by the community without creating anomie and homelessness. Whether cultural sustainability prevails is not primarily for the external observer to tell.

First and foremost, cultural sustainability and continuity is felt and experienced by community members themselves. Through interpretative techniques the researcher must analyse to what extent an observed continuity or discontinuity mirrors what is being perceived in the minds of community members intergenerationally.

In conclusion, then, our central thesis is that the minimal conditions for sustainability of coastal communities are closely linked to the nature of embeddedness, to the way economic activities, whether they take place ashore or offshore, are nested in social relationships and networks that provide direction to and basis for sustainability. This implies the hypothesis that there may well be communities that are embedded in the traditional way, but that are nevertheless economically or socially unsustainable. There may also be communities that are socially disembedded, but that are nevertheless economically, if not ecologically, sustainable. Re-embeddedness may under these circumstances offer the opportunities to break with the past in ways that may enhance all dimensions of sustainability in a way where one reinforces the other.

9

Modernization and Crises

The history of the fishery in Atlantic Canada and Norway, despite obvious differences in the postwar era, has a number of striking similarities. Foremost among these has been the adoption of the Fordist model in the fishery, the industrialization of fish processing, the strength of the offshore sector, and the emergence of single-industry communities. In each area, whole communities become reliant on large-scale, vertically integrated companies whose scale of operation dictated a Fordist style of plant management and a monocultural approach to the harvesting sector. However, in both areas these single-industry towns coexisted alongside other and more embedded fishing communities with a very different structure. These latter communities are characterized by small plants and an independent fleet of small boats. Plant operations are flexibly organized, and the harvesting sector is based on diversified production. In both the Norwegian and Canadian cases, we are interested in the fate of each fishery sector as a consequence of the transformations that have taken place in the postwar era. This chapter attempts to describe how the present structure of the industry, the Fordist model, came into existence in our two regions, and how the supply crises and globalization have affected the industry.

The Finnmark Fishing Industry – An Overview

The Fordist model incorporates structural rigidity, is very dependent on a large and steady supply, and is hence very vulnerable when crisis hits. In no other area in Norway did the fishing industry become as moulded within the Fordist framework as in Finnmark. Neither was there any other county as deeply affected by the resource crisis than Finnmark. Thus, in 1996, when market prices drove several fish plants out of business, Finnmark was hardest hit.

Finnmark is the northernmost county of Norway and by far the largest

(49,000 square kilometres). The population is sparse (76,000 in 1990) and settled in communities along a rugged coastline, where the fjords cut deep into the landscape, and in communities on the inland plain. The population is ethnically diverse with Norwegians, Saami, and naturalized Finns (Kvæn). The latter immigrated to Finnmark and northern Troms in the nineteenth century, while the Saami are indigenous. To the east Finnmark borders Finland and Russia, with whom historically there have been strong links. Until the 1917 revolution Finnmark had close trading relations with Russian communities along Beloye More – the Pomor trade – where Russian traders provided flour, wooden materials, and artisanal products in exchange for Norwegian fish. Thus, the recent fish business with Russia can be seen as a revival of traditional trading patterns, only this time fish go the other way.

War and Reconstruction

The Second World War left Finnmark in ruins. At the end of the war, to slow down the Soviet forces that were attacking from the east, the German troops, having evacuated the whole population, employed a scorched earth policy. Everything that could support the Russian troops was burned and destroyed. Thus, when the war ended, and the Russian forces left Finnmark, everything was in ashes, the entire infrastructure destroyed. With few exceptions, every fishing community and the entire fishing industry had to be rebuilt from scratch.

Plans to rebuild Finnmark were prepared by the Norwegian government while still in exile in England. With a 'tabula rasa' to start with, the aim was to concentrate the population into larger settlements. However, people disregarded the plans and hurried home as soon as they were allowed. Thus, by and large, fishing communities in Finnmark were re-established just as they had been prior to the war.

In the 1950s the government launched a major modernization program for the fishing industry. Through state initiative and support, fish processing was rebuilt as a modern, large-scale, freezing industry. Partly as a consequence of this industrialization process, many communities could not sustain their population base over a long period. Young people especially would seek out better opportunities in urban areas within and outside Finnmark. Also, if a school or a fish plant had to close, something that happened often because of frequent market crises during the 1960s and 1970s, communities were often abandoned. The government generally viewed this as positive and established a resettlement support scheme to facilitate further out-migration. Therefore, the coastal population in Finnmark gradually clustered, and today most people live in towns and larger settlements, where the fishing industry is the dominant employer.

Located adjacent to the Barents Sea, which is an area rich in cod, capelin, haddock, saithe, turbot, and shrimp, Finnmark became affluent with its fish resources. Many of these stocks, such as the cod and capelin, have a migration pattern that brings them close to shore, where they can be harvested by small vessels. This pattern has greatly influenced the fleet structure of Finnmark, which is predominantly of small and medium scale. However, it also made the fishing industry of Finnmark vulnerable to weather conditions and subject to seasonal variations, which can be extreme in an area that far north. This was seen as the major obstacle to the modernization process, and deep sea trawlers were thus regarded as crucial to ensuring a year-round supply. In spite of vociferous opposition from coastal fishers, trawlers were introduced during the 1950s in the Finnmark fishing industry and in other parts of north Norway, most often vertically integrated with fish-processing plants.

The Norwegian government came to play an important role in fish processing as the owner of Finotro (Finnmark and Northern Troms), a concern with branch plants in several communities in Finnmark. Finotro plants handled 22,000 tonnes of raw fish at the various plants. It possessed trawlers but was dependent on the coastal fleet (Hersoug and Leonardsen, 1979). Two other large concerns also became cornerstones in the fisheries of Finnmark. The Swiss-owned Nestlé (originally Findus) company in Hammerfest was the biggest single company. In 1960 Nestlé had close to 1,200 employees at its peak, produced more than 20,000 tonnes of raw fish, and had more than twenty big trawlers. The Aarsæther concern, like Finotro, had a number of subsidiaries throughout Finnmark. Aarsæther was well established in Finnmark (as well as in other parts of north Norway and the Ålseund area) before the war. Five of their plants – in Vadsø, Mehamn, Kjøllefjord, Ingøy, and Tufjord – were destroyed during the war. They were all rebuilt, but during the 1970s the plants in Vadsø, Vardø, Båtsfjord, and Kjøllefjord were given priority in the expansion and modernization program. In the late 1960s Aarsæther as a whole processed around 18,000 tonnes of raw fish, about 13 per cent of the Finnmark total, and employed close to 1,000 employees on two shifts. In the 1960s they invested heavily in deep sea trawlers, and in the late 1970s these trawlers provided half of the fish landed at the Aarsæther plants.

Organizational Structure and Public Policy

Despite massive investment in industrial fishing and fish processing, Finnmark maintained a second tier of small-scale, coastal vessels, and traditional, seasonally operated processing plants. At first they worked alongside the large-scale units, but later on they were either converted into freezing operations or they retreated to communities more specialized in small-scale fisheries, particularly to communities along the fjords. Here fishing was practised in combination with agriculture and the harvest of other natural resources (Eythorsson, 1991). This adaptation was particularly evident among coastal Saami.

This dual structure – that is, Fordist vs embedded models – developed partly in spite of government policies aimed at rationalization, partly because of them. Through grants, cheap financing, and tax releases the government aided the industrialization process. But in face of strong criticism from coastal areas and from the Norwegian Fishermen's Association, the government also imposed a series of restrictions on the industry. First, vertical integration was generally forbidden by the Midlertitid Lov om Eigedomsretten til Fiskefarkoster (Ownership Law) of 1954, as only fishers were allowed to own fishing vessels. This condition was partially lifted when fish-processing plants, in cooperation with municipal authorities, were granted permission to form separate 'ocean fishing companies' for the acquisition of trawlers. This meant that a processor was hindered from buying a trawler directly, but there was an alternative route through these companies. Second, on the basis of a previously enacted law on trawling operations, the Fishermen's Association was successful in pressuring the government to introduce a licensing system for the acquisition of new trawlers (Lov om Trålfiske or the Trawler Act of 1936). Officially, trawlers were legitimized as supplemental to, not a replacement of, the coastal fleet. The Raw Fish Act (1938) was also important. This law granted the fishers' sales organizations (MSOs) authority to fix minimum prices which a buyer was obliged to follow. Thus, the distribution of income and power because of these regulations changed in favour of fishers. These laws facilitated capital accumulation and reinvestment in vessels by independent fishers. In addition, a State Fishers' Bank provided subsidized loans and the Main Agreement for the Fisheries, between the government and the Norwegian Fishermen's Union, ensured income support, both contributing to a better financial standing of fishers.

While the small-scale coastal fleet prospered, the traditional small salting and drying plants disappeared in great numbers. For instance, the number of traditional plants increased from 20 in 1945 to 180 in 1955, from then on they started to decline, and by 1973 their number had halved. In contrast, freezing plants continued to increase at least up till the early 1970s, when they were thirty-seven in total.

This pattern led to an inherent flaw in the organizational structure of the Finnmark fishing industry, characterized by a 'mismatch' between fishing and fish processing at the local level (Aegisson, 1993). Large-scale freezing plants, which needed a steady resource flow, became dependent on a small- and medium-sized fleet in which, during the spring season, many vessels were irregular visitors from other counties. As a consequence, the capacity of the processing industry was never fully utilized, averaging only 50 per cent. The fishing industry always operated on the margin of profitability, if not at a loss, and was therefore unable to establish a sound financial footing.

In the 1950s there was also 'downstream' integration, as fishers established fish-processing cooperatives in great numbers. Finnmark became the stronghold of the fisheries cooperative movement in Norway. In the mid 1950s there were as many as thirty-one fishers' cooperatives in Finnmark, half of the total number in Norway at that time. Their popularity grew out of the situation that prevailed after the war. In many communities there were simply no other alternatives to fishers setting up their own fish plant. But as small plants, the cooperatives faced stiff competition from the private and state-owned industrial plants. When markets for traditional products collapsed in the 1960s, most of the small plants vanished (Otnes, 1980; Revold, 1980). Today, just a couple of them remain.

Stagnation and Decline

By 1970 expansion was over, and the number of processing plants and vessels in Finnmark had levelled out. The cod resource in the Barents Sea was showing some weaknesses, and with the then recent collapse of the Atlanto-Scandanavian herring stock fresh in mind, people began to realize that resources were not inexhaustible and that lack of fish could very well become a future constraint. No new freezing plants were built, even though established plants often continued to increase their capacity over the next decade. For a short period the 1976 Norwegian extension of the jurisdiction to 200 miles created new optimism, but realism was reasserted when the government imposed a quota cut on cod of 35 per cent in 1980.

By 1980 the numbers of both deep sea trawlers and coastal vessels had already started to decline. The total number of vessels larger than 10 metres dropped from 446 in 1983 to 365 in 1990. However, reduction was largely compensated by increasing technological efficiency. More mobility did much to increase the efficiency of the fleet, but not much to change the seasonal pattern. Aegisson (1993) maintained that it enabled the coastal fleet to increase their landings during the peak season.

The signs of decline that appeared during the late 1970s became even more transparent the following decade. A very profitable shrimp fishery collapsed. In Finnmark, 40,000 tonnes of shrimp were landed in 1985. Two years later landings had dropped to 10,000 tonnes, and many shrimp plants had to close. Over-

Modernization and Crises 251

fishing of the capelin resulted in a government-imposed moratorium in 1987 that lasted for four years. This affected not only the fish-oil sector, but also the white-fish plants that during the season had a lucrative business producing roe for the Japanese market.

From 1982 onwards the cod fishery was subject to strict regulations. Later, also, the seal invasion scared away the few fish that remained. Thus, fishers experienced great difficulties in catching their quotas, quotas that were record low to begin with. Unavoidably, the landings of the Finnmark fleet fell drastically, from 48,000 tonnes in 1986 to 17,000 tonnes in 1990. Another consequence was that the southern fleet that used to visit Finnmark every spring, and that normally accounted for as much as 50 per cent of the landings, chose to stay away. Some processors counteracted by importing fish from as far away as Alaska and Canada, but they found themselves in a losing battle. In 1981 there were 102 traditional and freezing plants in Finnmark. By the end of the decade their number had been halved. By 1989 the entire fish-processing industry, with the exception of three plants, was bankrupt or had negotiated debt financing. When later that year the cod crisis was officially declared, Finnmark already had serious problems.

Crisis and Recovery

Bad turned to worse. In 1990 landings dropped to 40 per cent of the 1986–8 average (Jentoft, 1993). No other county was as seriously affected. In the fishing municipalities, unemployment sky-rocketed to an average of 23 per cent, which for Norway is extremely high. Worst off was the fishing municipality of Måsøy, with 33 per cent unemployment. First, fishers started to lose their boats, and then their homes, which they often used as collateral for the boat loan. Many left the industry for good – not a new thing, though, but it had not previously been in such great numbers. In 1992 the number of registered fishers in Finnmark was only 80 per cent of the average number of 1985 to 1988. From 1988 to 1992 the number of plant workers decreased by a third, from 3,000 to 2,000.

Surprisingly, people did not leave Finnmark as they had done before. Higher unemployment than normal in other parts of the country prevented that from happening. The news media, however, were filled with dramatic stories of people in despair. Most sensational was an advertisement in an Oslo newspaper in which the whole population of Bugøynes in the Varangerfjord pleaded for a community down south that would accept them as a group. This had a strong symbolic effect, reminding people of the evacuation that had happened forty-six years before. (The full story of this incident will be presented in Chapter 13.)

The following year, 1991, was slightly better, and things continued to improve the year after, as landings gradually increased. From 40 per cent of the 1986–8 average in 1990, landings rose to 75 per cent in 1992, and then growth lingered. Russian fish accounts for much of this, about half of total landings in 1993. However, Norwegian cod quotas of Finnmark vessels also grew, from 12,000 tonnes in 1990 to 63,000 tonnes in 1993. Communities in east Finnmark were certainly salvaged by Russian landings. For instance, in 1992 more than 93 per cent of the fish landed in Vadsø, 77 per cent in Vardø, and 69 per cent in Båtsfjord were Russian. 'Without Russian landings east Finnmark would have been laid in ruins,' a fisher-processor claimed in an interview with these authors. Even Nestlé in Hammerfest, which used to be self-sufficient with their own deep sea trawlers, got more than half their supplies from Russian vessels in 1992 (Arbo and Hersoug, 1994).

The cod stock recovered much faster than anyone had predicted in 1989. Reports by marine biologists on fish populations in the Barents Sea in 1992 were as surprising as they had been in 1989, only this time the message was optimistic. Not only cod, but also haddock and saithe seemed to be in good shape. The prognosis had rarely been better. 'Now it is full speed ahead,' said the fisheries minister. And, indeed, it was. In 1990 depression reigned in Finnmark, and the future looked bleak. Four years later, the fishery was booming. The fleet of coastal vessels coming from other provinces for the spring fishery returned after several years of absence, and the processing industry was working full-time. Partly this has occurred as a consequence of changes in the quota system. In 1994 regulations changed from a whole vessel quota system to a system of vessel quotas combined at the beginning of the year followed by a competitive quota from 5 April. With the recovery of northern cod, hundreds of vessels from southern counties would take up their traditional habit of travelling to Finnmark in order to fish on the competitive quota.

Run-down buildings and wharfs were being renovated, and old-fashioned machinery was replaced by the most sophisticated technology there is, computerized flow-lines, high pressure water cutters, individual quick frozen (IQF) freezers, and the like. The fishing industry, as it presented itself in the spring of 1994 when we toured the coastal communities of Finnmark, was much much leaner than it used to be, but it was certainly a more self-confident one.

Ownership and the Local Interest

The mismatch between what the local community can offer and what the fish-processing industry needs to survive has always been a concern of plant managers in Finnmark. The problem itself, however, is an indication of the abundance

Modernization and Crises 253

of the Barents Sea resource base, on which so much of the economy of Finnmark rests. This situation should then provide a golden opportunity for growth not only within the community, but also by the community.

But the during postwar reconstruction, local entrepreneurs were few, and private capital was always scant. Thus, growth and development had to be generated from outside. Without government support, the ambitious plans of rebuilding could not have been realized. Consequently, the state itself created the Finotro plants. It also encouraged Swiss involvement in Hammerfest and Nestlé. Southern Norwegian investors, particularly from Bergen and Ålesund, became heavily involved in Finnmark, of which the Aarsæther concern is the most prominent example.

This structure made Finnmark special in comparison with other Norwegian counties and created the impression of Finnmark as a 'colony' (Brox, 1984). Local ownership was the exception, and not the rule, and was mainly concentrated in the traditional, embedded sector. Although this solved the critical problem of job creation and lack of landing facilities, frictions at the community level often occurred. The community was totally dependent on owners the local people had never learned to know and fully trust. Community members could not be sure of the future plans of the companies, on which so much of their existence relied. As Gerrard (1975) pointed out, in a study of a Finnmark fishing community, for workers and fishers alike, the fish plant often presented itself as an 'obscure' decision-maker. People feared that profits would be dissipated out of their community, and local fishers often claimed that they were less favoured than company trawlers or outside vessels. In many communities dissatisfaction with the foreign-owned plants led to the formation of producer cooperatives, but these cooperatives often had a very short life.

The Nova Scotia Fishing Industry – An Overview

Nova Scotia is one of Canada's easternmost provinces. It is a peninsula of some 55,490 square kilometres that juts out into the Atlantic Ocean and is surrounded by 96,000 square kilometres of some of the world's richest fishing grounds. At its peak, in the mid-1980s, the Atlantic region accounted for 80 per cent of fish landed in Canada and two-thirds of Canada's total fish exports, which in 1988 were valued at Cdn $1.8 billion. The Food and Agriculture Organization estimated that 3 to 4 per cent of the entire world catch came from this region.

Nova Scotia's population reached one million in 1991, and it is heavily split between an urban population of 53.5 per cent located primarily in one centre, Halifax (population 300,000), and 46.5 per cent rural and in small towns. Between 1981 and 1991 the latter population actually increased by nearly 2 per cent of the total as a consequence of counter-urbanization within the greater metropolitan region of Halifax. Bollman (1992) observed that Nova Scotia is only one of three provinces in Canada with this level of its population located in rural communities and small towns. This change is reflected in the recent efforts by the provincial government to incorporate four moderately sized cities and a larger number of rural hamlets into one super-city. In Chapter 15 some of these pressures are described in the case of one fishing village.

The province of Nova Scotia is one of the most diverse rural areas in Canada (Barrett, 1993). This stems from two factors: first, it is one of the most ethnically diverse regions in eastern Canada, with sizable settler and indigenous groups outside the dominant white Anglophone majority. These groups are particularly interesting since they have a predominantly rural base that dates back some three hundred years. Second, Nova Scotia has a diverse industrial structure that is heavily resource based, but also characterized by a substantial secondary manufacturing sector that was indigenously owned at the turn of the century, but that now has heavy levels of outside ownership.

In the past rural producers and their communities in Nova Scotia were pluriactive, but this pattern has declined substantially after the Second World War as a consequence of modernization and professionalization in agriculture, forestry, and the fishery. Predictably this process also coincided with dramatic reductions in overall employment in these sectors as a consequence of mechanization, government policies, and financial exigencies. Rural employment shifts to manufacturing and the service sector since 1971 reflect the opposite side of this structural shift in the structure of employment.

Nova Scotia, along with the three other Atlantic provinces of Canada – Newfoundland, Prince Edward Island, and New Brunswick – has the dubious distinction of being one of the most underdeveloped regions of Canada. The 1994 unemployment rate in Nova Scotia was 13.3 per cent, while the Canadian average was 10.3 per cent. Average weekly earnings averaged 11.2 per cent below the Canadian average in 1993, and average annual household income was 10.5 per cent below the Canadian average. To maintain a standard of living that approximates that in the rest of Canada, the province has been dependent on transfer payments from the wealthier provinces in the form of development grants and subsidies, social security programs for the unemployed, and targeted programs transfers for education and health care.

The fishery in Nova Scotia is arguably the most diverse in Canada, with approximately thirty species of fish with commercial significance.

Nova Scotia was Canada's largest fish-producing province in the mid- to late 1980s. At one-half billion dollars, the landed value of fish exceeded that of agriculture and mineral products. Approximately two-thirds of the total landed

Modernization and Crises 255

value of fish products in the mid-1980s was accounted for by the groundfishery – cod, haddock, redfish, pollock, hake, and others – 18 per cent by the pelagic fishery – herring and mackerel – and 16 per cent by the shellfishery – lobster, scallops, crab, and others. The fishery is also characterized by a substantial degree of social and industrial differentiation that stems back to well before the Second World War (Apostle and Barrett, 1992). In the postwar period the government embarked on a modernization plan that involved the transformation of the technology of both the harvesting and processing sectors, as well as a regulatory regime (Barrett, 1984). The older traditional sector represented by salt fish processing, lobster canning, and the small inshore coastal fleets was transformed into a modern competitive sector. The fish processors tended to specialize in fresh fish, salt fish, and the live lobster trade with the United States.

An important division seems to characterize this subsector industry between those processors who rely almost exclusively on the fish supplied by an independent coastal zone fleet and those who are detached from such local ties and pursue flexible input-sourcing in broader ways (ibid.). In effect, such relations establish a mutual chain of dependence between processors and fishers by which a flexible imperative is set in motion for all. They rely on a fleet of small, independently owned vessels, which by themselves, need to be flexible and mobile. The fleet, in turn, is vulnerable to fluctuations in the resource base on which it exists. Niche production is the hallmark of small-scale manufacturing in competitive markets. In the fish-processing industry this is no less true. By the mid-1980s, we estimate that this subsector accounted for upwards of 70 per cent of total plants and 62 per cent of vessels. Most of these firms were located in the 750 small communities that are scattered along the coastline of Nova Scotia.

Paralleling this competitive and small-scale sector is a large-scale corporate sector in the industry. A number of local and multinational firms have taken advantage of generous subsidy schemes since the 1950s to expand their plant operations, mechanize their fishing fleets, and move into new market segments. On the whole one can say that during this period this subsector specialized in frozen fish, prepared fish, scallops, herring, and fish-meal. Since the late 1970s a new company has been able to make substantial inroads into the European lobster market as well. In contrast to the competitive sector, the hallmarks of this group of firms were twofold: they were multiplant operations with a substantial number of intrafirm transactions; second, their primary form of vertical integration was the ownership of large offshore fleets with which they had a captive pricing arrangement.

The two largest firms have engaged in a series of corporate takeovers and mergers. The crisis of the early 1980s crisis mainly affected these large firms.

This economic crisis was a consequence of overexpansion, and the impact of high interest rates on inventories and long-term debt in the early 1980s. The state, which was overly sympathetic to the corporate sector, stepped in with an unprecedented investigation into the entire industry (Barrett and Davis, 1984; Kirby, 1982). State restructuring of corporate debts followed, and this ushered in a period of high profits, landings, and buoyant markets in the mid-1980s. Enterprise allocations (EAs) were experimented with, allowing the corporate sector private control over a particular share of the resource and allowing them to reduce throughput inefficiencies. (EAs were also effective collateral with financial institutions.) This greatly eased their perennial problems of seasonal gluts and scarcities, which had been accentuated under competitive quotas. Now the companies could rely on a certain volume of raw material and leave it in the water until markets or prices were lucrative enough to process the fish.

Few foresaw the crisis that emerged in the late 1980s. Groundfish landings, particularly northern cod and, later, other cod stocks and groundfish species began to show dramatic signs of collapse. Quotas were slashed from year to year in an effort to reduce fishing pressures. Firms began to retrench and downsize. Large plants were closed temporarily, then mothballed, sold off, and torn down. The same fate faced the wetfish trawler fleet of the large companies.

Points of Comparison and Contrast

The fishing industries of north Norway and Nova Scotia share some common structural features that allow for comparative analysis. In both regions the fishing industry is primarily export oriented and dependent on competitive markets. To a large degree fish processors in north Norway and Nova Scotia also compete in the same markets. Furthermore, in both areas, the fishing industry employs Fordist models and traditional models, the first characterized by vertically integrated large-scale companies, the other, by small businesses, independently owned, and less vertically integrated. In both regions, fish processors within the traditional subsystem make up three-quarters of the processing firms (Apostle and Jentoft, 1991). Fish processors play a prominent role in their rural communities, where plants are most often embedded. Thus, they are key to fisheries as a way of life in coastal communities in both areas, as fish-plant work is often the only available employment outside the household. This is particularly so for women, as fishing is typically considered to be men's work. However, in Norway, 5 per cent of the registered fishermen in 1990 were actually women. In a comparative study of fishing households in north Norway and Nova Scotia, Thiessen et al. (1992) found that women are involved in a number of tasks related to the fishing enterprise, such as repair of gear, baiting trawl/seine,

Modernization and Crises 257

keeping the books, preparing income tax reports, arranging for credit, cleaning the boats, and preparing meals for the crew. On average, women are more involved in these activities in Nova Scotia than in north Norway, but the interest in increasing their participation is, by and large, more pronounced among north Norwegian fisher wives than among their Nova Scotia counterparts. In both areas, we identified a potential for job creation related to support functions of fishing.

There are, however, some notable differences between the two systems. In their comparative study of fish processing in the two regions, Apostle and Jentoft (1991) found that the Fordist plants in Nova Scotia are larger than their north Norwegian counterparts. (The median size of a large plant in Nova Scotia is 120 workers, as compared with only 80 in north Norway.) Moreover, traditional, small-scale fish processors in Nova Scotia exhibit more organizational diversity than do their north Norwegian counterparts. For instance, small firms operating solely as subcontractors or 'feeders' for other processors are much more common in Nova Scotia than in north Norway, where specialized exporters or consortia of fish processors handle the marketing function.

Another contrast is the smaller degree of vertical integration in north Norwegian fish processing. This can be attributed to legislation that has hindered fish processors from owning and operating fishing vessels. Apostle and Jentoft (1991) claimed that plant managers in both regions, particularly in small-scale plants, tend to lack general education or relevant training in management, food technology, or mechanical trades. In north Norway, they found that 60 per cent of all fish-plant managers do not have high school education. They argued that a close understanding of the unique features of the fishing industry, which the managers traditionally have acquired through practical experience, is crucial to success, but that the lack of formal education in many instances poses a real disadvantage both in the company's internal affairs and in their interactions with people they do business with, for instance, brokers.

The symbiosis between small firms in traditional resource industries like the fishery and rural communities is well known. This relationship is part mutual self-interest – often accentuated by adversity – and part altruism, rooted in traditions of common identity and kinship. The embeddedness of small fish processors has traditionally been strong. An expression of this is the fact that managers most often are recruited locally. According to Apostle and Jentoft (1991), 80 per cent of the managers of fish plants in north Norway are born and raised within the same municipality as where the plant is located. A similar pattern was found in Nova Scotia in the mid-1980s. Apostle and Barrett (1992: 88–9) found that 71 per cent had grown up within fifteen miles of the plant that they currently managed. This means that there will often be strong bonds of familiarity and often close friendship between managers, their workers, and the fishers that sell their fish to the plant.

In conclusion, then, the historical overview and structural comparisons of the two fisheries systems presented in this chapter demonstrate striking similarities and differences as to public policies, socioeconomic processes, and organizational patterns. The fact that the fishing industry of north Norway seems to have recovered from a crisis that Atlantic Canada is still deeply in, raises some interesting questions related to the possible future development in Canada.

The large infusion of outside capital into Finnmark in the postcrisis period is a direct consequence of the recovery of stock and the abundance of resources relative to elsewhere in the North Atlantic. As such, the greater mobility of capital engendered by globalization has only accelerated a trend that would have occurred to some degree anyway. As unlikely as it may seem at this point, if there was such a stock recovery in Atlantic Canada, we could expect the same kinds of pressures there. Will the crisis and the partly consequent globalization trend towards *laissez-faire* deregulation facilitate a renewed trend towards outside control, further concentration of the industry into single-industry communities, and modernization along Fordist lines? Or is there still a future for the traditional, embedded fishing communities that have the advantage of small-scale, flexible specialization and community support?

10

Traditionalism and Crisis: The Social Bases of Disembeddedness and Re-embeddedness

This chapter explores the nature of traditionalism in the coastal communities and fishing industries of Nova Scotia and north Norway. For many years the prevalent social science framework has been characterized by a dualistic view of the industry. A monolithic 'traditional' sector is thought to be comprised of small plants utilizing archaic techniques and organization, unskilled 'sticky' labour, and closely tied to an artisanal and undercapitalized fishing fleet. In the 1960s and 1970s this sector was blamed for inherent inefficiency and backwardness (Brox, 1966). Most recently this sector has been blamed for overexpansion since the declaration of the 200-mile exclusive economic zone (EEZ) in 1977 in Canada, and the source of the industry's overcapacity. Lately, the Fisheries Council of Canada has attacked it as a dead and dying social or welfare fishery that is a drain on the public purse (Fisheries Council of Canada, 1994).

The 'traditional' is counterposed to the 'modern' industrial corporate sector. The latter is comprised of vertically integrated firms that can take advantage of modern technical and organizational forms, and economies of scale. This modern sector is thought to represent the efficient way of promoting development in the fishing industry. Far from ever being blamed for any problems in the industry, at least in Canada, the privatization of the open-access resource is thought to provide everyone with the kinds of throughput efficiencies that characterize the modern sector under enterprise allocation or individual transferable quota schemes.

Elsewhere this dualistic notion of 'traditional' and 'modern' has been criticized on both empirical and analytical grounds (Apostle and Barrett, 1992). We wish here to explore these issues further, particularly since a fishing crisis is thought to be some kind of natural selector in an economic shakedown that some think should be left to play itself out. The focus in this chapter is the relationship between traditional fish plants and their communities and what, if any, are the prospects for community renewal based on traditional forms of adaptation to crises by these firms and communities.

The Embedded Fishing Community

Traditionally, fishery social science has conceptualized coastal communities as closed, coherent, if not homogeneous, systems. Fishing communities are constituted by three principal action, or decision-making, groups: households, fishing vessels, and processing plants. Each has the hypothetical freedom to act as sovereign entities in making decisions. Nevertheless, such actions are shaped and constrained by the embeddedness of individuals in wider associations of families and kin groups, neighbours, and communities. The decisions that people make as to what goals to pursue or which strategies to employ are both limited and enhanced by the collective obligations and memberships that they have.

In embedded fishing communities, the 'employment system' (Jentoft and Wadel, 1984) forms social networks, not markets or hierarchies. Transactions are voluntary, but nested in relationships that are personal, complex, and symmetric. Thus, a fishing community, unlike a formal organization, has no formal leadership that oversees the whole system and makes sure that it is kept in balance (Wadel, 1980). However, members possess a reservoir of moral sanctions that are employed to ensure conformity with community norms and values. With reference to Tönnies (1963 [1887]) and his famous distinction between 'Gemeinshaft' and 'Gesellschaft' (community and society/association), the embedded ideal type of fishing community is both: the latter is nested in the former. As to ecological ties, the embedded community is characterized by modest 'adapted' levels of technology that facilitate adaptability, flexibility, and low technological imperatives. As a system, this type establishes a sustainable relationship to its resource environment as a consequence of indirect, default, low-intensity exploitation.

We examined the position of small firms in this system to assess their potential for dynamism in a changing global market. While the traditional system may be said to have been sustainable in ecological terms, was it so in economic and social terms also?

Small-Scale Fish Plants and Embedded Communities

Unlike large firms, small fish processors do not have the capacity to handle large volumes of fish supply or to make a significant impact on their primary markets. They tend to be price takers and to avoid cost-related vulnerabilities associated with large overhead. Part of this strategy entails a *social investment* in ties with independent fishermen and to a lesser extent their plant workers.

Family firms utilize community-based social capital, just as they do kinship ties, to maximize 'functional flexibility.' As with family relations, however, these advantages can be a double-edged sword. Community 'free riders' are as much a problem for small businesses as are family 'free riders.' Relations of solidarity, in this case, create a climate of opinion, sympathy, or obligation that tempers cold hard business decisions. Again, such dilemmas reveal community embeddedness as both an enabling and a constraining structure (Portes and Sensenbrenner, 1993). For example, one manager in eastern Nova Scotia has tried unsuccessfully to convince one fisherman to be more careful with the quality of his fish. When asked why she wouldn't just put some more pressure on him, she replied, 'We can't do that, he went to school with my brother.' The brother, as it turned out, was part-owner of the company.

There is also evidence of embeddedness in the labour market. When there is a shortage of workers, functional flexibility is a necessity – workers must employ various qualifications and they must be able to handle different work tasks. Fishing communities like Vengsøy in the county of Troms, because of its isolation from the specialized labour and service market of the urban centre, are dependent on the knowledge and skills that can be provided locally. This is true both on the household as well as the community level. In other words, functional flexibility has been developed as a necessity for survival and provides an opportunity for further development, particularly in times of crisis. This is the crux of traditional small-firm embeddedness in rural communities. In the case of Vengsøy, the fish plant can draw on a core of skilled workers who are able to switch among work tasks. The variety of skills that can be mobilized within the community is also important. The fishers may be a valuable asset to the processing firm not only through the fish that they bring ashore, but also through other skills they have, for example, as carpenters.

A pluriactive workforce with strong ties to the informal sector is a result of the community not being intimately integrated into the industrial division of labour. Two traditional advantages stem from this: access to a wide range of skills and experience and the opportunity to participate in exchanges based on reciprocity. It is not so much that these are barter based, as that they involve elements of reciprocal expectations, support, loyalty, familiarity, and trust.

The case for embeddedness was also illustrated in the instance of a manager in Digby County, Nova Scotia, who bought extra fish to provide his workers with enough work to qualify for unemployment insurance. We also encountered evidence of a wider community-based worker solidarity, wherein workers shared scarce employment to qualify for unemployment insurance. In Barrington and Argyle municipalities, in Nova Scotia, a multiplicity of plants workers would move around among plants, taking days of work, or giving them up to someone else, in order to qualify for unemployment insurance.

Small family firms are most likely to finance their operations conservatively through retained earnings. Several firms testified to this: to their unwillingness to loan money, to a policy of gradual expansion, to a conservative business philosophy of earning before investing. They avoid investments in certain kinds of technology that carries with it long-term commitments regarding throughput volumes. This relates to large investments in vessels and expensive processing technology like computer flow-lines. As one owner-manager put it, 'The company always came first.' This insulates the processing firm from financial crises during a downturn, but it may also limit its manoeuvrability in responding to new opportunities. This conservatism in investment strategy has a lot to do with the degree of embeddedness that exists between the firm and the community in which it is located. The more trust there is, the greater the degree of freedom to innovate, utilizing functional forms of flexibility.

These factors relate to a distinction between *paternal* and *fraternal* traditional firms (Goffee and Scase, 1982). The smallest fraternal firms are only one step removed from their producers and workers: the managers work alongside both, indistinguishable in dress and demeanour from the others. Part of the reciprocal nature of small embedded communities, and *fraternal* small local businesses, is that there are greater normative constraints on the interaction between management and workers, for instance, governing the nature and extent of the exploitation of the workers.

In some fundamental respects, small fraternal firms are the lifeblood of the coastal zone fishing fleet, providing a competitive edge in port markets. Often the firms will have been established by fishers themselves to give them direct access to the market. This trend has been more prevalent in Nova Scotia than north Norway because of the ease of access and proximity to a lucrative market for fresh fish in New England. A good example is a fishing family of three generations in Sambro, Nova Scotia, that has recently gone into various kinds of independent fish processing and wholesaling.

Response to Crisis: Paternal versus Fraternal Firms

The fishing crisis appears to have increased subtle differences among firms in a process that has a distilling effect on the relationship between these firms and their traditional rural base. Traditional fish firms have retrenched, hunkering down, hoping that their time-tested traditional markets hold, because they are unwilling or unable to make the effort or investments necessary to reduce their dependence on one or two primary market segments. One manager commented, 'We can't make any bad moves. A few years ago, you could gamble.'

For most managers in the midst of crisis, a conservative philosophy has prevailed. This ranges from cynical fatalism to more cautious approaches to new opportunities. A manager at Forsølbruket, Finnmark, echoed these sentiments, saying that even in an improved economic climate, 'We will grow within these walls.'

More often than not, in the past this attitude was a function of conservativism. A plant that failed to reach a minimum technological threshold necessary to achieve flexibility would not be able to reduce vulnerability during crisis. However, in our estimation, it represents a significant differentiating feature between static and more dynamic small-scale firms.

Most small firms have to walk a fine line between maximizing flexibility and minimizing vulnerability. Often the two coincide and are a function of uncertainties inherent to the industry. As an illustration of this dilemma, one manager told us that his firm's chief problem in the current crisis stemmed from the debt incurred during the last boom, when the company invested Cdn $500,000 in a labour-saving filleting machine. This seemed like a prudent risk at the time, given an overabundance of fish and a shortage of labour in the local community. Two years later the fish supplies to sustain the costs of the equipment were not there. Paradoxically, however, a relatively small investment of a further 100,000 CAD in a freezer might have allowed the company to access supplies of frozen Russian fish, which in turn would have allowed it to put its filleting machine back to work and perhaps move into new markets. Such decisions are agonizing, since there is no way of predicting what the right move is. Even if one does everything right, it still may turn out to be wrong because of circumstances beyond one's control and prediction.

A common theme in our interviews was the problems the crisis presented for intergenerational succession. Firms that went into the crisis with older, traditional managers who had little formal education were especially vulnerable. One of the most prevalent reasons cited for bankruptcy and closure in the fish processing sector in north Norway during the crisis of 1989 was poor management. When pressed to be more specific a number of managers referred to problems of *paternalism*: managing firms in a personalized, rather dictatorial style, cultivating dependent relations with fishermen and workers, acting as patron with the community at large.

One manager in Honningvåg, Finnmark, remarked that traditional fish buyers '[came into the industry] from the ground up. They were the only employer in the area and inherited whole communities and the islands around.' Their production and marketing strategies were essentially conservative: overdependence on one main product or marketing channel, low reinvestment levels, and archaic technology.

A manager of a small plant in Eastern Passage, Nova Scotia, remarked on essentially the same problem there: The small plants that get into trouble were isolated, overspecialized in one main product, and too dependent on a single market.

The traditional response to crisis in this system is retrenchment, using social capital to absorb the costs associated with weak markets or poor fishing. In other words, passing on the social costs of the crisis to one's own and one's workers' families and households. Once the cyclical downturn was over, the traditional system could be expected to re-emerge relatively unchanged. In the past this strategy could work where plants were well established and carried no significant debt load that would make them vulnerable to banking institutions or government agencies. A lucrative social safety net would carry over the fishers and plant workers until times improved.

A manager from Cape Sable Island, Nova Scotia, observed that plants with debts on machinery cannot easily downsize in response to the crisis. One manager in eastern Nova Scotia quipped that the Atlantic Canada Opportunities Agency only gives firms like this 'help to get them into trouble.' Since the time that we conducted these interviews in Nova Scotia, in 1993, two of the small firms that had been trying to retrench but carried moderate debt loads have gone bankrupt.

The crisis has increased the rationale underlying functional flexibility. Where firms are in a position to hibernate until the crisis is over, production workers and office staff are laid off, and the manager's family becomes increasingly active. One plant on Cape Sable Island, Nova Scotia, employed 120 workers in 1991. This was reduced to 25 in 1992, and to 10 in 1993. Its main competitor down the road had gone from 70 workers to 6 over the same period. In one family firm in Shelburne county, Nova Scotia, the crisis had led the manager to reduce his office staff by laying off his wife. The manager was also driving the truck and trailer in 1993, whereas in the past he had had a full-time driver. The family had also reduced their self-consumption. By 1994 this firm was in receivership. 'We are not all running around with suits on. The days when there isn't a lot of work, we do it ourselves,' said another manager in eastern Nova Scotia.

Such flexible strategies are an option, especially where family firms operate on the livelihood principle rather than on the accumulation principle (Goffee and Scase, 1982). One cannot expect the same commitment from a manager with no kinship ties to the previous owner. Neither can one anticipate the similar willingness to make sacrifices, for instance, regarding other career opportunities and self-exploitation in terms of lower pay for more work, particularly as a way to buffer crisis.

Downsizing in the small plant is also significant as a strategy for reducing debt loads and operating costs – one manager in the Clare district of Nova Scotia remarked that his 'streamlined plant runs on what my competitor down the road pays his office staff' – and improving marketing leverage. This latter strategy was explained as follows: When it stayed open for twelve months, the plant produced a volume of salt fish that, in a soft market, was difficult to get a good price for. By cutting back production to a six-month operation, the manager was able to hold a smaller inventory and wait for the optimum price over the twelve-month period. He produced for half the year, but sold over the entire year.

We observed a dramatic increase in seasonality in the industry and a consonant rise in the importance of unemployment insurance for workers and fishermen. Managers of small plants, dependent on a small skilled workforce, were more conscious than ever of the importance of keeping these workers available. Great emphasis was placed on ensuring that their best workers got enough hours to qualify for UI. One manager even purchased four container loads of Icelandic pollock to provide his workers with the extra work time needed to qualify for UI benefits.

Typically, small plant managers give a rhetorical response in face of crisis that praises the long-standing virtues of the traditional way of doing things. Two small plant owners in southwestern Nova Scotia praised the virtues of hand-filleting, sundrying processes, and piece-work as cost-effective techniques that could be relied upon to cope with the crisis.

Such linkages also translate into another kind of discourse about the virtues of traditionalism, particularly in our interviews in Nova Scotia. Processors would express a lot of sympathy and solidarity with coastal zone fishers employing fixed gear and opposition to dragger technology and foreign fleets. All of this would usually be couched in terms of a sustainable fishery: 'If the fishery were managed on a sustainable basis, our access to fish would be assured, and we could go on supplying markets – and making a good living for ourselves and our fishermen – in the traditional way. The crisis is not of our making, we are victims of mismanagement and greed. Why should we be victimized for being unwilling to change?'

In Nova Scotia a sharper division separates small plants that rely strictly on a longline fleet, and those plants that accept 'dragger' fish. Given the absence of organizational cohesion among fishers in Nova Scotia, scapegoating has become a prevalent feature of an oppositional discourse surrounding the crisis. To a large degree the conflict between the two gear sectors is spilling over into a division between processors and even communities. For example, one plant on Cape Sable Island in Nova Scotia has been labelled a 'dragger' plant in a predominantly longline-based coastal community. The firm now has posted a sign that reads: 'Keep away from the property unless you work here!.' They have also installed an eight-foot steel fence and a closed-circuit TV security system.

To remain loyal members of their communities, processors are having to take sides in Nova Scotia. Several small processors in this region expressed a strong resentment of mobile gear. One said, 'I have never and will never buy dragger fish. It goes against my principles.' Commitments were often made to longline and handline fishers faced with an uncertain future that these processors would never abandon them by purchasing foreign fish, or if they had to, local fish would be treated preferentially. Such commitments are shrouded in a rhetoric of loyalty and trust: one manager swore that he would never buy 'dragger' fish, that it would violate his obligation to buy from the small boat, 'hook and line' fleet.

The crisis in Atlantic Canada is undermining much of the cooperation that has traditionally characterized the competitive sector of the fish-processing industry (Apostle and Barrett, 1992). Debt-laden plants, experiencing shrinking markets and nervous bank managers, are resorting to distress sales and dubious transactions to survive. A manager from the Acadian area of southwestern Nova Scotia remarked that 'plants are undercutting each other out of desperation to the point that they are even hurting the financially secure plants.'

Contrary to the Darwinian assumptions that underlie neo-conservative economic thought of late, one Nova Scotian manager observed that 'the strongest are not surviving,' and it is not a consequence of too much state subsidization. Another manager from Clark's Harbour, Nova Scotia, pleaded for the need for trust: 'We have to work it out among ourselves, otherwise the banks will get theirs first.' With reference to the banks and the Fishermen's Loan Board, he said, 'They are running scared. They want cash now rather than extend credit for a bigger return in the future.'

All this bears testimony to the value of social capital to these firms, but if the strategy is to respond to the crisis by retrenching and passing on the costs to the community to absorb, their 'social credits' in the community will become 'social debits' rather quickly. They ultimately run the risk of seeing their embedded communities turn against them or even disintegrate to the point where their firms can no longer function.

Disembeddedness and Re-embeddedness

It is worth reflecting on the implications of the preceding discussion for sustainable development and re-embedding coastal communities. 'Skipper-tak' is Norwegian for an 'all-out effort'; etymologically it means the skipper gives his crew a hand – usually the skipper stays at the wheel. Skipper-tak culture in embedded coastal communities can be seen to have provided the social capital underlying numerical and functional flexibility. Managers gain access to a compliant and readily available workforce on a highly erratic schedule. Workers in return can expect small preferences for themselves and their families, congenial work conditions, and a substantial amount of autonomy (Apostle and Barrett, 1992; Midré and Solberg, 1980).

Clientism, however, is a well-known aspect of paternal capitalism. While reciprocity has often been identified underlying these transactions, inequality, hierarchy, and systematic exploitation are also commonly described in such situations. Subservience, servitude, docility, and passivity are the hallmarks of defensive responses among the underclass. In work environments that lack a demand or reward structure for education, skill, and innovation, little of these qualities come to characterize the workforce. These are hardly the qualities needed by small *dynamic* firms to cultivate quality circles, worker innovation, and synergism on the shopfloor. One quality control manager said as much in referring to the cultural problems firms in Finnmark encounter in hiring Tamil immigrants to work in fish plants. She stated that their traditions of status deference, fear of being held responsible for mistakes – and losing one's job because of it – not only made these workers extremely reticent, but created dangerous situations on the shopfloor. For instance, workers would always reply to instructions in the affirmative, even if they had not understood what the instructions were, for fear of giving offence.

Sweatshops are not hotbeds of technological change. Communities that are strongly conditioned by these economic structures are clearly embedded but highly vulnerable at the same time. They lack resilience, diversity, initiative, and self-reliance.

In the case of *fraternal* firms, the transition to a more *dynamic* type is more likely. The day-to-day involvement and interaction of the owner–manager with the workers and fishers creates the opportunity of shopfloor innovation. Open lines of communication, social egalitarianism, organizational flexibility, and problem solving characterize such 'flat' organizational structures. The movement beyond the traditional in this case is more a consequence of interactive or agency factors, whereas the obstacles to this occurring in the *paternal* cases are more structurally rooted. Individuals have to be willing to try new ideas, be accepting of newcomers, take risks, travel, and learn.

In our estimation the dilemmas of intergenerational succession represent a fundamental difference between *static* and *dynamic* small firms in the fishing industry. This is a difference between a future orientation and a past orientation, particularly when it comes to meeting the challenges posed by a crisis. Many of the fish plants included in our study have recently gone through a generational succession, as a son or a daughter of the manager was taking over, sometimes with the father still watching from the wings. Interestingly, the generational succession is a process where both change and stability are generated; the younger generation tends to have more formal education than the older. This process does not always lead to a qualitative break with the past. One manager quipped, 'You need a birth certificate to take over the firm.'

Sons and daughters have been sent to university to qualify in business management and food technology in order to ensure the viability of the firm as a family enterprise. Formal education gives the manager more latitude for decision-making and innovation. In one Norwegian fish plant, the son came back with skills in computer science, which he immediately put into practice in the plant. In addition to his university education, he had worked seven years in a fish export consortium, which had provided him with detailed knowledge of fish marketing. Being raised in a small community, university also brings the heir in contact with a new and more cosmopolitan environment. Said a fourth-generation fish processor: 'My four years in university did not so much teach me to run this fish plant as it made me more mature and more confident in relating to the outside world. I became aware that it was a greater world out there.' However, it did not reduce his attachment to rural life or change his attitudes to what constituted a sound business strategy. Like his father, he was pledged to the principle of spending not more than you earn. This 'conservative' attitude, he argued, was particularly helpful in times of crisis. 'With no debt, we can lock the door of this plant any day and go home without losing any money.'

Conclusion

Individual sources of dynamism are not sufficient for coastal communities to reorient themselves on a sustainable footing. For traditional firms to really become transformative they need to act as catalysts in the re-embedding of communities of which they are a part. One cannot succeed without the other. As we found in the case studies that follow, re-embeddedness is very much dependent upon the reassertion of a collective spirit on the part of workers, fishers, and community members at large in a range of social settings and interactions that reinforce and fuel the kinds of individual entrepreneurship that are all too often the sole focus of economic studies. This theme is taken up in the material on post-Fordism that follows.

Nevertheless, community decline and disintegration is also a common result of the crisis. Individuals come to the point where they have to make decisions in their own and their family's interests which, when compounded by others doing the same thing, undermine the ability of the community to respond in a coherent collective fashion to the crisis with new and innovative strategies. Instead of re-embeddedness, a process of disembedding occurs. For example, by 1995 in Nova Scotia, the embedded rhetoric of solidarity between small processors and longline fishers was wearing thin. In a trend that echoes the experience of the longline fleet in communities like Vardø in Finnmark, fishers were seeking new and more diverse opportunities such as targeting species and landing fish outside their home communities. Fisher wives were becoming even more instrumental in their role as brokers negotiating sales and landings of their husbands' fish on a highly speculative supply-starved market. Processors were having to chase down fish wherever they could buy them. Most often they had to go out of the community for fish, sometimes they were even forced to go abroad.

New technology, particularly cellular telephones and fax machines, are greatly accelerating the need for spontaneous flexibility on the part of plant managers. These kinds of developments, which especially affect the most embedded sectors of the industry – the coastal (inshore/midshore) fleet and small plants – represent a crucial problem for coastal communities. These developments signal problems for communities re-embedding in response to new opportunities now or in the future. If it is left to individuals to respond in an atomized way to new global opportunities, communities will not only lose the initiative, but, as we argue below, communities lose the opportunity to establish socially sustainable systems in the long term.

11

Fordism, Neo-Fordism, and Community Disembeddedness

Almost a hundred years have passed since the time of homogeneously embedded coastal fishing communities along the coasts of Canada or Norway (Barrett, 1992). As in forestry, mining, and agriculture, the Fordist model of industrialization has prevailed in the fisheries sector. Coastal communities assumed a single-industry character with distinctive sociological features. The fisheries crises of the 1990s greatly affected these communities, and this process has been compounded by the general crisis within the Fordist system.

In this chapter we examine Fordism in coastal fisheries communities and explore some issues related to the transformation of this type of firm following the fisheries crisis. The Fordist model of industrialism in the fishery is unsustainable. There has been restructuring of the industry along what we term 'neo-Fordist' lines. However, this has created what amounts to a facade of economic sustainability to accommodate globalization by deregulation. Specific forms of disembeddedness have become more acute than ever, and the regional fishery has exported its ecological unsustainability to other jurisdictions.

The Fordist, Single-Industry Town

Fordism requires the fragmentation and differentiation of the traditional embedded community. Households and friendship revolve around work and occupation. Social class issues largely define personal identity and community culture. One's embeddedness, to the extent that it is remains at all, is mediated through a reconstituted occupational community that segments space, neighbourhoods, and patterns of association and sociability (Bulmer, 1975). Particularism and moral economy are replaced by the rule of law, contracts, and bureaucracy. The hierarchical social structure is reflected differing standards of living, status, and power relationships.

Fordism strips the economic 'heartblood' of the community away from encumberments such as cultural or social demands or considerations of the interests and concerns of the local community as a whole. Decision-making, whether on the part of managers and professionals, workers, fishers, or householders, becomes unfettered and occurs in what, for the most part, is an atomized climate of free choice. In the fishery, for instance, the vessel skipper makes unilateral decisions as to fish targeted or to whom fish are sold. His crew is drawn from a casual labour market and defined as subordinate wage labour more than as co-adventurers' and equal team members. The manager of the fish plant makes decisions concerning fish purchases, employment in the plant, and product configurations based on 'cool and hard' economic realities and bottom-line profit margins.

Interpersonal trust may develop between these actors, but under Fordism only as a function resulting from a deliberate strategy of reducing 'transaction costs' (Granovetter, 1985). Embeddedness, in the sense of collective obligation and loyalty or responsibility *vis-à-vis* the local community, is either entirely absent or subordinated to utilitarian self-interest.

At a formal level, input, output, and demand linkages continue to characterize the economy, but, by and large, they are 'internalized' within the corporate organizational structure (O.E. Williamson, 1975). Technical and economic imperatives are such under Fordism that an independent, local, small-scale fishing fleet becomes insufficient to provide a stable flow of resources. To ensure stability and security, processors acquire vessels that are technologically capable of fulfilling the demands of the industrialized production system. Only as long as economic conditions make such transactions prudent, will the processor purchase fish from the independent local fleet, recruit and employ local labour in the plants, and utilize local services.

Sustained and sustainable relationships between the Fordist economic system and the local ecology are accidental or non-existent. More often than not, such balances fall victim to industrial agendas associated with technological change and quasi-urban landscapes. When firms reach a technological threshold, certain cost-driven imperatives are unleashed. Flexibility and adaptation are replaced by the rigid and unrelenting needs of industrialized fish-processing plants. In this circumstance, human relations and culture have to adapt to the industrial discipline demanded by the machine and the clock. The extraction of resources must be directed from ashore to fulfil the demands of the Fordist industrial process, promoting vertical integration from processing through to harvesting.

In the fishing industry the fully articulated Fordist model emerged out of the ruins of the Second World War, when governments were promoting both modernity and an end to traditionalism. In north Norway and Atlantic Canada whole communities, thus, become reliant on large, vertically integrated companies, whose scale of operation dictated a Fordist style of plant management and a monocultural approach to the harvesting sector.

Sources of Disembeddedness

The fishing crises of the 1980s and 1990s affected the Fordist firms in both countries in fundamentally the same ways that crises have affected other sectors of the economy in the western world after 1970 (Murray, 1983). Corporate capital was forced to retrench and restructure in ways that would allow it to improve cost efficiencies in the short run, and productivity in the long run. The outcome is a substantially leaner and meaner industry comprised of downsized capital that is far more responsive to global opportunities at both the input and the output ends.

Unsustainability is a hallmark of the Fordist system. In the fisheries sector, this is most clear in the perennial mismatch between what rural communities can offer and what the fish-processing industry needs to survive. This contradiction has always been a concern of plant managers in Finnmark, where fishing has never been a self-sustaining industry. It relied on external inputs such as the supply of an additional fleet from some other region, seasonal labour recruited from other parts of Norway and even from neighbouring countries, government capital and private investment from the south, or, as in the case of Nestlé, from abroad. For the external entrepreneur, the lavishness of the resources of the Barents Sea provided great opportunity, while for Finnmark the outside contribution compensated for scarcity that prevailed at the local level, partly as a result of the Second World War. Without government support, the ambitious plans for rebuilding could not have been realized. Consequently, the state itself created the Finotro plants, and corporate capital from outside the region both domestic and foreign moved in.

Finnmark was an 'outpost' and 'frontier' that attracted outsiders – who rarely stayed for long. In the peak season, the local labour force was far from adequate to fit demand. Thus, labour had to be recruited from outside, often from southern counties. The workers would stay for the season, work full-time in the plant, and live in dormitories owned by the fish-processing firm. In the 1980s many Finns came for work in the fish plants, along with refugees from Sri Lanka. In most cases these workers would form separate groups and live apart from each other and from the community. The Finns had language difficulties, while in the Sri Lankan case there were also great cultural barriers.

Although Nova Scotia has had its share of foreign investment in the fish-processing sector, such investment has generally been restricted to single-plant, single-location subsidiaries of British or American capital. Large-scale capital in Nova Scotia has primarily been of the home-grown variety. National Sea Products (NSP) and H.B. Nickerson and Sons, and now Clearwater Fine Foods and Comeau Seafoods, represent large Fordist operations that have flourished in the postwar period under state modernization programs.

NSP is by far the largest and best illustration of this relationship. By the mid-1980s it had grown to be a multinational operation with plants in Paraguay, Australia, France, Portugal, and the United States. While not an overly successful vertically integrated firm, NSP organized its plants and fishing fleets into divisions that had a high degree of internal coordination and exchange. Such external control based on corporate interests is and was the root of various disembedding effects in coastal communities.

Disembeddedness at the community level was most obviously felt by trawler crews and their families, because fishing vessels target fish and land their catches according to centrally controlled decision-making within the organization. The local resource was irrelevant to the operations and decisions of the plants, and the vessels were run and crewed as small industrial factories on the ocean. These vessels operated with little regard for, or ties to the local ports where they were based. Work conditions were bad, turnover high, and the firm was regularly recruiting labour from distant communities and Newfoundland (Binkley and Thiessen, 1988).

A third form of disembeddedness stemmed from the high labour turnover that these types of assembly-line plants suffered from. Organization of the plant was geared to the mass production of low-grade frozen and cooked products, primarily for the U.S. institutional market. Throughout the 1950s, 1960s, and 1970s, quantity rather than quality was the primary focus of these operations. Thus, routinized and cheap labour was the backbone of the plant operations. Rigid productivity quotas, coupled with arduous and monotonous work conditions, led to chronic labour instability (Connelly and MacDonald, 1990a, 1990b).

In a study of this sector in the 1980s, Apostle and Barrett (1992) discovered that the plants could only function well in a rural environment with high levels of un- and underemployment. In some locations the plants were 'using up' available labour supplies – particularly of young women – and having to extend out from the communities in concentric circles to recruit workers. One plant in Canso, Nova Scotia, established a commuter bus service to bring in workers on a daily basis from as far as sixty miles away. The fish plant in Lockeport also recruited a large percentage of its workforce from various outlying communities up to thirty miles away. The firms involved, rather than being pleased, were actually exasperated at the labour costs associated with high turnover rates and distant recruitment. They would have preferred to have 'locals' working in their plants, but such work was a last line of desperation for local people, particularly if they had any kind of alternative. In a number of communities, in fact, small independent plants would steal away National Sea's best older workers, forcing the large plants into external recruitment efforts.

Crisis and Neo-Fordist Restructuring

The restructuring of corporate capital in the fishing industry from the mid 1980s onward pursued a distinctive trajectory. Following theorists in the literature on flexible specialization (Murray, 1983), we term this restructuring 'neo-Fordism.'

Neo-Fordist firms are distinctive in at least four different respects. First, they focus on the high-end, value-added market represented by such things as individually quick frozen (IQF), portion-controlled fish products used by restaurants and sold in supermarkets. Just-in-time delivery systems and a demand-driven operation reduce inventory costs. Second, production is more mechanized than ever with computer-controlled grading, flow-lines, and IQF tunnel freezers. This technology gives managers an unprecedented ability to reduce waste and maximize productivity. They are able to monitor and measure individual worker productivities on a minute-by-minute basis with great detail. Third, global communications and improved access to transportation have facilitated the global inputting of raw material supply. Value-added production allows the firms to substitute fish from one part of the globe for that from another without undermining market demand. Such purchases are closely tailored to product demand, providing improved throughput efficiencies in the use of processing and storage technology. Fourth, the entire system is increasingly driven by smaller management teams comprised of young technically trained individuals with a background in commerce, technology, and/or science. With the sophisticated computer technology employed in the steering and control of production flow and with a more professional management and labour force, these plants are up to date with international trends. The focus of management has shifted from the local to the international level as the market management approach, with its stronger emphasis on quality, has replaced the traditional orientation on supply volume as the source of profit (Johansen, 1994).

The neo-Fordist version of fish processing needs tight and extensive connections with the outside world. In what follows, we illustrate these trends with material from north Norway and Nova Scotia.

Neo-Fordism and Disembeddedness

The fishing crisis hit Finnmark harder than any other county in Norway and brought the industry to the brink of collapse. The government came to the rescue and saved many boat owners and processing plants from devastation. So also did foreign capital, which, in many cases, obtained ownership of bankrupt plants at very low prices. The county government, with the Finnvest affair (see below), had hoped to counteract the tendency to increased foreign control, but failed. Thus, today outsider presence in the Finnmark fishing industry is more conspicuous than before, with Danish, British, and Russian capital, in addition to the Swiss.

The question of ownership has not only been a local concern, but also an issue in county politics. During the crisis there were examples in Mehamn and Bugøynes of bankrupt plants being sold to outside people for an exceptionally low price – and then stripped of all their equipment. This triggered loud local protest and created a general fear that this could happen to other communities. What little remained of local control in the fishing industry was at risk.

In Finnmark, one counterinitiative was the formation of the Finnvest holding company. In 1989 the Ålesund-controlled (south Norway) Aarsæther concern was in economic trouble. With plants in five communities and with seven deep sea trawlers, it was a cornerstone of the Finnmark fishing industry. Bankruptcy would clearly cause serious disruption. There was the possibility of plant closure. Another, and even more worrisome prospect, was the risk of losing the trawlers, and with them their licences. For county politicians this was a daunting scenario. Thus, the county council decided to form the Finnvest holding company with the purpose of acquiring the Aarsæther concern, and thus bring it under Finnmark control.

But a government is not usually equipped to run a business enterprise, and things did not go as planned. Without any experience in running deep sea trawlers, this operation was contracted out to the Ålesund company Westfish - a subsidiary of the Aarsæther family. In 1990 and 1991 Finnvest had great losses and decided to go public to expand its capital base. Then Westfish came in as a major owner, with 40 percent of the shares and later in 1994, at the expense of a Finnmark shareholder, with 90 percent. And so, the Aarsæther family was back in again. The outcome, therefore, is somewhat discouraging from the point of view of those in Finnmark who envisioned a regional ownership and a re-embedding of the fishing industry.

Today, therefore, many would argue that the county government initiative was a total failure. Nevertheless, there were positive accomplishments. Finnvest rescued all the Aarsæther plants from closure, jobs were saved, the concern was kept together as one entity, and, importantly, trawlers and trawling licences remained in Finnmark.

This did not happen with another major player in the Finnmark processing industry - the Frionor-Polar Group Ltd (formerly Finotro). When this company went bankrupt, the county authorities refrained from interfering, and the company was sold in pieces to private owners. Only in two cases, in Båtsfjord (Havprodukter A/S) and in Honningsvåg (Storebukt Fiskeindustri A/S) did local people take over.

Outside ownership has been common in Finnmark since the Second World War. The fisheries crisis, however, seems to have consolidated and even augmented this pattern. Arbo and Hersoug (1994) concluded that since the crisis, most of the key processing companies in Finnmark are controlled from outside the county and that only a few small traditional plants still belong to local owners.

A new player appeared in the Finnmark fishing industry, the Danish multinational company Foodmark, that bought out the local owners of the Mehamn and Gamvik plants, and formed Nordkyn Products Ltd. In these two communities people feared that the new owners would utilize the facilities only as a landing station for transporting fresh fish to the Foodmark plant in Denmark, thus eliminating local jobs. However, the company invested heavily in renewal. In Gamvik a new plant was constructed after the old one had burned down. Foodmark accepted heavy losses (Nkr 80 million in two years) in their two Finnmark plants.

Nordkyn Products was already bankrupt, but with four new managers brought in from Denmark in 1993, and substantial economic support from the county government, Foodmark had hoped for a brighter future. This was not to be, however, and the firm closed by 1995, which put workers out of job and fishers out a buyer.

Outsiders lack loyalty to local communities and their people as compared with locally recruited owners and managers. At the same time they are isolated from community social capital. Where their plants are branch operations of a larger conglomerate, the decisions of these owners are even further externalized in their orientation.

The social unsustainability of these large-scale plants is exemplified not only by their effects on local labour relations, but in their reliance on outside labour. In Finnmark, for instance, Tamil refugees from Sri Lanka were working there before the fisheries crisis, and they still represent a highly visible group in Finnmark's fishing communities. In contrast to other guest workers, they bring their families. Although statistics to confirm this are unavailable, it is our impression that there is a broader Nordic representation in Finnmark than there used to be. Partly this is a consequence of fisheries crises in other Nordic countries, especially in Denmark and the Faeroe Islands. New Nordic owners also bring workers with them. Russians are a recent group of guest workers in

Finnmark fishing communities, though their numbers are not great. Given the dire economic circumstances in Russia, however, they may well become more numerous in the years to come.

According to the production manager of Havprodukter Ltd in Båtsfjord, quality management requires better communication between management and workforce. His experience is that this can be difficult with an international workforce. Language and cultural barriers create misunderstandings of intentions and misinterpretation of information. Said the Havprodukter production manager, himself a Dane: 'We cannot afford to make mistakes in the production process, and the risk of misinterpretation of orders given is a constant concern.'

The Aarsæther Vardø company now requires that foreign workers take a Norwegian language course. The management has contacted the local school to make sure that workers know the language necessary to function properly at the workplace. The plant also has an information bulletin which, to improve the Norwegian workers' understanding of Tamils, ran a series of articles on Tamil culture. Twenty-eight Tamils are working in the plant. But communication across cultural boundaries remains difficult. The Tamil workers criticized the articles as misrepresenting their heritage. While all managers think that foreign workers fit well into the work schedule, they argue that there is a limit to how many nationalities they themselves can relate to. In many plants more than 50 per cent of the labour force are from other countries, and in some this is as high as 70 per cent.

Whether Finnmark fish processing will be able to attract more local people to the industry, and reduce the turnover of workers, remains to be seen. Many fish workers moved south during the crisis. To limit the effects of the crisis, the government offered unemployed workers training courses to qualify for jobs in other industries. According to the personnel manager in the Aarsæther Vardø plant, of those who took the courses many have now moved away from the community and most likely they will not return. For the locals with alternatives these neo-Fordist plants are even more likely to be an employment of last resort.

In Nova Scotia a common response of large-scale plants to crisis is downsizing, disinvestment, and production slowdowns. As in north Norway, a 50 per cent rule seems to prevail: plants, employment, and fishing fleets are all down by one-half over precrisis levels.

National Sea Products provides insight into the neo-Fordist transformation in the fishing industry in Nova Scotia. The crisis meant that the firm lost more than 50 per cent of its normal supply of groundfish. From a high of fifty wet-fish trawlers, fifteen plants, and employment of 8,000 in the mid-1980s, the firm downsized its operations to three plants, ten trawlers, and under 1,000 employees by 1993. As with other firms in Nova Scotia and north Norway, National Sea Products was experiencing a 50 to 60 per cent cut in overall capacity.

Neo-Fordist companies respond to crisis by retreating into a technologically driven cost-efficiency model, reducing inventories and unprofitable product lines, and specializing in value-added product, such as individually quick frozen (IQF) fish. This kind of policy was implemented in the Mehamn- and Gamvik-based Nordkyn Products Ltd. For cod alone, their inventory included more than seventy different varieties. The new management decided to concentrate on cod blocks and fresh fish and only produce prepared food like fish fingers and battered fillets on order.

In a pattern that seems to be mirroring the agricultural side of the 'food chain' (Winson, 1988), global marketing and intrafirm transactions in the neo-Fordist corporation are strongly affected by a counterweight at the food chain's other end, retail supermarket chains. Both National Sea Products in Canada and Nordkyn and Aarsæther in Norway are (or, in the case of Nordkyn, were) carefully nurturing joint venture agreements with large chains to preserve the traditional market shares they held with their own brand names. Often this took the form of no-name or retail brand labels.

One of the concerns of management is how to mesh quality standards of the European Union – which was the ISO and HACCP standards – with other customer countries such as the United States and Canada. The International Standards Organization's (ISO) standards are in use worldwide and across a great variety of industries, including the fisheries. In August 1994 only two fish processors in Norway were certified according to the ISO-9002 standard which was the one relevant for fish processing. Many plants, however, are now working to become certified. The Norwegian Fisheries Association of fish processors and exporters has initiated a project called 'total quality management' that comprises twenty-seven fish plants from north to south, of which fourteen are in Finnmark. A part of the project is for these plants to become certified according to the various ISO standards for final product control, quality control, and quality management. To become certified, the firm must document and formalize every step of the production process in terms of quality handbooks and plans. An internal quality coordinator must be appointed to overlook the process and to implement and enforce quality measures.

The Hazard Analysis Critical Control Points (HACCP) originated in Canada. It has a narrower focus and is based on the principle of self-control. The HACCP standards are now being introduced by the Norwegian government as part of their licensing requirements. If plants do not fulfil the HACCP requirements within two years, the Directorate of Fisheries may withdraw their licence to buy fish.

According to the Fisheries Directorate, within a couple of years these requirements must be fulfilled to get access to international markets. In all the Norwegian plants we visited, steps are being taken to become certified in accordance with the quality standards.

Qualification of workers is only one of several demands. For instance, to comply with hygiene specifications plants invest in new equipment and renovate their production facilities. They have to establish working routines and quality control measures, which they have to present in a handbook for inspection. This, of course, is a new burden on managers who already face a heavy workload. Still, managers have a positive attitude and think that the ISO certificate will give the industry a competitive advantage. But they prefer to draw on their own knowledge and experience and not spend too much money on external consultants.

The effort of the industry itself to professionalize fish-processing work in Norway through various educational programs is very promising, but it is still too early to make any final judgment. The typically pyramidal organizational structure of neo-Fordist plants contradicts the open style of relations between management and labour that is needed to generate synergisms. One large firm in Finnmark had a sizable number of employees participating in the educating program, but the manager indicated that the training was of little practical use, as the workers would be returning to their old jobs where the new skills and knowledge were not relevant.

Mass collective unions are often caught up in the perpetuation of these industrial systems (Freudenburg, 1992; Murray, 1983). The routinization of work and the defensive resistance of workers to protect wage levels, seniority, and work conditions has led to certain 'bargained rigidities' in the organization of work in neo-Fordist plants. Managers attempting to restructure such operations and introduce greater flexibility are faced with intransigence and institutionalized inefficiencies. For example, the plant manager in Stofi, Honningsvåg, told of how, because of the scarcity of fish during the crisis, seniority-defined lay-offs meant that the youngest, best trained, and most energetic workers were the first to be laid off. Another manager in Vardø observed that the union had become captive to a 'best worker bias.' The top pieceworkers in the plant, earning the equivalent of Cdn $15 an hour, resisted attempts to negotiate a payment schedule that would reduce performance pressures on the workers. Such situations aggravate community decay by failing to replace community embeddedness with a cohesive class-based 'occupational community' (Bulmer, 1975). In this circumstance, workers become far less concerned with the overall needs of the community, focus on job-related issues, and increasingly restrict their interaction patterns exclusively to other workers. Rural communities undergo a process of fragmentation when this happens, as class replaces community in terms of people's identity and sense of belonging.

Social capital is based in reciprocal transactions. There is a limit in terms of how far externally controlled firms can go in capturing these 'externalities,' since they are inherently less flexible, have less autonomy, and are less responsive to local needs and resources over the long term. These factors become especially apparent the moment the subsidiary firms have locally owned competition. Thus, there are important constraints and limits on the degree of re-embeddedness one can expect to see from a neo-Fordist restructuring.

Lockeport, Nova Scotia, is an interesting case in point. Corporate expansion into two communities on the South Shore was carried out through mergers. Because knowledge of fish processing and fishing is so idiosyncratic, the parent company remains dependent on the manager of the fish plant and the local workforce. To utilize this knowledge and to the benefit of the corporation as a whole, the local manager must have independence, and the company must rely on his knowledge and personal relations with the workforce. Such relationships mark an important attempt by Fordist plants to capture some of the benefits of the flexible specialization model (see below), particularly the advantages offered by social capital.

As to fish landings, the Russian connection definitely helped to ease the crisis, but it has now apparently become a lasting link both in Finnmark and in Nova Scotia. Even though there is in Finnmark a preference for Norwegian fish, the industry has become dependent on the Russian supply, at least as a buffer against variable Norwegian landings. For instance, for Nestlé in Hammerfest we were told that its own trawlers have the highest priority, but Russian supply ranks second, ahead of even the independent Norwegian coastal fleet. The contracts Nestlé has with the Russian vessel owners make them very reliable. Nestlé had planned to continue buying Russian fish for the next couple of years at least, but how long and how much would depend on the quota situation. Most of the plants we visited had invested heavily in new lines of productions, and to efficiently utilize this capacity they need to expand the volume of landings. They reckoned they would accomplish this at least in part using Russian supplies.

Large-scale firms seem more able to diversify sources of supply, albeit under the more desperate circumstances associated with high operating costs and high debt loads. The impact of global fish supply is particularly strong here. In 1993 National Sea Products in Nova Scotia was only operating with Russian fish; its twenty trawlers were tied up without any domestic quotas. A subsidiary division was established in Europe specifically to coordinate global input sourcing for its resource-starved Canadian plants. In neo-Fordist plants, the dictates of throughput efficiency define these strategies. The administrative manager for Nestlé in Hammerfest told us that Russian fish during the crisis were 20 to 30 per cent more costly than their own trawler fish when all factors related to uncertainty, production scheduling, and the like were taken into account. Coastal vessel fish were a further 10 per cent more expensive than Russian fish. He observed, 'If we had to buy fish from other sources [than our own trawlers], we think that we would reduce our possibility to control our operation and it would be more expensive.'

In Nova Scotia, where upstream integration is not legally restricted as it is in Norway, a large firm built and purchased large automated longliners in a deliberate attempt to improve the value of its product by improving the quality of its raw material. This was done in conjunction with a policy of buying up as many other types of licences as possible. This company can now target swordfish, tuna, offshore clams, and offshore lobster. In another case, a processor paid a fisherman Cdn $70,000 for a shrimp licence. This strategy has accelerated a concentration process already under way in the 'dragger' sector in Nova Scotia: individuals with capital buy up quota shares in the fishery. In the middle of a crisis, with uncertainty and debt payments rising, the prospect of distress sales of quotas mounts. One small processor remarked: 'IQs are a rich man's game.' A manager for National Sea Products in Lunenburg revealed that such quota transfers had allowed his company to actually increase its landings of cod, perch, and pollock in 1993 – at the very time that there were drastic cuts in allocations.

Conclusion

With increasing imports of both fish and labour used to compensate for the loss of local input, and with the stronger presence of foreign ownership, the fisheries system loses much of its foundation in the local community. Also, the links between the three principal action groups at the local level, the household, the vessel, and the processing plant, grow weaker and weaker. When transactional relations get weaker, the foundation for cooperation – solidarity and trust – is eroded. For an increasing number of local participants, the community becomes like a 'camping ground,' not a place to settle and live in, but one to stay at for a limited time. Fewer people have roots locally, and their external relationships are more important than the ones they have within the community. For the functioning of the fisheries industry, the community itself becomes increasingly irrelevant, and for the local residents the industry becomes a stranger in town.

There is a close interrelationship between outside ownership, large-scale plants, a disarticulated economic system, and community decay. Certain fishing communities seem to be host or enclave economies with an external orientation that bears little relationship to locally embedded traditions, or, in our estimation, a sustainable fishery system. Expatriate southern professionals dominate the managerial and professional ranks in many communities in Finnmark. One manager observed that the doctor, lawyer, teachers, and even priests were from the south. In the schools, courses related to or about the fisheries had been dropped, as these teachers lack the expertise or fail to appreciate the relevance of such course material for the local community. Another manager accused teachers of using the fishing industry as a threat for failure. He said that teachers encouraged students to seek professional careers away from the community. A fishery officer commented about the disintegrating levels of communication between workers in the fishing industry and local municipal authorities as the economic impetus shifted outside the community. A 'culture of decay' is plainly visible in these communities. Wharfs are in a state of disrepair, the harbour and tidal flats are littered with garbage, local hotels and retail stores are run-down, and public drunkenness abounds.

In Finnmark we see the contours of an ideal type of community, what, with apologies to Marshall McLuhan (1960), we may call 'the global fishing village.' Many of the traits of Fordism remain, but in a more advanced form. Globalization has advantages that people in Finnmark can appreciate. Foreign investors are better than no investors. They help maintain employment and tax revenues. They provide the economic base for a flourishing service economy and derivative employment. To some extent, such development can be empowering: in export the locals can be released from the traditional bonds of dependence. But globalization also has a price. The small fishing village, even if global in its orientation, is dependent on a fishing industry that cares, that is aware of its critical role in the community, and that feels a social responsibility. Faced with the choice, the community prefers the local owner, from which members expect greater indulgencies, better communication, and more security.

The implications of neo-Fordism for locally disembedded communities are mixed. In some cases where plants have been modernized and upgraded, their competitiveness has greatly improved, and employment, while down overall, has stabilized. In other instances, modernization has not prevented plant closure. Where there is only one fish plant in the community, its closure has led to unemployment figures similar to those during the resource crisis.

Other negative consequences of neo-Fordism can be detected as well. First, one can point to the worsening of managerial disembeddedness from the local community. Managers of neo-Fordist plants are less likely to be local, less likely to have grown up in the fishery, and less likely to have much autonomy in how they run their plants. On two different occasions in north Norway, we were told by managers that they were discouraged by their head office from forming ties or arrangements with other local fish companies. Second, the alienation of labour worsens except for the few who are young, fast, and good, and can take advantage of improved productivity incentives. Overall employment and pay scales are greatly reduced, even if unions remain in the plants. Third, under the new quota regimes, company-owned fleets of trawlers are reduced or eliminated altogether. From a resource sustainability perspective this might be said to be a positive outcome, however, in the globalization of input sourcing, these plants are merely exporting the way they relate to the resource to other jurisdictions. Their over capacity continues to drive an industrial approach to the resource in an unsustainable fashion, even if it is in another jurisdiction.

12

Post-Fordism and Re-embedding Coastal Communities

In the previous chapter we described the fish-processing industry in Finnmark as having become more disembedded as a consequence of outside take-overs in the wake of the 1989 fisheries crisis. These developments are consistent with neo-Fordism. There is, however, an alternative model of development for crisis-ridden communities: our term for it is 'post-Fordism.'

The post-Fordist model is characterized by an attempt to move beyond the limitations of traditional embeddedness, without destroying the firm's ability to accumulate social capital. In the process, communities can re-embed their resources on a more sustainable basis. The focus in this chapter is on a range of distinctive principles relating to technology, forms of organisation, and styles of management. Here we encounter a solution to many of the problems that result from disembeddedness – organizational rigidities, poor working environment, and high labour turnover – that have affected the Fordist model and the many coastal communities in which this model has been employed. Here we also encounter dynamic strategies of networking among small firms that assist small isolated and traditionally embedded communities in meeting the challenges and opportunities of globalization head on.

Localized, Global Marketing

In Norway recent institutional reforms have made it possible, and necessary, to pursue a market-oriented strategy. Until a few years ago, export legislation prevented fish processors from integrating forwards into fish processing. The 1955 Export Act protected fish export as a specialized trade, which effectively restrained processors from becoming exporters of their own products (Hallenstvedt, 1982). Within traditional production, a three-tiered structure prevailed. Fish would be sold from independent fishers to fish buyers residing (permanently or temporarily) in the local community, the buyers would process (dry or salt) the fish and sell it to independent exporters, usually in Bergen or Ålesund. In frozen products, fish processors collectively established the Frionor and the Nordic Group companies to export the fish for them. Thus, the functions of export marketing would be handled centrally from Oslo and Trondheim, and not by the fish-processing firm itself. In many ways this was a positive step, for it did solve some of the fish processors' problems that stemmed from small scale and remote location. However, the processors were also thereby barred from direct contact with the market, from accumulating knowledge of their customers, and from acquiring competence in marketing (Jentoft, 1993).

Export legislation was altered in 1990 to allow fish processors to freely establish their own export business. Certainly, this was an important institutional change for fish processors. And while it is too early to know all the practical results of it, some of its effects in north Norway are easily detected. At the level of the processing firm the organizational pattern for the export of frozen products has largely been maintained. For traditional products, such as dried fish, however, there has been a clear tendency for firms individually or in cooperation with other processors in the local area to move into direct export. For several years now, fish processors have been frequent visitors to countries in southern Europe in search of business partners, and the communities of Finnmark have been receiving numerous foreign guests with similar intentions.

An example of this new trend was observed in the management of fish processing in Nordkapp Seafood Ltd, in the fishing village of Nordvågen, a few miles outside Honningsvåg. The plant had been idle for a couple years after a series of bankruptcies, but new owners took it over in 1992. After a short period of producing frozen fish, the plant switched to salt fillets and fresh fish targeted at Spain, Italy, and Greece. The manager considered direct links with the customer to be a must, and concrete steps in that direction were taken. A link was established with a customer in Spain. The customer had very specific demands not only with respect to the product itself, but also the production process. For plant management, this was a new experience. Previously, management at the fish-processing plant had been totally dependent on market information as provided by the exporter, who was found to be very selective as to what he would tell them, including prices. The plant management's ambition was to become totally self-reliant as a fish exporter, but they realized that they needed to move ahead gradually and learn as they went. The processing firm maintained its relationship with the exporter, but signed a new contract stipulating better provision of market information than had been the case previously.

Nordkapp Seafood was not alone in experimenting with export marketing. Although hardly a panacea for survival – it did not prevent Nordkapp from a later bankruptcy – independent marketing has become a more common strategy among fish processors. In all the north Norwegian fish plants we visited, a more self-reliant marketing strategy was in the works. For plant management this is a demanding challenge. It will transform fish plants into open systems, and also put them at lower levels of the processing organization. As well as challenging the capabilities and style of top management, independent marketing will put new demands on middle management. Export marketing requires knowledge and skills that small-scale fish-processing firms typically do not have. For the time being this is a bottleneck. In the long run, however, we may expect a better educated and more staffed management. The top manager, and the managerial staff, need to acquire expert marketing knowledge. Middle management must also become more involved in the overall day-to-day external interactions of the firm.

Niche production, the hallmark of small-scale manufacturing in competitive markets, has increased. We observed a tendency towards a more narrow product mix. With the old marketing system, even a small fish plant would keep a broad range of fish products, sometimes in the hundreds. This not only required several production lines, but also a large marketing management staff which only the large, specialized marketing organizations like the Oslo-based Frionor could provide and maintain. With greater responsibility for marketing accruing at the firm level, specialization into fewer products becomes indispensable. The production manager of Forsølbruket said, 'We cannot be best on all products.' Last year this plant reduced the number of its products by 25 per cent. 'For instance, we used to produce flounder in many varieties. Now, we only do fresh flounder.' The firm produced fifty different kinds of fish products, and they thought this was still too many.

As in north Norway, many small firms in Nova Scotia have ventured out to develop new opportunities and product lines. Modest investments in intermediate technology – small freezers, driers, deep-well salt water pounds, computerized graders, salmon and mussel farms – coupled with a new market focus, improve the viability of these firms to the point that one manager bragged, 'The worse the crisis gets, the better it is for me.' In another case a manager in Pugwash invested in an innovative salt-water holding facility for lobster. It is supplied with salt water from a deep well at a constant 42-degree temperature which chills the plant in summer and warms it in winter. He can now produce year-round and has reached a capacity threshold to secure a steady export business. Some individuals are now devoting their entire research and development budget to sales promotion. One manager of a small plant in Eastern Passage regularly sets up sales booths in food shows in Asia, the United States, and Europe. Taking the Japanese lead, managers are concerned with new outlets, especially the retail and restaurant trade, with or without traditional wholesalers. There is a keen emphasis on customer servicing, and market-driven, small-batch, 'just-in-time' production schedules. The Pugwash plant hauls lobster in ships live from Halifax to Amsterdam twice a week. These shipments are based on tailoring exports to orders with forty-eight hours notice. (The cancellation of KLM's direct flights from Halifax to Amsterdam in the spring of 1996 will dramatically affect this firm's ability to market its product in Europe. It remains to be seen if the proposed Icelandair service will be able to meet the needs of local processors as well.)

Functional Flexibility

Crisis in the fisheries has increased appreciation for the rationale underlying functional flexibility. Traditional firms hibernate until the crisis is over: production workers and office staff are laid off and the manager's family becomes increasingly more involved. Some firms attempt to innovate and develop novel product lines or new technology, with emphasis – again reminiscent of the Japanese experience – on job rotation and versatility.

One Norwegian manager pointed out the task orientation of his unionized workforce. The union understood the importance of flexibility in the production processes in small plants, and it had negotiated a flat wage irrespective of task. The manager argued that job rotation functioned to broaden training and experience, as well as to distribute the most difficult or arduous jobs more evenly among the workforce. The firm was already noticing various improvements. Health-related problems associated with overconcentration on one task (or boredom) were reduced. Morale and interest were up, while absenteeism was reduced. Another manager remarked, 'Workers understand rather than being told.' In the new regime, workers were being drawn into the problem-solving process, both formally through government-sponsored upgrading and professionalization courses such as Norway's Reform '94, and informally through worker innovation. Individuals were encouraged to think about new ways of doing things, or they could be called upon if they had special skills such as carpentry.

Such improvements were a function of two factors. First, the small size of the firm facilitated personal interaction, informal communication, an absence of hierarchy, and, most of all, interpersonal trust. Second, the firm had a minimum technical capacity that was flexible in its applications and configurations. One manager remarked, 'Often it is ideas we get in the middle of the night. We can try them because we are so small. Simple changes ... adding or removing one thing, or changing the direction of the conveyor ... take ten minutes.'

Apart from the benefits of synergism and shopfloor innovation, one manager in Honningsvåg described the very real benefits from improved trust with his workers, what Gouldner (1964) referred to as the 'indulgency pattern.' Workers could be depended on without his close supervision. The manager remarked that for the first time in years he was able to spend time with his family. Thus, there are tangible benefits from being integrated into a tightly knit local community: 'We know what everybody is like, we know what they can do and what they cannot do. We know that we can trust them. Therefore we can let them go to work on their own not needing supervision. They all have keys to the plant, so that they can let themselves in and start working when they arrive in the morning.'

Quality Management

A market-oriented business approach involves emphasis on product quality. All of the fish processors that we interviewed, expressed a positive attitude on this matter, and in all plants concrete steps were being taken to improve quality performance. A good example of this is the Båtsfjordbruket plant which was the 1994 Norwegian 'fish plant of the year.' In presenting the award, Fisheries Director Mr Viggo Jan Olsen said: 'Today, Båtsfjordbruket offers itself as a modern and rational fish plant where quality control and management have been decisive in the choice of solutions. The plant's premises are a model for handling raw fish and the systematic enforcement of hygienic standards, orderliness, building standards and product quality.' Furthermore, he drew attention to the successful campaign for quality improvement that Båtsfjordbruket had launched towards Norwegian and Russian vessel standards (*Fiskeribladet*, 11 August 1994).

Plant managers realize that a market approach puts demands on the final product, the whole production flow, and, in fact, the total organization, including relations with workers. Thus, learning is an integral part of the process. In every Norwegian plant we were in, workers were encouraged to become certified fish workers. For this purpose, an educational program has been developed and standardized by the state for the entire fish-processing industry. It consists of three courses, one in microbiology and hygiene, one on issues of health, security, and the work environment, and a third on production and quality factors. The course work includes papers and a written exam. To enrol students are required to have at least two years of practical experience.

For the workers the certificate means more job security, a better understanding of the production process, and, thus, higher motivation. For management a more qualified workforce is a means of achieving higher quality performance.

The manager of Storebukt Fiskeindustri Ltd found that workers' training had reduced sick leave. Now, workers were more willing to work overtime, and they showed a more participatory attitude in improving work routines. Eight of the twenty-five workers in this plant were enrolled in the program. Storebukt's manager pointed to increased trust between the two parties, in large measure attributable to the course program for quality enhancement: 'My continual presence is not needed as much as before. Now I can go home and have free time, in the afternoon and on Saturdays, with full confidence that middle management and the workers will run affairs with competence and commitment.'

Others expressed similar thoughts. The production manager of Forsølbruket, a plant with one certified worker and twelve in the program, said: 'When workers acquire more knowledge and higher motivation they demand more from us, but so can we from them.' The manager of Nordkapp Seafood was hopeful that the certification program would give fish work more status and thus make it more attractive to local people, particularly the young who today regard this work negatively.

Quality management requires learning not only at the individual, but also the organizational, level. Thus, organizational development is part of the new concept of quality management. Quality circles that involve workers in a participatory manner are part of this strategy. In Storebukt Fiskeindustri, the quality circle had eight members who had been recruited from all levels of the organization. The owner–manager preferred to be an ordinary member, while a production foreman is chair. The group makes decisions regarding new equipment, ways to improve the work environment, and ways to enhance product and process quality. Thus, workers now have a channel of influence in addition to the plant branch of the labour union.

Relations to the Fishing Fleet

The kinds of relationships that exist between management and the workforce also extend to partners outside the firm. In contrast to the vertical integration strategy of Fordist and neo-Fordist firms, post-Fordist firms establish informal but cooperative relationships. One firm in Sommarøy, Troms, went so far during the crisis as to divest itself of its interest in a trawler, and it encouraged the local independent fleet to modernize and introduce new forms of fishing and fish storage. The inherent flexibility of the coastal zone fishery makes this possible. The local fleet responded, and it increased the capacity of its catch to more than compensate for the loss of the trawler. Faced by a quota reduction, this fleet became more modern and more efficient through close ties to one particular processor.

Unlike large firms, small fish processors do not have the capacity to handle large volumes of fish supply or to make a significant impact on their primary markets. They tend to be price takers, and to avoid key kinds of cost-related vulnerabilities associated with substantial overhead costs. Part of this strategy entails a *social investment* in ties with independent fishers. The return on this investment is in the form of social capital.

Social responsibility is an issue of constant concern among embedded fishers and fish processors. Since the end of the 1989 crisis, Norwegian fishers have complained that the availability of Russian cod has made processors less interested in the local supply. They also dislike prices, which are low because of the Russian landings, and the longer period before delivery. Fishers claim that fish processors prefer cod to other, less valuable, species such as saithe, when saithe is an important resource for fishers during the summer season.

In 1995 there were reports of processors rejecting fish from the coastal fleet altogether as they said they had enough with the Russian fish. To ease the problem, Russian and German factory freezing trawlers have been contracted by the Norwegian Raw Fish Organization for the possibility for sale over the side. In the spring of 1994 there were reports that processors broke a tacit agreement to buy the entire catch. According to *Fiskeribladet* (31 May 1994), processors apparently would only buy cod and reject other species.

When access to resource supplies declines for the coastal fleet, as we see happening now in Nova Scotia, the system of cooperation between fish processors and the local fishing fleet begins to break down. Processors start to see the costs of servicing their fishers outweighing the benefits of this form of input sourcing. One such small processor in Eastern Nova Scotia remarked to us that he could no longer justify the cost of icing boats or providing other free services such as baiting space and refrigeration when, under circumstances of local fish shortages and reduced size and yield of local fish, fishers were demanding higher prices. At this point the opportunity costs of contributing to make social investments came to be outweighed by alternative sources of supply such as frozen Russian groundfish, which is cheaper and has higher yields.

The question of maintaining ties to the fishing fleet represents a very real distinction between embedded and non-embedded firms, whereas in most other respects both are fairly dynamic. This distinction revolves around the decision to use alternative sources of supply even while local fish supplies are available. This is a significant dilemma for small firms in the midst of a crisis and faced with new opportunities. What effect the increasing distancing of these firms from their communities has on the re-embedding process remains an open question. In the case of Bugøynes, the community itself seized the initiative and established broad-based participatory institutions to facilitate the ongoing accumulation of collective social capital. In other instances, the re-embedding process will be stillborn as the gradual break up of these coastal communities begins to assume a neo-Fordist profile.

The difference between embedded and non-embedded firms stems from the degree of personal embeddedness of the manager. Non-embedded business decisions are far easier to reach where individuals are essentially outsiders, and/or where these individuals can use external constraints as a rationalization for their 'disloyalty.' These people, while dynamic and innovative entrepreneurs, are far more likely to be 'mavericks' and operate outside interfirm networks, except when it is in their immediate interest to do so. We will return to this distinction in the concluding chapter.

Interfirm Networking

For the more embedded post-Fordist firm in times of crisis, establishing close cooperative ties with external partner firms may be both a defensive and an offensive strategy. In defensive terms, firms that cooperate may save costs, exchange information, and provide support for each other, also at the individual level. Crisis can give rise to personal trauma that family, friends, and colleagues may help one cope with. Cooperation can also be an offensive approach, for instance, it may be an integrated part of a new marketing strategy. Rather than being involved in cut-throat competition that will eventually hurt everyone, fish processors may gain from collective action such as forming a marketing consortium, joint ventures, and technology transfer. In several communities that we encountered in Finnmark interfirm networking has increased, initially in response to crisis, but lately as a more permanent innovation.

The most institutionalized interfirm network in the Finnmark fishing industry is in Båtsfjord. In 1994 the local association of fish buyers had fifteen members – all fish processors and firms related to the fishery such as the local ship repair shop, the marine electronics supplier, and the local bank. They meet twenty times a year to discuss issues of mutual concern. Notably, not only matters concerning the industry are on the agenda, but also community issues. The association sponsors local sports activities, and it works for the improvement of the local infrastructure. A recent issue was the improvement of the airport which the government had threatened to close. Also, the association financed a search dog for the local sheriff for tracing narcotics which have become an increasing problem with the presence of Russian fishing vessels.

Projects directly related to the fishery are, nonetheless, the association's main concern. The plants jointly purchase various inputs such as boxes, containers, spare parts, freezing forms, and trucks. Fish too large for the filleting machines are sold to one member plant that has specialized in salt fish production. When this plant went bankrupt during the 1989 crisis, three other Båtsfjord plants rescued it to prevent outside buyers from coming in. With outside owners there was a fear that the plant could be used merely for transshipment of raw fish out of the community. The Båtsfjord firms were afraid that an uncooperative and 'hostile' outsider would ruin the informal norms that underlie their cooperation. For instance, the five fish plants in Båtsfjord practice the rule that they offer each other surplus fish before they sell it out of the community.

Collectively, as joint ventures, the Båtsfjord plants have established three subsidiaries. The first is the Miljøprosess, a firm that serves as the outlet for fish offal from all member plants originally (it was just another fish plant that went bankrupt in 1992). Silage and meal are produced at Miljøprosess. The second joint venture is a freezing and storage facility, the largest in north Norway, which also supplies local plants with ice. As was the case in the Hammerfest region, the Båtsfjord firms have formed an organization that runs courses for upgrading workers' knowledge. Better facilities are needed, and plans are ready for the expansion of the Fishers' Welfare Station for this purpose. One other example of the cooperative spirit among Båtsfjord fish processors is that they present themselves as a group at the annual fisheries exhibition (Nor-fishing) in Trondheim.

The association's most recent project was the formation of an export company - North Cape Export. For every kilo the member firms buy of Russian fish, they transfer ten øre (one tenth of a Norwegian kroner) to North Cape Export for the purpose of improving relations with Russia. For example, this money has financed the stay of children of Murmansk business partners in Båtsfjord. In addition to attending school, they work one day a week in the fish plants. The main purpose in forming the North Cape Export company is to facilitate marketing of niche products, and possibly also value-added products. It has taken more time than had been anticipated to get the company going. One problem has been the shortage of people qualified in export marketing. Another was the anticipated negative reaction of Frionor, through which most of North Cape Export's frozen products were exported.

The cooperative ties among fish plants make Båtsfjord unique in the Norwegian fisheries. Cooperation is multidimensional, inclusive, stable, and innovative. The plant managers are firmly convinced that as a group they are more effective this way both politically and economically. Together they can have an influence not only in the community, but also in government and on the industry as a whole. Many of the collective projects mentioned were initiated before the 1989 crisis and contributed to their survival through it. Of only three

Finnmark fish plants that did not go bankrupt during the 1989 crisis, two were in Båtsfjord.

Apparently, the crisis made the Båtsfjord processors more aware of the benefits of cooperation. Said the owner-manager of Båtsfjorbruket: 'Each of us is too small to solve all our problems. Through cooperation we can support each other. Cooperation also creates trust both outside the community and among ourselves. The Båtsfjord community wouldn't have survived if we hadn't worked together. In the future, we hope that other plants in the region will join our group.'

Solidarity is vital to interfirm networking and necessary if transaction costs are to be overcome. Transaction costs are particularly high in the fishery because of the uncertainties associated with resource supply, perishability, and competitive markets. Solidarity is also crucial at the community level because of the variability of production schedules. One manager said: 'We all need each other.'

A good illustration of the value of solidarity at the community level was demonstrated to us in Berlevåg. A young manager of one of the fish plants there was called by the neighbouring fish plant to fill in for their manager, who had just suffered a heart attack. In underwriting business transactions, the more informal the ties, the more crucial is trust. In Båtsfjord, one of the managers was especially explicit about this: '[By acting co-operatively through their buyers group] there is greater responsibility for each other and the community, than there would be if we were acting independently. We are too small to solve all our problems on our own. We need each other. Open-mindedness builds trust.'

In Finnmark, and elsewhere in Norway, interfirm networking has become part of the postcrisis business ethos, and marketing is one of several tasks suitable for collaboration. For some firms, like those located in the Honningsvåg area, interfirm cooperation is still in the initial phases, but managers are hopeful for the future. For example, they have established the Nordkapp Fiskerigruppe (North Cape Fisheries Group) for informal discussion of matters of mutual concern.

Some tangible results have already been obtained. The manager of Storebukt Fiskeindustri in Honningsvåg, a most zealous promoter of interfirm cooperation, and currently leader, of the managerial group said, 'If I have a skilled employee who can help some of the others out, they 'borrow' him for a few days.' To free themselves from brokers in Oslo and Stockholm, Honningsvåg firms have looked at the idea of forming a joint venture for Russian fish imports. A problem, according to Storebukt's manager, is that the other manager are not owners, but employees. Therefore, they have less latitude to discuss and commit themselves to cooperative projects than he himself has as owner-manager. This, in other words, is yet another characteristic of non-embeddedness.

The Flexibility of Embeddedness

In a welfare state like Norway, government institutions have replaced the paternalistic fish merchant – in Norwegian 'nessekonge' – as the provider of community services. Despite institutions like the Raw Fish Act, the municipal authorities, and the local fisheries adviser, there is an institutional void in the community. In the absence of deliberate planning and coordination, a well-functioning fisheries system requires mutual responsibility, local knowledge, familiarity, and trust among its members. To reduce uncertainty, each must understand each other's work conditions, demands, and future aspirations. A local owner is better situated to identify the needs of the system than is a foreign manager or an 'absentee' owner. Between the non-embedded owner-manager and the local population, myths and stereotyped images of the other easily develop. For members of the community outside owners appear occult, they fear that the foreigner has a 'hidden agenda.' Minimal support and enthusiasm coming from the community appear as indifference and indolence. The outside manager considers that importing labour is a solution, not a problem.

A local owner will have stronger commitments to the community than a foreign, 'absentee' owner. This is to be expected. From the perspective of the community, local ownership provides some guarantee of stability and predictability. From the perspective of the firm, this may be felt as a restriction that reduces its flexibility and ability to meet market demands. However, the production manager of Havprodukter also saw some advantages in local ownership:

Decisions can be made quickly. They don't need endorsement higher up, and this gives us flexibility. Also, a local owner-manager knows the community well and how to interact with local authorities. He must live with his decisions, which is good. It makes him more careful, for instance, regarding the working environment. We are here also to take care of the community. But hard decisions can be harder if you are a local guy – for instance, when closing down something that should be closed down.

Flexibility increases with social capital that is accumulated through embedded relations that are basically personal, reciprocal, and characterized by a degree of social solidarity. The production manager of Havprodukter was cognizant of these relationships. Referring to the owner-manager of his plant, he felt that being a member of the community implied a moral commitment: 'There is more than business [to consider] when he takes his decisions. We are also here to take care of our community. We are not sitting in Oslo.' He felt that it was beneficial to have to face everyday problems and the ramifications of business decisions, because this provided extra motivation to avoid problems in the first place.

Conclusion

The Finnmark experience offers an intriguing glimpse at a constructive future for coastal communities. In Båtsfjord and Honningsvåg cases globalization and recovery of the fish stock have revitalized the small-plant sector, the independent fleet, and the small coastal communities. Entrepreneurship is revitalized, and there is a greater sense of local independence, pride, and initiative pervading these communities. In Båtsfjord, globalization has been most noticeable in strengthening interfirm networks and cooperation in the areas of product specialization, marketing, quality control, and worker professionalisation. But the 1989 crisis has also made Båtsfjord fish processors more conscious of the importance of working together. Supportive state funding programs and university-based research links have also encouraged intermediate-scale expansion.

Although there has been plenty of evidence that the small-scale fishing sector in Nova Scotia shares many of the same characteristics as its north Norwegian counterpart, the dynamism that is apparent in the Finnmark case is not yet so evident in Nova Scotia. In interviews conducted in Nova Scotia in the summer of 1993, the same kinds of intergenerational succession were evident. However, great difficulties were faced by managers in modernizing technology to the degree necessary to take advantage of what few opportunities did exist. The resource crisis was the obvious factor underlying these problems, but isolation from either state or university research and public funding support was of greater significance. For instance, interfirm networking among small-scale fish processors in Nova Scotia is not part of prevailing business ethos and practice, as is true in some communities in Finnmark. In contrast to the large-scale fishing sector in Nova Scotia, small plants and their coastal communities have not been able to create a political climate that would legitimize a coherent program of state support.

Given what has been said in previous chapters of this part about the complicity of public policy in the disembedding process, what can be said about a more adequate response by the state to the re-embedding of small-scale fisheries and coastal communities? What can government most effectively do to improve the conditions for sustainable communities? Support for educational and training opportunities that make individuals and firms more flexible and able to diversify is paramount. Interfirm network support, particularly in the areas of business services, international marketing, and transportation would be well received, and, finally, support for improved self-reliance through community organizations and strategies of economic diversification is very important.

But is the new post-Fordist firm able to turn around the process of community decline? While there seemed to be a clear connection between neo-Fordist firms and community disembedding, the connection between the post-Fordist firm and the process of community re-embeddedness is far more complex. These questions take us much further afield. For example, certain communities are advantageously placed regarding transportation networks and markets, and they have become sub-agglomeration points within the wider rural landscape, while other communities continue to wither (Robinson, 1990). This point is also argued in the flexible specialization literature to be a necessary factor (Scott, 1988). In the context of a general economic recession in nearby urban centres in Troms, for example, we found evidence of a return rural migration among the most mobile segment of the workforce – young school-leavers. This development is qualitatively different from that which has characterized community restructuring based on Fordist growth poles and resettlement in the past. That trend was based in the transformation of large segments of the rural countryside based on numerical flexibility and a centralized Fordist production regime. The current trend is based on independent small processors operating in a competitive market and structuring production relations around functional flexibility. This is a process far more in keeping with the integrity of traditionally embedded patterns, and, in our estimation, far more sustainable.

The next chapters present three fishing communities in north Norway, Newfoundland, and Nova Scotia: Bugøynes, Fogo Island, and Sambro. We focus on how the fisheries crisis has affected each of these communities, and how they have struggled to ride out the storm. In the Norwegian case, Bugøynes, the crisis seemed to be over until it struck again. In the meantime the community had changed in some fundamental ways. As the fisheries crisis is still lingering on in Canada, the fate of Fogo Island and Sambro remains uncertain. Despite this, the two communities are struggling on, and they display impressive fortitude and resilience.

13

Bugøynes: A Case Study of Community Resistance[1]

During the cod crisis few fishing communities in Norway caught the attention of the press as did Bugøynes. For some days in August 1989 Bugøynes was headline news, and not only in the Norwegian media. Reporters from Finland and Sweden went there. *Le Monde* of Paris sent a journalist. A radio reporter even came from across the Atlantic, from Canada. And, the Bugøynes case was discussed in the Norwegian Parliament. How did Bugøynes come to deserve this fame? It was created by an advertisement placed in a leading Oslo newspaper, *Dagbladet*, headlined with a simple but rather peculiar question: 'Will Someone Accept Us?'

Is there a place in Norway that will welcome an increase in population of about 300 people? We ask as citizens of the fishing community Bugøynes in eastern Finnmark who now are fed up. The last few years have been a constant struggle to maintain the settlement. The reason is a fisheries policy that failed. Among other things, it led to the bankruptcy of the fish plant – our cornerstone firm – in 1987. In the two years since then no one has been able to get the plant started again. This is because of bureaucratic clutter and lack of will among bureaucrats and politicians. (In fact, they are still arguing about who owns the plant.) Now we feel it is time to put everything behind us, and start again somewhere else. We want to avoid becoming burned out and worn out in our struggle for existence for no purpose. We want to use our strength in a community where we can work for a future for ourselves and our children. We want to move together as a group – solidarity among the people of Bugøynes is strong!

The adult part of the population has a mixed professional background. With our competence and go-ahead spirit we have much to give. We would be bringing 50 children.

---

[1] Oddvar K. Nilsen helped to collect the data presented in this chapter. He is completing his Master's thesis at the Institute of Social Science, University of Tromsø.

We are interested in moving south of Trøndelag. Even if there are difficulties in providing jobs there too, we won't let that scare us. We can help in creating new jobs.
Correspondence should be sent to 'Action Bugøynes, 9935 Bugøynes.'

The message was extraordinary, but clear. Even for a crisis-ridden region, this was special. A whole community wanting to leave voluntarily, and as a group, was a novelty well suited for media coverage. 'The story of the Bugøynes community is the chronicle of a group of people that feel betrayed by the authorities and businessmen from the south,' the *Dagbladet* wrote on 17 August, a couple of days after the ad had appeared. What led to this situation? What caused the people of Bugøynes to publicly announce their desperation? Let us go back into the history of this community and describe some of its more recent developments.

Short History of Bugøynes

The community is situated on the eastern shores of the Varangerfjord, almost as far east as one can get in Norway. On a clear winter night one can faintly see the lights in Vadsø, the county capital of Finnmark, across the fjord. Before Bugøynes was connected to the highway by road in 1962, Vadsø was the link to the outside world. Afterwards Kirkenes became a more important connection. The mining industry in Kirkenes also had jobs to offer, and with the new road it became possible for some of the residents of Bugøynes to get employment there.

From Bugøynes to Kirkenes, the municipal centre, is 100 kilometres, which is too far for commuting, at least on a daily basis. Thus, for most people, a job in Kirkenes meant moving away from Bugøynes. Despite that, the population was fairly stable at a level of 420 people until 1980. In the following decade Bugøynes lost 18 per cent of its inhabitants. In particular, young people left.

There is no other fishing community in Norway as far east as Bugøynes, and the distance to the Russian border is not great. But Finland is even closer (60 kilometres), and this connection has made a much stronger impact on the community. From the 1860s to the early 1880s Bugøynes was settled by Finns who fled to find refuge from hunger and poverty. Before that the community was mainly used seasonally as a fishing station for Norwegian and Saami fishers. Even today Bugøynes is considered a Finnish community, and it is popularly called 'Pikku Suomi' (Little Finland). Most of its inhabitants are descendents of the early immigrants, and Finnish is still spoken daily among 60 to 70 per cent of the people, which is higher than in any other community in north Norway where Finnish immigrants have settled. The Finnish heritage can also be observed in the architecture, as Bugøynes was one of the very few communities not razed when the Germans retreated in 1944.

The Fishery

The fishery was always the main employer in Bugøynes. The Varangerfjord is abundant with fish, and the Bugøynes' fishers did not need to travel far. This can be seen in the fleet structure. In the 1960s, there were fourteen or fifteen boats around thirty-five feet, but from then on fishers switched to smaller boats more suitable for a home-based inshore fishery. These boats go as far as the Russian border, starting with gillnets for cod in January until May, when the salmon trap fishery opens. In the fall cod is the main species and handline the main gear. The salmon fishery has traditionally been very important economically as well as culturally. The fishers of Bugøynes attribute much of their identity to this particular fishery. For generations, during the summer months they have fished from cabins that they possessed farther east in the Varangerfjord.

The larger boats did not participate in the salmon fishery. They would also go after species other than cod, like shrimp, which has been a major source of income for the fishers as well as the local fish plant. These vessels would also target halibut and haddock, particularly in the fall season. The fish plant, however, would also rely on outside vessels and buy capelin in late winter and whale meat during the summer, but cod still accounted for 80 per cent of its purchase. The local fleet concentrated 75 per cent of its landings of cod in April and May, which meant that other supplies were essential to a steady production at the plant year-around. For this purpose, in 1976 the plant acquired the trawler *Bugøyfisk*.

Crewed by non-locals only, the trawler played a crucial role in sustaining the fish plant. In a normal year, it supplied 60 per cent of total landings, and provided stable employment, particularly for local women, for whom the plant was the only job alternative outside the household. In the late 1970s, 87 persons worked in the plant, 57 of them full-time. The plant was owned by a local until it went bankrupt in 1960. Since then, and until recently, there have been outside owners. External control of the fish plant potentially meant tensions in relation to the community, and these tensions eventually reached a climax with the newspaper advertisement in 1989 when unemployment was at a peak 50 per cent.

In the late 1970s prospects looked good for Bugøynes. The fishery was going well, the plant was running full-time, and the number of inhabitants had never been higher – 425 people in 1980. People were optimistic.

Bugøynes belonged to the most successful of the Finnmark communities. But quickly and unexpectedly things turned worse, and in the 1980s they became disastrous. In 1978 the first signs of a seal invasion appeared, and Bugøynes was among the very first places that noticed its effects (Eikeland, 1993). The seals arrived in March. Some fishers say they came with the capelin, others say it was with the pollock. Seals got caught in the gillnets. The larger vessels snared up to 120 seals in one hauling. Then, after a few weeks the seals disappeared as swiftly as they had arrived, only to return the next year, and for many years thereafter.

For the fishers, the seals created severe damage: they destroyed the gillnets, ate the fish in the nets, and scared the fish into deeper waters. A coastal vessel that used to catch sixty to seventy tonnes of cod during the winter and spring fishery, now only captured two to three tonnes. The small boats had a similar experience. After a while, the cod disappeared totally. Fortunately, the shrimp fishery boomed and became an alternative, particularly for the larger vessels. The small boats changed to tusk. Some even bought stronger nets and started to catch seals for sale to the Norwegian Raw Fish Organization. A few also left for the Lofoten fishery, something uncommon for Bugøynes boats. But most boats tied up at the wharf waiting for better times. According to Eikeland (1993), 67 per cent of all registered boats eight metres or more were not in use in 1981, with 28 per cent fishing less than thirty weeks, and only 6 per cent more than thirty weeks.

Bankruptcy

The trawler *Bugøyfisk* prevented the seal invasion from having major repercussions on the fish plant because it still landed shrimp. Nevertheless, the company that owned the plant went bankrupt in 1985, and then problems really started for Bugøynes. Several bidders, who were more or less serious, showed interest in the plant. A company in Måsøy, a fishing community in the western part of Finnmark, won the bid, and thus took over the plant including the trawler. There was a requirement that the plant and the trawler had to be sold as one entity, but the latter was apparently the most attractive item of the two.

The *Bugøyfisk* was one of the rare vessels with a cod licence as well as a shrimp licence. According to the newspaper *Sør-Varanger Avis* (13 June 1985), the new owner paid approximately Nkr 18 million for the trawler and around 2 million for the plant.

After a couple of years, the plant was bankrupt again, which came as a surprise to the people of Bugøynes. The previous owner managed to retain ownership of *Bugøyfisk*, as the Ministry of Fisheries allowed the trawler to be separated from the plant on the condition that the vessel would continue to land its fish in Bugøynes. The fish plant, however, went up for auction. For three years there was almost no activity at the plant. The delay was partly because the trawler owner refused to fulfil the obligation to land in Bugøynes, while he disputed the interpretation of the clause within the contract. In the meantime the bankruptcy administrators and the municipal authorities worked hard to find a new plant owner. But with the uncertainty related to the trawler, they unavoidably failed. For this reason the Frionor Polar Group, among others, rejected the offer to become the new owner. The municipality of Sør-Varanger also declined the opportunity.

For the people of Bugøynes the situation became increasingly frustrating. 'Shall Bugøynes be allowed to live?,' the Bugøynes Labour party members asked in an open letter to the Minister of Fisheries (*Sør-Varanger Avis*, 14 April 1988). 'If the government says no to Bugøynes, it must say it loudly.' People felt that they were being left in the dark and claimed that they were excluded from information on what was going on, that negotiations were kept secret, and that there seemed to be a tacit agreement among politicians that Bugøynes should die. The mayor of Sør-Varanger declared that the municipality was powerless.

Things remained unsettled until the summer of 1988. Then a fruit company owner in Fredrikstad, southeast of Oslo, took over. With no experience in fish processing, he wanted to use the plant to produce salt fish. To compensate for the lack of local supplies, he intended to buy fish from Iceland and Greenland. However, the new company appeared to be a castle in the air: The plant was in operation only a few months before the employees were laid off, and the owner started to sell off the processing machinery.

The people of Bugøynes feared that the plant would be stripped of all its equipment and that they would then be out of work permanently. When the truck came to pick up the newest machine, people decided to act. They blocked the road, and after a couple of days the truck had to leave empty. Again, the company went up for auction. The fruit company owner was again the highest bidder, but this time the State Fishers's Bank intervened and accepted an offer from three local bidders.

Collective Action

With the development committee Action Bugøynes, which sponsored the advertisement in the *Dagbladet*, the community had speakers and a pressure group. The development committee also took part as a mediator in the effort to realize the local takeover.

Action Bugøynes was formed in 1987 by local residents 'with the purpose of fostering industry and the well-being and prosperity of the people.' Prior to this four local people got together and wrote a letter to all out-migrated citizens of

Bugøynes and asked them for ideas on how to initiate community development. In a public meeting with representatives of municipal authorities present, the committee was formed, according to its charter, with the aim of improving 'the utilization of resources and realization of potential within the fishery, aquaculture, manufacturing and craft industries, tourism, and other existing and potential industries in Bugøynes,' in collaboration with local enterprises, government agencies, and other community organizations. Thus, in 1989 the committee was involved in making a development plan for the community of Bugøynes.

The initiative for the notice in the *Dagbladet* came from leading figures in Action Bugøynes. They did it, but without consulting the membership. Some of the members were sceptical when the news came out, but many gave it hearty support. 'If this is what it takes to wake up the government, then let's do it.' The mayor of Sør-Varanger perceived the ad as 'a cry for help,' and told the newspaper *Sør-Varanger Avis* (17 July 1989): 'As I know the people of Bugøynes feel good about their community and don't want to leave.' The journalist who covered the story for *Dagbladet* called up county prefects all around Norway, and found that the people of Bugøynes would be welcome everywhere. However, as time would show, nobody in Bugøynes accepted their invitation. Instead, the people of Bugøynes decided to stay and ride out their difficulties.

It took another year before the fish plant was on its feet again, but since then things have improved sensationally. 'Look at Bugøynes,' county government officials stated, when we interviewed them in Vadsø in May 1994. By then news reporters could attest to the optimism prevailing in the community.

The plant was running again full-time, and the advertisement in *Dagbladet* that put Bugøynes on the map, was attracting many tourists. Two hundred tourists visited Bugøynes in 1989, and in 1994 – 2000! Bugøynes received visitors from Finland, as seems natural. A small café, cabins for rent, and a site for campers were established. Even an authentic Finnish sauna was built for the visitors. People opened their homes to tourists. Organized boat trips for bird-watchers became a summer tourist attraction. A seminar of Finnish artists was held in the community, apparently with great success.

Bugøynes has tried to capitalize on its Finnish heritage, which people in the community see as holding great potential for the growth of heritage-based tourism. In the development plan of 1989 a great number of project ideas were listed, but the natural environment is vulnerable, and the people of Bugøynes feel it is important to keep tourism under control.

Diversification

Industrial diversification has taken place in Bugøynes. At the time Bugøynes hit the wall, in 1989, its industry was already fairly mixed. It had a small shipyard – Varanger Maritime Ltd – that had originally grown out of a local boat-building tradition initiated by the Finnish settlers. Today, with six people employed, its main activities are boat repair and the development of new boat designs, in addition to mechanical products for the fishing industry. Even during the seal invasion, the yard was fully occupied, and thus it compensated somewhat for the fishery in decline.

Another plant, Varanger Salmon and Game, established in the early 1980s, employs fifteen people. This firm purchases reindeer and salmon for value-added production aimed mainly at the Finnish market, and it remained largely unaffected by the resource crisis. Thus, it could maintain jobs during that difficult period. A new firm, Varanger Steinindustri Ltd, established in 1989 by two newcomers to Bugøynes, exploits local mineral resources for various stone products as building material designed for the European market. Five people are employed there, although only one from Bugøynes.

The school and kindergarten together employ eighteen people. With twenty-five people working at the fish plant, twenty full-time fishers plus a few part-timers, and two people employed at the food store, four at the fish farm, one taxi-driver, a husband and wife running a small company called Sea Safety that sells building equipment, one free-lance journalist, three health care workers, and two people the cleaning business, the Bugøynes labour market is diverse. There is a shortage of jobs for women, particularly because the fish plant now employs fewer people than when the production of cod fillets peaked in the early 1980s.

What happened to the fish company after the local owners took over is an interesting story, but also a sad one. The plant has been renovated physically, partly with support from the Norwegian Raw Fish Organization, and has all standard government quality requirements. It has participated in the ISO-9002 quality management program. The wharf has been repaired, and the baiting facilities have been modernized. As an organization, the new company consists of three independent subunits: (a) Arctic Property, which owns all facilities; (b) Arctic Seafarm, which runs the fish farm; and (c) Arctic Products, which processes the fish. With this organizational form the new owners had hoped to avoid the possibility that any future bankruptcy would cause a domino effect, bringing all the community's economic activities to a halt. Unfortunately, as we will see, things did not turn out as hoped when a new crisis struck in the winter of 1996.

Total sales of Arctic Products in 1993 amounted to Nkr 30 million. When they took over, the new local owners were already involved in salmon farming as well as grouse drying. They were met with some scepticism in the community in the beginning regarding local ownership, but doubts faded when things started to work well for the new company. Some of the machinery that had been sold was replaced, and the plant got a filleting and a salt-fish line. The production profile changed from frozen fish blocks to greater emphasis on fresh products, in line with the general trend in Finnmark fish processing.

The new management followed a strategy of wide product mix. 'We want to become a "total dealer" of fish,' the manager said to *Sør-Varanger Avis* on 2 February 1990. 'The customer who asks for fish also wants shrimp, redfish, and halibut.' Futhermore, the marketing strategy did not stop with the wholesaler. 'For us, the sale is not completed before our customers have sold the product further on.' The firm succeeded in establishing an export business. Here, they exploited their Finnish background and language skills, as Finland became the most important market.

But the management was careful not to overexpand. Consolidation of prevailing activities had a higher priority than hasty growth, even if the supply of fish looked good at the time.

Globalization

Markets for Bugøynes products included all the Nordic countries, Germany, even destinations as distant as Hong Kong, and once a week, a shipment of smoked salmon went to Miami. Besides salmon, the main products were cod, redfish, shrimp, and crab. Crab is a new product and a recent species in Finnmark waters originating from a Russian research project in ocean ranching in the Kolafjord in the 1960s. Particularly numerous in the Varangerfjord, the 'kamtschatka' – or king – crab was a main reason for the plant's success, and Arctic Products was the only firm in Norway with a producer licence.

Restaurants in Oslo, Finland, Belgium, and Germany tested this new product, and England and Hong Kong expressed interest in it. With a price of Nkr 150 per kilo, it has the potential for becoming an important product supplement for Bugøynes and possibly other fishing communities in the area (if properly managed). Norway and Russia share the crab quota, and Arctic Products has signed a contract with the Russians to buy a portion of their catch. Two Bugøynes vessels were participating in this fishery, and they were using offal from the local salmon farm as bait.

The trawler *Bugøyfisk* was lost in the last bankruptcy, but the plant bought shares in another trawler based in Båtsfjord in order to obtain a steadier supply. However, in 1995 salmon played a major role. The plant was in fact almost self-sufficient, as 250 tonnes came from the company's own farm, and another 50 tonnes were purchased from individual fish farmers in the area. Salmon was exported either fresh, filleted, smoked, or marinated, that is, as 'gravlax.' 'As much value added as possible is our policy,' the managers said, and 'all filleting is done by hand. This policy applies to all fish species, and the main reason is to create maximum employment within the limits of profitability. We provide what the customers want, and if we don't have the fish ourselves, we buy from other plants.'

Arctic Products had a contract with a Båtsfjord plant to buy redfish. One specialty product was a delicacy 'gift package' that included smoked and marinated salmon (gravlax), mustard sauce, shrimp, and crab.

All in all, prospects for the plant looked good. The sales manager of Arctic Products, also a native of the community, told us: 'Now, Bugøynes is blooming as never before. At times, we have difficulties in providing enough labour to the fish plant. If things develop as we hope and believe, the bad times of Bugøynes should be gone for ever (*Vårt Blad*, September 1994).' His optimism was shared by most people of Bugøynes.

The state of the fish supplies was not the only positive factor. The plant was diversified, self-supplied, and flexible. Because of the ethnic closeness and physical proximity to Finland, it had its special market niche. The special structure of the firm should also have provided more resilience from internal and external turbulence. The local ownership ensured a management attached and committed to the community and to its future development. Together with a cultural cohesiveness that exceeds most other fishing communities in Finnmark, Bugøynes stands out as a community where economic activities are closely embedded within the community.

Bugøynes was in a better position than many other fisheries-dependent communities in Finnmark to withstand another resource crisis. First, the supply of cod no longer had the importance it used to have. Second, the community had jobs outside the fishery, in boat building and repair, the stone industry, and tourism. As long as these activities could be sustained, there would be private and public service jobs such as within the school, the day-care centre, and the post office. There would also be a demand for the grocery stores and café. In the 1989 development plan, several projects were identified that would make Bugøynes a better community to live in. To mention a few, the planners saw a need for a petrol station, a bakery, a hairdresser, an accounting bureau, a service station for fishers, a home for the elderly, and a community hall. Thus, for Action Bugøynes and its recently hired full-time manager, there were many things to become involved in.

The turnaround of Bugøynes did not come easily. The community was not simply rescued by the improvement in resources that helped so many other communities in coastal Finnmark from 1992 onwards. The people of Bugøynes had to fight for their survival. The advertisement in *Dagbladet* was a clever idea. After that, nobody could ignore their demands. But it also served as an eye-opener to the Bugøynes residents themselves. It made them realize that the destiny of the community was in their own hands. They spoke up and struggled, they organized and diversified, they were wise enough to draw on their human and natural resources.

Bankrupt Again

In March 1996 news was released that Arctic Products, Arctic Seafarm, and their mother company, Arctic Property, had gone into receivership. 'Unbelievably regrettable,' the leader of Action Bugøynes said in reaction. To the people of Bugøynes it came as a surprise, even though there had been lay-offs the previous fall. There was fear that a standstill would lead to a loss of customers, and that this would be a problem for a new firm. Crab was still good business and had a profitable future, and prices on salmon products were on the rise. Extremely low salmon prices, together with an accident at the fish farm, were two of the factors that led to the bankruptcies. With considerable liquid assets still in the firm, such as the live salmon in the pens, the situation did not look all bleak. Therefore, the committee responsible for the bankrupt firm decided to keep it running until new owners and management could be found. Thus, in the interim, at least some of the employees kept their jobs. Rather than locking the door and leaving the community, as Foodmark did in Mehamn, the manager hung on, fully committed to solving the plant's problems. The union representative in Arctic Products told the local paper (*Sør-Varanger Avis*, 16 March): 'Jarle [the manager] is pushing forward and works hard day and night for the rest of us. He has tried and never given up.'

This time there will not be an advertisement in *Dagbladet*. 'Now we must hope for the best and think positively,' the leader of Action Bugøynes said.

14

Fogo Island: A Case Study of Cooperativism

The islands off the northeast and east coast of Newfoundland offer access to rich fishing grounds, and hence these were the first areas of Newfoundland to be settled by English (and, briefly, French) in the seventeenth and eighteenth centuries. Fogo Island, and its town of Fogo, was one of the centres of the Newfoundland fishery for more than three hundred years (see map).

'Fish' means cod in the Newfoundland vernacular. The northern cod stocks that range from Labrador to the Grand Banks of Newfoundland and overwinter and spawn off the coast of southern Labrador were the centre and mainstay of Fogo Island's fishery. But the development of viable communities depended on the existence of a more diversified economy (Head, 1976), where cod fishers could also fish for salmon and lobster, and for herring, mackerel, and capelin bait fishes; where they could go into the woods in the winter months – when the bays and coastal waters of Newfoundland and Labrador freeze up, and the fishery ends – and log for their own use or for lumber and pulp and paper companies; where they could participate in the spring seal fishery. Families also tended subsistence gardens and livestock, and when it was necessary or possible people went off to the United States or Canada to work. Birth rates were high and so was out-migration. People did come back, however, and they could do so, because access to the fisheries was open. Accordingly, for almost a century, the population achieved something like a steady state (McCay, 1976).

This rocky, barren outcropping provides a case study in the social construction of a sustainable fishery-dependent system, and in the problematics of the resilience of such a system to crisis. The term 'problematics' is appropriate because it is not at all clear whether or not the Fogo Island system will make it through the current crisis.

Fogo Island is the site of ten coastal settlements organized into seven municipalities or rural development districts. The population hovered around 4,000 for a century. However, it had gone down to an estimated 3,500 by 1995 because of the decline of the northern cod fisheries, which began in the late 1980s and resulted in a closure of all fishing for that stock, as well as many others, in 1992. About twelve kilometres off the mainland of Newfoundland, an unreliable ferry ride, and about 120 kilometres from any sizable town (and hence alternatives), Fogo Islanders are almost totally dependent on coastal fishing and fish-plant work. This dependency, together with an exceptional degree of embeddedness and particular constellations of leadership, vision, and opportunity, have given this island community its shape and the directions its members are taking in the context of extreme ecological and economic distress.

Fax, phone, computer lines, cable TV, trucks, ferries, and airplanes link Fogo Island to the rest of the world. So do unemployment, marginality, and despair, particularly in the wake of the disappearance of the mainstay, northern cod. And crisis it is. The options of diversification and mobility that worked so well in the past (see McCay, 1978) are no longer available, at least not in the same forms as before. Diversification once meant fishing for different species, with different gears. Now it has to include product diversification within seafood processing, as well as the development of income-producing opportunities beyond the fisheries. Mobility once meant moving to other fishing grounds, as well as migrating elsewhere for work. Now the option of finding other fishing grounds is narrowed by government-limited access programs and by the fact that failure of the fishery is widespread. Opportunities in urban areas, as well as other sectors of Canada's economy, for people with little formal education and few skills beyond fishing are diminished by nationwide and global recession.

An Earlier Crisis and Development of Present Institutions

Only in the 1950s and 1960s did places like Fogo Island come to seem marginal to the economy of Newfoundland. With the building of roads trucks began to replace ships as critical links of commerce, and islanders became isolated and increasingly vulnerable. Moreover, competition for fish changed to include foreign fleets of longliners and trawlers, technological light years beyond the simple, small-scale, but effective inshore fishing technology of the locals – effective as long as the northern cod migrated inshore in substantial numbers, which they stopped doing because of heavy fishing offshore. By the late 1960s stocks of inshore fish (which still meant cod) were unreliable and increasingly scarce. It was at that time that Fogo Islanders came into the public limelight and developed institutions critical to their survival then and now.

The islanders became famous for resisting the temptation to accept support from a government program to resettle, choosing to redevelop instead. The

Fogo Island: Cooperativism 309

resettlement program of Newfoundland was a dramatic and often wrenching challenge to rural Newfoundlanders (Copes, 1972a; Matthews, 1976). How the Fogo Islanders resisted it is a long, complex story, involving education extension workers, clergy, local merchants, politicians, and fishers. It also involved, and was recorded by, innovative documentary filmmakers who used the 'Fogo process' as an experiment in film as a medium for communication and hence rural development (Wadel, 1969; H.A. Williamson, 1975).

Resistance depended on the embeddedness of economic activities within the culture and social relations. Fogo Islanders tend to define themselves in terms of community, by which is meant shared experiences and memories, kinship and genealogies, and place. The decision not to resettle required expanding the boundaries of community, which had been limited to the individual small settlements, or outports, as they are known in Newfoundland, and even neighbourhoods within the outports, riven as they often are by differences in religious, ethnic, and other sources of identity and affiliation (DeWitt, 1969). Accordingly, there is now a sense of Fogo Island as a community, enhanced by the creation of a central high school in the early 1970s, which has since been expanded to all grades.

The decision to redevelop instead of resettle was also predicated on the creation of an islandwide fishing cooperative, which became another instantiation of the newly defined community. Withdrawal of the resident fish merchants in the 1960s, which worsened the crisis of that time, opened the door to the creation of a large-scale cooperative. This was not entirely accidental or fortuitous: Fogo Islanders had a long if chequered history of experimentation with fisheries, consumer, and credit cooperatives, periodically challenging the monopoly position of embedded and non-embedded fish merchants over both production and consumption (McCay, 1976; Carter, 1988). The islanders finally had their chance in 1967, when they formed the Fogo Island Producers and Shipbuilding Cooperative.

Once scheduled for resettlement, the island's fishery economy turned around in the late 1970s and the 1980s. The success story for Fogo Island – like other eastern and northeastern Newfoundland fishing communities – lies in diversification as well as cooperation.

The cooperative began by building 'longliners,' 35-foot to 55-foot gillnetting vessels to enable more extended fishing seasons on this patently seasonal coast (local and Arctic ice often lasts until May or June, and the fishery is usually forced by storms to end in October or November). The longliner fishery began with cod, but it diversified to other species as soon as the new cooperative could find markets. Cod was the major species landed by volume, but as early as the late 1960s, with the development of the longliner fishery, islanders had begun to bring in other groundfish species, such as halibut, turbot (Greenland halibut), yellowtail flounder, and American plaice. Moreover, the island's economy also included lobsters, salmon, and, in the winter and spring, seals. Pot fishing for snow crabs (sometimes called Queen crab) emerged as important in the 1980s. Even the inshore fishery experienced some diversification, with the emergence of good markets for pelagics such as herring and capelin and for lumpfish roe, which is salted in barrels for the caviar industry.

The Crisis of the 1990s

Today, however, there is very little opportunity for diversification, and the staple has shifted from northern cod to snow crab. The seal fishery (for harp seals) continues, but it is not very profitable because antisealing activists have destroyed most markets. The commercial fishery for salmon in Newfoundland ended in the late spring of 1992, because of pressure from sports anglers and environmentalists. This immediately preceded the decision by the Canadian Minister of Fisheries to close the northern cod fishery in a moratorium that was to last two years and is now scheduled to continue for an unspecified time, possibly into the next millennium.

None of the other groundfish species is abundant enough in local or regional waters to support a fishery. In the summer of 1994 the only fisheries taking place by islanders were a small-boat inshore gillnet fishery for lumpfish (for the roe, made into caviar), a small-boat inshore fishery for lobsters, and a large-boat offshore fishery for snow crab. The crab fishery was the only one where people made money. Lumpfish and lobsters were very scarce. A hoped-for fishery for small capelin never took place because the fish did not arrive in sufficient numbers and sizes. The offshore gillnet fishery for turbot had only a few participants because of fish scarcity and management measures. The crab fishery lasted for only five weeks, because of quota limits. The lobster fishing season ended 15 July. From that time on, the harbours of Fogo Island have been very still. The groundfish plant, run by the Fogo Island Fisheries Cooperative, did not open and the cooperative's crab plant provided only seven weeks of work.

Islanders are receiving some financial support from the government. A small group – 35 boats, about 175 fishers – are licensed for the crab fishery, and they participated in 1994, making good money – enough to qualify for fishers' unemployment insurance and more – but only for four to five weeks. About 300 people worked at the crab plant in 1994, but only for five to seven weeks, not enough to qualify for unemployment insurance. In this highly seasonal economy, wintertime unemployment compensation has become crucial, and the majority of people – excluding the crab fishers – did not qualify. However, virtually all island households include men or women, or both, who are getting compensation for not being able to fish for or work in fish plants with northern cod because of the moratorium. This money is locally called 'the moratorium,' or 'the package.'

The current moratorium program, begun May 1994, is called the Atlantic Groundfish Strategy (TAGS). It is intended to provide support for participants in the fishery (fishers and fish-plant workers) for up to five years. It follows the previous fisheries support program, which was to last for two years with expectations that the cod fishery would resume. The programs are similarly structured, indexed to previous unemployment compensation earnings, and with provisions and requirements for education and training. Both are intended to help individuals adjust to decline in the fisheries in ways that will result in a downsizing of Newfoundland's fishing industry. Most public statements refer to a downsizing from about 30,000 people to about 6,000 or 7,000 people in Newfoundland.

Fogo Islanders are thus 'getting by' with little or no work to do on the water and in the fish plants, and in the context of support that carries with it the message that many will not be in the fishery in the future. An important question is how islanders perceive and are preparing for that future. Will there be adequate individual, household, and collective responses to the challenge?

Marginality

Fogo Island is somewhat marginal to the fisheries and community development policy, being small, distant from the provincial capital, and non-unionized. Fogo Island is exceptional in that its fisheries are largely controlled by a cooperative, rather than private enterprise, with the result that neither fishers nor fish-plant workers are unionized.

Fogo Island's political isolation is countered in several ways. First, fishery cooperatives have status, as such, on various government advisory committees and similar fora and thus have a voice. (The Fogo Island cooperative is part of an organization of fishery cooperatives in the province – there are three – which is in turn part of a general cooperative organization.) The fish cooperatives have smaller roles and less power than the Fishermen, Food, and Allied Workers (FFAW) union. On the other hand, they have more flexibility because they are not hamstrung by the need to coordinate policy within a centralized union. Second, islanders can use their political representatives in the provincial and federal legislatures to help them accomplish their goals. Third, some Fogo Islanders have emerged as effective leaders and spokespersons within the region and beyond, in particular Bernadette Dwyer, who has developed an international reputation in community and fisheries development.

Marginality is not all bad. The fact that Fogo Island is not part of the unionized system has also meant that experiments in the inshore fishery could more easily take place, and that it could experiment with innovations in processing. For example, Fogo Island has been the site of an experimental program in dockside grading of fish, as well as more specialized programs of quality enhancement, intended to provide incentives for higher quality handling of fish by harvesters and processors. The FFAW union has long negotiated off-vessel landing prices on what is known as the 'tal-qual' basis, whereby there is no grading of the fish, it is taken 'as is' and for one price. Although this system does away with the problem of cheating by cullers – a long-standing issue in the fisheries – it has reinforced the tendency for Newfoundland to produce low-quality, high-volume fish products, mostly notably frozen cod blocks for the U.S. market. Gaining more value locally and, for the future, re-entering the groundfish trade will require a shift from a volume orientation to a value orientation. This is a central argument of the Fogo Island cooperative in its own planning for the future, and key to it is the ability to engage in dockside grading, to this point impossible for unionized plants.

The Cooperative and the Community

The cooperative has two personae, two lives as it were: as a business and as the economic underpinning of a community. This has been shown clearly in the management of labour at the cooperative's fish and crab plants, where a productivity-based system of hiring, firing, and compensation exists in uneasy tension with and in contradiction to practices – backed by strong community ethos – favouring the needs of workers for unemployment compensation, which in this context means the needs of the larger community for survival (Carter, 1988; McCay, 1988). In other words, while the managers of the cooperative would like to hire and fire on the basis of productivity (with a measure of seniority), they must also acknowledge the needs of workers to reach their minimum requirements to qualify for unemployment. Beyond that they must deal with strong pressures for 'job-sharing,' letting one person go, once that person qualifies for unemployment, so that another may be hired on to be given a chance to qualify.

Thus, the embeddedness is two-way: neither the business nor the community-welfare components of the cooperative can be free of the demands, pressures, subterfuges, and complaints of the other. Those who demand their 'rights' to unemployment benefits must acknowledge the fact that if the cooperative is to provide jobs for them at all, it must try not to lose money, let alone make a profit. And the business-oriented managers know, from their board of directors and from the many people who call them on the telephone, that they cannot run a profitable business without the cooperation of the community.

One of the important institutional constraints for the Fogo Island cooperative is the principle of voluntary participation. In the Western 'Rochdale' tradition, a cooperative cannot force people to participate. Another related constraint is the principle of 'equal treatment,' a powerful statement about what equity means generally, and especially in a cooperative. These two institutional principles have made it difficult for the Fogo Island cooperative to stop members from defecting to other fish/shellfish buyers, for whatever reason, and, as important, to offer special terms (such as free ice), to entice them back unless it can offer those terms to everyone.

Ironically – given the exceptionally deep embeddedness of the Fogo Island cooperative in the larger community – one of the benefits of the cooperative form of fish buying and processing in Newfoundland is that it cuts through and across the paternalism and kinship as the basis of the traditional mercantile system, which lives on in the numerous small family-owned fish-processing firms that have arisen in the past two decades in Canada.

The first manager of the Fogo Island cooperative was a former fish merchant who ran the cooperative in a paternalistic fashion, getting it woefully in debt. Subsequently, pains have been taken to find managers who are more business-like, in the abstract, capitalist sense, and in any case to get the cooperative out of the business of providing major lines of credit to its members. Succession to positions of influence and power in the cooperative has little or nothing to do with kinship or even the community connection. That contrasts markedly with the norm and reality of everyday life on Fogo Island, where kinship and community are, ultimately, all that count. Kinship is both enabling and constraining. In any case, the board of directors has taken great pains to hire managers 'from away' and to impress upon the rest the importance of avoiding signs of favouritism.

Another special feature of cooperatives in the context of crisis is that there may be less pressure – or different kinds of pressure – to take a conservative stance about matters such as investment and training than there might be for private firms. Cooperatives may have linkages and buffers by virtue of their organization that enable them to take more risks and a longer term view of the future, and in this sense, they are more like unions than private firms. In Newfoundland the fisheries cooperatives have some protection from government agencies, and the Fogo Island cooperative, in particular, has been able to negotiate loans, grants, and access to particular government programs that have helped its leaders implement far-reaching planning.

On Fogo Island there is a lot of tension about the cooperative fish plant's demands to come to work at a moment's notice, when a boat comes in, and to work when the fish are cut and frozen, not when the clock reaches a certain point. It is not so easy to say that the cultural ethic is that you should do this, that there is, what has already been described as a 'skipper-tak' mentality; it is rather that this is a reasonable position for debate.

The situation intensified after the moratorium in 1992. From that point on, people called in to work have had very little assurance that their 'call to duty' will result in enough hours to help them meet the goal of qualifying for unemployment insurance, upon which survival in rural coastal Newfoundland has come to depend. Compounding the problem is the moratorium option of relying on the government assistance programs (which are based on precrisis unemployment insurance qualifications) rather than going to work. It is not really an option: if called to work, one must appear. But it is a major question that informs the behaviour of people: why should I go to work if I get less for it (given my travel and child care expenses) than I would if I stayed home and received my government assistance? The 'skipper-tak' art of it on Fogo Island is that people can still accept that there is another argument; there is not, however, widespread agreement, as of the summer of 1994, that this argument is a good one. Nonetheless, most people were going to work if called, afraid of being completely cut off from even the option of working in the fish plants in an uncertain but possibly much changed future.

15

Sambro: A Case Study of Participatory Development[1]

Sambro is a fishing community of four hundred that is located approximately twenty kilometres from Halifax, Nova Scotia, a metropolitan centre of 300,000. Sambro has been the site of an active fishery since the founding of Halifax by the British in 1758. However, it was not until the 1820s that a significant influx of loyalist, white Protestant refugees established the contemporary religious and demographic character of the community (Loucks, 1993). Thirty to forty years ago three local merchants ran the local grocery stores, gas pumps, and the post office. The local merchants and fisher families were intimately connected through credit and debt relationships spanning generations. Fisher families were heavily involved in self-provisioning and pursued a pluriactive lifestyle to supplement their meagre earnings in the salt fish trade. Kitchen gardens were tended by fisher wives, and families kept a cow, chicken, or pig. Owing to poor soil and drainage conditions, gardens and hay fields were shared among kin groups (ibid.). From the mid-1900s fishers gained the right of access to four hundred acres of Crown land which surrounds the village and comprises what are known as the 'backlands' of the Chebucto Peninsula (ibid.). This region has been used by generations since for hunting and trapping, fishing, forest resources, and leisure (Manuel, 1992). It has become an intimate feature of the community identity of Sambro.

Because of climatic and locational advantages the community has always been able to fish all year. One local fish processor introduced the Sambro fishery as follows: 'We have had a sustainable system here built on the inshore.' The fleet is independently owned and includes thirty or so mid-sized boats: fif-

---

[1] Laura Loucks participated in the data collection for this case study. She completed a master's thesis in Atlantic Canada Studies at St Mary's University (Loucks, 1995), assisted in part through support from this project.

teen use handline gear, fifteen longline, and there are two Danish seiners and two groundfish draggers. This fleet represents seventy to a hundred fishers. The longline and handline fleets are renowned for their aggressiveness in the nearshore fishery. Stories abound about Sambro fishers going to the Flemish Cap or Anticosti Island in their 44-foot boats. There fleets are highly diversified, fishing for cod, haddock, swordfish, halibut, and lobster. Groundfish is sold locally to two small plants from whom they receive a certain number of services such as bait-freezing space, ice, and fuel. Swordfish were fished by three vessels using longline gear and twenty vessels using harpoons. Unlike the fixed gear fleet in southwest Nova Scotia, the Sambro fleet does not have a lucrative winter lobster fishery to fall back on.

Two groundfish draggers have been on individual transferable quotes (ITQs) since 1989. We were informed that both the local fish plants consider their fish to be inferior in quality, and both have refused to buy it. The fishers have therefore had to cultivate markets outside the community for their catch.

The two Danish seiners, while classified as part of the mobile gear fleet by government regulations and on ITQs, are highly respected in the community, based on their personal commitment to sustainable fishing practices. They use large mesh sizes in their nets, and they keep discards to a minimum. Unlike the groundfish draggers, these boats are fairly diversified in their approach, and reputedly they produce a high-quality product.

Two fish-processing firms operate in a closely 'competitive but cooperative' atmosphere in Sambro. One firm boasts a long lineage in Sambro, going back through one owner's family several generations. The other company is a new venture since 1987, but it is located on a spot that has housed a fish plant for over a hundred years. For the older firm, peak employment in the mid-1980s was fourteen full-time and ten part-time workers. This is mainly a fresh-fish operation selling whole round or dressed fish to a processing operation in Halifax (40 per cent) or directly to the U.S. market (60 per cent). In 1993 cod and haddock represented 70 per cent of their business. Sales averaged Cdn $5 to $6 million from 1990 to 1992.

Both firms are strongly embedded in the community. The manager of one said he feels a strong 'loyalty to fishermen and employees ... you can't cheat people and live next door to them. We go out of our way to help people.' The firms regularly lend money or make advances to fishers, and they have undertaken an impressive list of community-oriented activities. They have sponsored fast-pitch softball and two minor hockey teams in the past. They provide donations on a regular basis to the local volunteer fire department and a scholarship fund at J.L. Ilsley High School. They also donate fish to church suppers and local community fairs.

The firms have stuck by a commitment they made to longline and handline fishers a few years ago that they would not displace local fish with imported Russian frozen fish. They continue to buy from approximately thirty regular boats. The co-owner and manager of one of the firms is a nephew of the company founder. He is in his late thirties and has a university degree in commerce. Until the mid-1980s he had worked for a former Sambro fish processor. This manager has a close personal history with the owner of the other major fish plant in the community. Their friendship has carried over into business and community activities, including both the Fishers' Association and the Community Development Association. The companies borrow ice, forklifts, and tote boxes from each other, ship in each other's trucks, and avoid competing for each other's fishers. Each firm has a more-or-less distinct market niche, so they also avoid open competition in the marketplace.

The second firm was established in 1987 through a partnership between a local contractor and a newcomer. They purchased an older plant and embarked on an ambitious modernization program. Initially 80 per cent of the plant's output was round, dressed fresh fish for the U.S. market and 20 per cent salt fish for the Ontario market. A diversification plan was launched the following year under strict new quality control guidelines designed to help the company break into the Portuguese market. From a concentration on only groundfish, the firm moved into swordfish, lobster, and halibut. Within a couple of years the Portuguese segment came to represent one-third of the firm's sales. The key to diversification in a small company, according to the owner, is flexibility. 'In twenty seconds we can choose what markets and products to do. It takes juggernauts like National Sea six weeks if they decide not to do "lite tonite."'

Workforce and Community

There is a long-standing social separation between fisher families and plant-worker families reflecting subtle differences in status. Unlike fishing, one respondent remarked that fish-plant work has always been an employment of last resort: 'People are not born to be a plant worker.' She went on, 'When I was going to school it was the girls who left school and had a baby and needed money, or guys who couldn't fish due to seasickness who worked at the fish plant.' In some cases, she pointed out, it was less by choice than circumstance that fishers' wives did not work in the plants. The older generation of men often had the attitude, 'No wife of mine is going to work in no fish plant. They would be expected to stay home, raise kids, cook, clean, and pace the floors.' Fish-plant workers in Sambro generally come from outside the community, and they tend to have strong kinship ties and intergenerational ties to each other.

The people of Sambro have always had an ambivalent attitude about the outside world. They have been quick to take advantage of the opportunities that presented themselves by virtue of their close proximity to Halifax. However, these advantages have not always been extracted without a price. The main coastal road to Sambro from Halifax was paved in 1947. A second inland road connecting Sambro to a Halifax suburb was paved thirteen years later. These improvements have made commuting to Halifax much easier, particularly for women seeking educational or work-related opportunities. However, the roads also brought an increasing number of newcomers to Sambro. In the past twenty years the pressures of suburbanisation have been strongly felt by the residents of Sambro. While the core population of the village declined marginally to 405 in 1991, the population of two new suburbs increased dramatically. In the past fifteen years outsiders, mostly from Halifax, have started moving directly into the core nucleus of the community. The value of property with access to the sea or seascapes has increased dramatically. This led to an attempt by newcomers to preserve the more 'idyllic' qualities of the village by changing local property codes. Local people strongly resisted these proposed changes. One fish processor commented, 'How are you going to conduct a business, or if your children want to do it, how are they going to do it?'

It is instructive to attempt to distinguish between outsiders who remain aliens in the community, and outsiders who become 'newcomers' by successfully integrating themselves into the community. As anywhere, outsiders can either represent a threat or can bring fresh ideas, new skills, expertise, capital, and opportunities to a rural community. A number of newcomers have established crafts and marine repair businesses in Sambro. They also play an active role in the local church, and a number of the leading activists in the Community Development Association are non-locals. In fact, the leading community entrepreneur and youth worker over the years is 'from away.'

Given the new ease of access to the city, urban-centred recreational pursuits have displaced traditional patterns. For example, one mother of three remarked that she takes her children into Halifax practically on a daily basis for music, dance, majorettes, and skating lessons. Other children in the community are involved in urban-centred organized sports such as softball, soccer, and minor hockey. While this integration acts to break down intercommunity rivalries and hostility among youth attending consolidated junior high and high schools, it points to a glaring need for a local recreation centre.

The majority of the residents of the community are members of the United Church. Traditionally, the United Church is tied symbolically to the fishery through the annual 'blessing of the fleet' ceremony and the 'fishers memorial service.' The clergywoman, however, laments the general decline in church involvement by the young, and especially by fishers and their families.

Another respondent feels that interpersonal trust and community solidarity are diminishing as a consequence of the declining influence of the church among the young: 'Strong, influential families in the community and powerful matriarchs at one time led the direction for community rules and standards. At one time the church was a major part of this. All fund raising was for the church and people were looked down upon if they didn't attend.'

The Fisheries Crisis

For the smaller of the two fish plants in Sambro, sales were down 50 per cent in the first six months of 1993. By 1995 swordfish was representing 30 to 50 per cent of the firm's business, up from 10 per cent a few years before. Apart from this move, however, the firm seems to have responded to the crisis by retrenching. One of the owners is of the clear opinion that the crisis is cyclical and that the smart thing to do is to wait it out. It would seem that the firm may be in a position to do so because it is relatively free of debt.

As a consequence of the crisis, the staff of one of the fish plants is at best skeletal. In 1993 it was reduced to two full-time and six part-time workers. The main ones to suffer in the initial downsizing were high school students working after school, on weekends, and in the summer.

By 1990 the bottom had fallen out of the other fish company's business as well: locally sourced landings were down by 50 per cent, and sales had declined by 30 per cent. (The owner remarked that had they not diversified outside groundfish their landings would have been down by 70 per cent.) While he too has maintained a commitment to local fishers, he has aggressively sought out global sources of fish supply – particularly Russian groundfish, Norwegian cod, and Icelandic cod and haddock. It has reached the point where other small processors in the area want to subcontract through this firm for global fish. In a reaction to the crisis which is reminiscent of the plants that survived the crisis in north Norway, the larger one of the two firms has instituted new efficiency standards in its plant operations to maximize return on volume and minimize costs and waste. ISO standards are being used as a model. This fish firm has also embarked on an entirely new venture in fin-fish aquaculture.

A marine biologist was hired in 1994 to evaluate and assess the prospects in Sambro harbour for aquaculture and to develop an experimental hatchery program for winter flounder. After an initial evaluation, the firm called a public meeting of all interested fishermen in February 1995. Detailed results of the experiment and others elsewhere, market analyses, and technical information on the local ecology and equipment costs were presented. The owner described his intention as selling the idea to the community by way of a demonstration effect. 'We could go it alone, but we have had a good relationship to the community and fishermen, and we want to involve them.' The fishers were presented with a number of options: individual ventures with which the firm would subcontract, a collective joint venture with the company, or a fishers' cooperative venture which would establish arm's length relations with the private firm.

Most significantly, perhaps, the owner of this firm has become an ardent spokesperson for small coastal communities, firms, and fishers in the conflict that has come to characterize an industry in crisis. This activism stemmed from a disenchantment in the early years of the crisis with the fact that the small players in the industry were in the dark as to the regulations and how allocations were being made. To protect their interests they needed representation. His appointment to the Fisheries Resource Conservation Council (FRCC) in 1992 was the upshot of these efforts. He felt that it was much more effective to seek representation than to take on the government in the courts. He finds that he is one of the few inshore fixed-gear voices on the council.

The fisheries crisis has led to a substantial decrease in wages for the people of Sambro. They used to earn Cdn $50,000 to $60,000 a year. The crisis has cut this to the Cdn $30,000 or $35,000 range. A total moratorium would reduce this to real distress levels of Cdn $10,000 a year. Only handline fishers collected unemployment insurance benefits up until the late 1980s. By 1993 all fishers, including the largest longline fishers, collected UI.

Intensification in 'self-exploitation' is a common adaptation: fishers have reduced crew sizes from four to three or three to two. They bait their own trawl rather than hire students or local women to do it. This has cut down on fishing time – two days to fish and two days to bait. Fishers are also diversifying, targeting a more diverse range of species, particularly swordfish and blue shark. A bidding system has evolved, wherein fishers' wives take bids on their husbands' catch from various buyers prior to landing. This fish can be landed anywhere along the Atlantic coast. The larger boats are also getting more into halibut, and the bidding system is also starting to characterize that fishery. Fishers bait on board and go out for twelve to fourteen days. As a consequence, they are less tied to local plants.

More people are having to work outside the community, particularly women. One woman remarked that since the crisis began, 'I don't know of one fisherman's wife who is not doing it.' Most have taken retail and clerical jobs. In 1993 one respondent estimated that 25 per cent of Sambro's local labour force had jobs in Halifax. It makes for a long day commuting (80 to 120 minutes). 'People only have enough time to look after their own needs.' A local bus service has been initiated. The driver offers two morning trips to Halifax and two return trips using extended minivans for Cdn $5.00 per day. 'Mornings, he is completely booked up.' With so many women commuting to wage jobs in Halifax, neighbours and relatives are picking up the slack in terms of child care. One local woman, with a history of local community involvement remarked, 'I'm more isolated, people are becoming more isolated.' More self-provisioning is also done, for example, hunting and gardening.

We were told that out-migration is not a great problem in Sambro, particularly among the young. Nevertheless, high school and university graduates are not, for the most part, returning to Sambro to work or start their own businesses. One respondent has four grown-up children. Two work in Halifax, and the other two attend university there. All want to continue to live in Sambro, but none of them expect to work there. While post-secondary education costs now average Cdn $10,000 or more per year, the attitude is 'you can't get enough education these days.' For those few who choose to return to Sambro to work, underemployment is an unappealing prospect.

A long-standing intergenerational, and gender-based recruitment tradition characterized fisher families in communities like Sambro, along the southwest coast of Nova Scotia. Sons would leave school as soon as it was legally possible to do so and go fishing with their fathers or uncles. Daughters would stay in school until they were needed at home to look after infirm relatives, or until they got married and had children of their own. The current generation, particularly with the uncertainty in the fishery, is not so rigid. Fathers are ambivalent about their sons following in their footsteps, and 'fishermen's daughters don't want "that" life for their kids.' Girls are getting more education now in order to go outside Sambro to work.

A preliminary review of intergenerational kinship patterns seems to indicate gender-based out-migration and educational achievement biases. Young women with higher levels of post-secondary education are marrying outside the community and leaving permanently.

Community Responses

Based on the foregoing analysis, one would not be too far mistaken in thinking that here is a 'typical' rural community going through the throes of disintegration as a consequence of modernization and the fishery crisis. Both individually and collectively, however, the people of Sambro are responding in ways that question this picture. They are demonstrating that they are not passive victims. Two interesting initiatives in recent years give rise to hope that atomization has not completely destroyed community cohesion: the Community Development Association and a community fishery co-management scheme.

In 1994 the United Church minister invited the residents of Sambro to a meeting. She posted notices around the village that read: 'How are we going to get through the winter with no fish?.' Forty people attended the meeting, and the Community Development Association committee was struck. Three of the eight-person committee were newcomers to Sambro. Their mandate was to establish linkages to the community and explore diversification strategies. Within a few months, however, it became clear that a local–outsider division was creating problems for the organization. This led to a six-month struggle before local control and direction of the organization could be reasserted. The main problem seems to have been an outsider expertise bias such that the opinions of outside experts were given more weight than local knowledge and opinions. More insidious perhaps was the formal organizational process that effectively intimidated and gagged local voices. Following an attempt to reorient the committee's efforts to community needs, one outsider, and a few other local professionals who had been active, splintered off to form their own organization to promote tourism.

A revivified local development association emerged from the next mass meeting attended by over sixty locals. A commitment was made to run meetings based on informality and consensus in order to empower local voices. Executive meetings were scheduled for every two weeks, and two community workshops on economic development were planned.

The first workshop was held in the spring of 1994, and it was attended by thirty people. Five areas of concern were established for action: development of the beach, aquaculture (in conjunction with a local initiative by one local fish plant), tourism, the arts, and highway infrastructure. Of these, aquaculture was the most contentious as fishers felt that any development of aquaculture would conflict with lobstering activities. More local representation and participation had to be encouraged in order to build trust. To that end a report on the workshop was published in the local paper.

A second community workshop was called for the fall of 1994. This endeavour was initiated by the newly created Halifax County West Economic Development Commission. This represented a second battle against the forces of outside control by the development association. Essentially, the meeting that was called was part of a larger 'top-down' approach which cast professional planners in the role of experts. Their approach was heavily criticized by members of the local development association as being hierarchical, destructive, irrelevant, and non-consensual. It was perceived as an effort to highjack local initiative and stultify participation at the grass roots.

In 1995 three areas were receiving priority in the activities of the local development association. First, in cooperation with the provincial Department of Economic Development and the Halifax County Industrial Commission, the association is sponsoring a study of local needs that will attempt to mesh local needs with local resources. For example, residents of the community may need help with literacy training, child care, domestic and marriage counselling, or financial management. In addition, they hope to focus on youth services, small business development, and educational retraining needs. Second, under TAGS, and in partnership with Parks Nova Scotia, the community of Sambro applied for a beach development initiative grant which would train twenty people to work for twenty-six weeks in the area of conservation and community-based resource management. From this core program it is hoped funding can be raised for a beach facility and a community centre. Third, the association is attempting to coordinate community-based development planning with the local fishers' association community quota scheme and provide a link to the coastal communities network. The immediate plan of action in this regard is a meeting of community residents concerning the report by the Fisheries Council of Canada, entitled 'A Vision for the Atlantic Fisheries.' A community meeting was called for in late February of 1996, and posters were distributed around the community with the slogan: 'Fishing Community For Sale. ... The Fisheries Council of Canada May Sell Your Community.' (Community members were stunned to learn of the Bugøynes advertisement, 'Will Someone Accept Us?' when we told them about it after they had drawn up their own advertisement.)

Collective Action

The efforts by Sambro fishers to regain control over their resources can be traced to the beginning of the crisis in 1992, when the government started to cut groundfish quotas. Sambro fishers felt they were being disadvantaged in two ways: first, they were at a locational disadvantage in terms of access to the resources. Consequently, in a competitive system they suffered from an unfair handicap. (The relative importance of groundfish was also greater in Sambro because of the weak lobster fishery.) Second, the competitive system led to gluts and scarcities on the port market and therefore low prices. Such supply-driven tendencies disadvantaged fishers and processors alike in a highly differentiated food chain.

The Halifax West Commercial Fishermen's Association, affiliated with the Eastern Fishermen's Federation, decided to calculate historical catch trends for eastern-shore fishers to determine exactly how disadvantaged they were under the competitive quota. From 1986 to 1993 their proportion of the overall catch was stable at 10 per cent cod, 10 per cent haddock, and 4 per cent pollock landings in a rich fishery district of the southwest called 4X. (While all communities from Pennant to Canso were lumped into these averages, Sambro fishers claimed to account for 99 per cent of these landings.) Under the competitive system the community had landings that were a fraction of this historical average. Trip limits were coming down, and the community was facing economic disaster.

Through their association, fishers made a presentation to the federal government asking for an experimental 'community quota' allocation to protect their historical claims in 4X. The vast majority of local fishers wholeheartedly supported the effort. In the fall of 1990 the Department of Fisheries and Oceans agreed to an experimental community-managed quota on haddock. Of a total allowable catch (TAC) of 1,900 tonnes of haddock in 4X, 45 tonnes was allocated to fishers through their association. The fishers, through their association, administered it themselves, applying trip limits. The system allowed them to rationalize landings according to the market. Throughput efficiencies and improved market responsiveness allowed processors to realize greater unit profitability, even though overall landing volumes were down because of quota cuts over previous years. These benefits were passed along to fishers in the form of higher prices. Fishers in Sambro discovered, in fact, that they ended the year with a 25-tonne surplus which they reallocated to fixed-gear fishers in southwest Nova Scotia. One processor remarked, 'It all boils down to greed, and we are not being greedy about this.'

Based on the perceived success of this experiment, DFO tentatively gave the green light to a broader 'community quota' that would encompass all groundfish in 4X. A flexible management regime was put into place that gave the fishers the right to manage and allocate trip limits and the maximum number of trips per week within a broad range of types of vessels under sixty-five feet. Self-policing was all important. Fishers had to show that they could be both responsible and accountable. One local observer summed it up this way: 'The eyes of the world are on them. The concept has never been tried before.'

The 'community quota' system seems to avoid the frenzied approach to fishing that comes out of competitive fleet quotas. This small amount of quota is intended only to compensate the fishermen for their costs in fuel and labour. They are now expected to participate voluntarily in scientific work for the good of their group. This management framework that allows latent conservation mechanisms associated with fixed-gear technology – for instance, selectivity and switching – to be realized as fishers are given a modest security of tenure over the resource, and they can rationally plan and allocate their time and effort on both an individual and a collective basis. In other words, the community quota does not seem to suffer from the two central drawbacks of the ITQ system that characterize the mobile-gear fleet: high-grading and a rich fisher bias. In other words, community quotas provide middle-income, moderately successful fishers with some security of tenure and income at a time when the fisheries crisis and changing government regulations are leading to a 'disappearing middle' in the fishery.

The next and final chapter of this section explores a number of complexities that are represented for the study by the ambiguous nature of globalization and the state. We are concerned to present an integrated picture of fish-processing firms and coastal communities in such a way that sustainable policy options are clear both for the communities themselves and for policy-makers.

16

Re-embedding Coastal Communities: Towards a Localized Globalization?

The central theoretical perspective in this section assesses the issue of community disembeddedness as a consequence of the fisheries resource crisis within an increasingly globalized economy. As we have seen, by disembedding we mean a process through which the economic activities, rational, and social relations are 'lifted' from local contexts of interaction (Giddens, 1990). What the consequences of such a process are for coastal communities is a theoretical focus of our work. For instance, globalization may be seen both as a threat to coastal fishing communities that are already weakened by resource crises, but it may also provide new opportunities for economic diversification and growth. We argue that disembeddedness does not necessarily have to follow in the wake of globalization, but that it is possible for communities to stay embedded, while becoming globally oriented at the same time. Our three case studies offered some useful insights into the matter of globalization and embeddedness.

Globalization and Culture

As Robertson (1993) pointed out, globalization is not simply a matter of economic penetration and incorporation of local economies. It also involves cultural and subjective matters. The global world is 'an imagined community' that is symbolically constructed. But so also is the local community (Cohen, 1985). Globalization, however, is a process that makes the local and the global world become one imagined whole. The impact of globalization on coastal communities can be assessed in terms of the institutional processes involved – economic, cultural, political, and so on – the state of community organization and embeddedness – marginalized or strong – and, finally, in terms of its impact on different segments of the population – for instance, youth or women.

Globalization is never complete, but rests in a pluralized life world (Berger et al., 1974). In the embedded community, the life world is one, while in the Fordist community different life worlds coexist according to class and occupation. Globalization adds new life worlds of a larger scale to community life, and it needs to find space within, and be balanced off, the already existing life worlds. Globalization takes place within the private as well as the public spheres, it manifests itself within the daily life of community members, within the family, at the workplace and other community arenas, and within the social as well as the economic domains. With Cohen (1985), we submit that the multiplicity of life worlds makes community a device for social aggregation rather than integration. Social solidarity, coherence, and capacity for tolerating deviance come to the fore. As Cohen (1985) argued: 'If the members of a community come to feel that they have less in common with each other than they have with members of some other community then, clearly, the boundaries [of the community] have become anomalous and the integrity of the "community" they enclose has been severely impugned' (p. 20).

Analytically, globalization should be seen as a property of the context for social interaction among community members, young and old, fishers and fish processors, males and females, local workers and guest workers, employers and employees, teachers and students, tourists and local people, the government and community clients (Barth, 1978). It brings more complexity into the daily life of community members as they are becoming increasingly interwoven with social networks of a larger scale. Focus must be directed at the social process of 'cognitive bargaining' (Berger et al., 1974) that occurs between the parties involved and the extent to which they are capable of striking compromise between the local and the global life worlds. Partly, also, globalization should be regarded as a process on the level of consciousness, something that gets under the skin of local inhabitants. Thus, globalization is likely to influence the rationales that individuals and collectivities pursue and the social values that they are derived from.

Paulgård (1993) discussed the complexities of cultural globalization in her case study of youth culture in Honningsvåg. Like Marshall McLuhan, she regarded the media as a driving force. 'Youths that grow up in Finnmark today do not only become exposed to experiences they encounter in their local environment, they also are influenced from the outside to a larger extent than before' (1993: 41). This, she maintained, leads to a reduced interest in traditional community values and the opportunities that the fishing industry can offer. In a parallel fashion, young people in Honningsvåg prefer service jobs, for instance, in tourism, to jobs in the fishing industry. Thus, they become increasingly marginalized in their home community and *vis-à-vis* their parents' generation.

Siri Gerrard (1992), in a study of a number of other Finnmark communities (Loppa, Sørvær, Havøysund, and Skarsvåg), noted a similar trend:

> It is precisely the tight coupling between households, local community, and industry that has been one of the hallmarks of the fisheries culture in Finnmark. One of the consequences of the new situation has been a decoupling of people between household and industry, something that among other things, has had an impact on the socialization in the fishing industry. [Traditionally], socialization in the fishing industry has been tied to the household and the local community. Through daily life activities crucial knowledge has been transmitted, particularly on the cultural and industrial levels. If these connections are broken down, we also risk a breakdown in the ways knowledge is transmitted. The question is how these connections can be replaced, and what kind of fishing industry and fisheries culture we then will have in the future. (1992: 229)

A fisheries system more loosely coupled to the community level, where its individual constituents have their primary linkages to the external world rather than to each other, risks segregation. Such a community provides neither organic nor mechanical solidarity (Durkheim, 1964 [1893]). At best, communities become entangled in 'a global Gesellschaft' in the economic sphere and a 'global Gemeinschaft' (Robertson, 1993) in the social and cultural spheres. At worst, globalization erodes social solidarity and cohesiveness and creates alienation and anomie.

The fact that the fish processors of Båtsfjord collectively and by their own initiative financed a dog for narcotics control of Russian fishing vessels is a sign of collective responsibility and re-embeddedness. But the drug problem in the community is in itself a sign of social disruption, where globalization has been undermining traditional mechanisms of social control such as stigma (Christie, 1982). With Finnmark fishing communities in mind, the Norwegian sociologist Ottar Brox warned against such a prospect already in the mid-1960s (Brox, 1966).

Resistance to Crisis

Another issue is the extent to which globalization makes communities more resistant to crisis. As we have seen with many marginalized and neo-Fordist communities, globalization may accelerate the erosion of community solidarity. In this instance, if crisis survival requires collective action by community members, globalization may have a weakening effect. However, atomization can allow individuals to survive in spite of community decline. For example, the systemic nature of the fishing industry at the local level brings collective vulnerability to environmental change. Problems in one link of the production chain easily spill over to others. If fishermen fail, processors do too, and vice versa. In Norwegian fishing communities, a domino effect has been demonstrated. Once decline has started, it tends to accelerate (Jentoft, 1984). Thus, less interdependence gives more individual leverage and reduces the risk of system wreckage. Individualized external linkages give a greater repertoire for crisis resistance, and, hence, system resilience. They provide lifelines for individuals when times are bad.

Following Paine (1972), we distinguish between two types of entrepreneur: the 'free enterpriser' who is not part of community, who does not care what community members think, and the 'free-holder' who is more integrated into the community, considers good relations with the community vital, and exercises self-restraint in terms of ostentatious display.

Our conceptualization parallels one made recently by Jentoft and Davis (1993) who distinguished between rugged individualism – independence as a value in itself, rooted in community, a strong sense of belonging, independence, and self-sufficiency as an embedded ethic – and utilitarian individualism – where individuals pursue accumulation and growth for their intrinsic value, abstracted from any particular social or cultural context.

Crisis may disrupt the commonality of interest between community and the 'free-holding' entrepreneur. If the bonds between the two are relatively loose, crisis will probably make them looser, transforming the 'free-holder' into an all-out 'free enterpriser.' As Cohen (1985) argued, however, crisis can draw communities together. In this instance, if the bonds between the 'free-holder' and the community are strong, the crisis will make them stronger.

For us the question is when is such an 'every man for himself' circumstance reached. Another question is what are the effects on the community, since fish-processing firms accumulate 'social capital,' and they often cannot afford to do things in the short run that undermine it in the long run. Put another way, how long can a small fish plant survive in a dying community? We are interested in the parameters that structure such individual decisions within a broader collectivity framed by the community. How much tolerance is there? Our study of post-Fordist communities and our community studies show that an open approach to the opportunities presented by globalization, often embodied in particular entrepreneurial personalities, is a common feature of community renewal and re-embeddedness.

Globalization does not manifest itself only as external pressure originating from the centre and spreading to the periphery. It is also manifest within the community itself as members struggle to acquire self-reliance and security. The Russian connection in Norway was established through a local initiative. The shift to a market-oriented approach to business management typically is a mixture of both a local and global enterprise. The same is true for tourism, which is now a burgeoning industry in all the rural areas we examined. The natural environment of coastal communities is extremely attractive and a potentially lucrative tourism product. But the tourism industry is also a likely competitor for the traditional industries, as demonstrated by Paulgård (1993) in the case of Honningsvåg. Tourism and its derivative service industry have a homogenization effect typical of globalization. The question remains whether such economic advantages always come at the expense of cultural authenticity (Cohen, 1979; Jentoft, 1993; MacCannell, 1973; Urry, 1995). Hammerfest may have fewer restaurants than Oslo or Toronto, but the quality of their menus reflects a growing cosmopolitan, it not global, identity. The pizza franchise in Vardø could be situated anywhere in the western world. In Nova Scotian fishing communities, Chinese restaurants are commonplace.

Ambiguities of Globalization

Although the major themes in rural sociology centre on community decline and marginality (Cohen, 1975; Gaventa et al., 1990; Perrucci et al., 1988; Robinson, 1990; Wadel, 1973; Young and Lyson, 1993), a number of other studies address cases where community development initiatives have been successful (Barnes and Hayter, 1992; Flora et al., 1992; Henderson and Francis, 1993; Lind, 1995). We see an intersection of three lines of inquiry in the latter literature: one deals with the nature of strategies of community organization (MacLeod, 1986; Nozick, 1992; Perry, 1987; Rahman, 1993), a second takes a distinctly postmodern view of cultural indigenization (Allen and Dillman, 1994; Appadurai, 1990; Friedman, 1990), while a third examines the local effects of decentred post-Fordism (Barrett, 1993; Brusco, 1982; Scott, 1988). We situate ourselves more eclectically, drawing upon some key theoretical perspectives within what has been coined as the 'new' economic sociology (Etzioni, 1988, Granovetter, 1985, 1990; Holton, 1992; Sayer and Walker, 1992; and Swedberg, 1991).

Following Polanyi (1944), our view is that traditional structures and norms in rural communities embed individuals within collectivities and also encapsulate institutions within broader processes. It is this two-edged action-based and structural perspective that defines for us the dimensions of Coleman's (1988) conception of 'social capital.' Various factors, both internal to rural communities and as part of the external environment – the capitalist mode of production, modernization, and the bureaucratic state – erode traditional embeddedness through the atomization of individuals from the collective, through the factionalization and stratification of cohesive communities, and through the functional specialization of institutions.

In this work we cannot explore adequately the 'dark side' of traditional 'moral' economies. At least three factors need to be considered in this regard: sexism and the nature of patriarchy in kinship and work relationships, clientism and the nature of 'captive' exploitation, and, finally, bigotry, racism, and moral panics that often stem from the social construction of community identity based on oppositionism and xenophobia.

Community decline and disintegration are neither universal, nor are they irreversible, linear processes. Our concern is to capture the range of diversity that characterizes rural fishing communities in the modern age and their responses to external processes such as resource crises and globalization. We wish to illustrate the parameters within which resistance, restructuring, or accommodation occur, the nature of the external–internal interface, and the nature of the individual–collective dynamic.

It is not by accident that we found ourselves drawn to write about the three particular cases described. They represent places that stand out as positive examples of community renewal in the face of crisis. In their different ways, they all demonstrate deliberate and ingenious efforts on the part of community members to re-embed their fisheries systems within their local communities.

Each, in its own way, is an interstitial case. Bugøynes is a rural community with a distinct ethnic heritage and composition *vis-à-vis* the surrounding areas. Fogo Island stands out as a community with a history of cooperativism and independence *vis-à-vis* dominant traditions of free enterprise and trade unionism in the fishery of Newfoundland. Sambro is a rural community coming to grips with the mixed blessings of being within commuting radius of a fast-growing metropolitan area. As such these communities have been confronted with challenges and opportunities and have produced responses that do *not* characterize the vast majority of rural fishing communities in their regions.

Classifying coastal communities simply as marginal glosses over the nature of social change. One risks portraying a static situation with no eye to social processes. Then in so-called marginal communities there is little room for human agency in both its individual and collective forms. As Fredrik Barth (1981) argued, structural classifications contain in themselves no explanatory value even if described in the most precise fashion. To explain a structural pattern, therefore, one needs to describe and model the generative forces of social processes.

Coastal fishing communities in our regions are not passive victims of external structural forces such as resource crises or global influences. For better or worse, people make choices, both individually and as groups within communities. It is the task of the social researcher to investigate what these choices are, what motivates them, and what aggregate effects they may have for the community as a whole. A key perspective advocated here is that choices are never made in a social vacuum, but reflect the ongoing relations that people have within social networks and the cultural values they share as members of a community. In other words, we see choices and decisions – individual and collective – as embedded in wider social groups to which individuals are attached (Granovetter and Swedberg, 1992). We argue, however, that resource crises and globalization are forces that may well disarticulate these relationships unless they are effectively addressed by community members cooperatively. This, again, calls our attention to community institutions and people's capacities, traditions, and motivation for organizational formation and concerted action.

The three case studies demonstrated clearly the capacity for collective action in face of stagnation. In spite of crisis, they also told a story of optimism and hope.

We contend that there are some important lessons to be learned from these communities. First, each case presents us with an example of communities that have come to the edge of an abyss and taken up the challenge of community survival. Through hard work, the people of Bugøynes have been able to capitalize on a common ethnic identity in order to build a stock of 'social capital' based on bounded solidarity (Coleman, 1988; Portes and Sensenbrenner, 1993). Fogo Islanders are poised to utilize a past repertoire of independence and self-reliance that has evolved through a cooperative tradition. The community of Sambro is moulding a long tradition of rugged individualism and individual self-reliance into innovative collective efforts at community reconstruction.

The responses of these communities are not simply the reinvention of past traditions in some kind of nostalgic renaissance (Cohen, 1985). In each case, the distresses of the fishery crises have been met with responses that draw on a combination of new and innovative ideas and opportunities that are cast and shaped in ways that are culturally and socially in keeping with embedded traditions. This is very much a process of re-embedding and a vivid illustration of the 'indigenization' of globalization, or 'global localization' (Appadurai, 1990; Cooke et al., 1992; Friedman, 1994; Giddens, 1990; Holton, 1992; Pocius, 1991).

The role of entrepreneurial newcomers who themselves become 'localized' seems particularly interesting. It would seem that they have built up enough trust and goodwill on the part of local people that when crisis forces communities into a collective catharsis, new ideas and opportunities can be put forward by them in a non-threatening way. Small successes and experiences then reinforce this trust, breaking down xenophobia, while at the same time new skills and knowledge are diffused to the local population.

For local people who assume new leadership positions, this challenge is particularly great. These individuals almost have to be above reproach. Lack of communication or misunderstandings can easily break down trust. Under the traditional patronage system, opportunism and self-interested behaviour were expected to be the hallmark of community leaders, so people often wait for an 'iron law of oligarchy' to sour things when someone does step forward to assert some leadership.

Government Policies and Globalization

The state is not a neutral player in the events described in this book. The state is active in the particular design, implementation, and enforcement of resource regulations and regional planning processes that have decided ideological biases. It has been an active player through limited entry regulations in the fishery, municipal and regional planning, and a myriad of other state policies that have had the effect of disembedding local fishers and communities.

The vertical linkages between individual citizens and a centralized bureaucratic state are promoted over local horizontal ties among fishers and community members. Local control is lost and former cooperative and symbiotic relations are transformed into competitive, atomized, and dependent relations with government. The compounding effect of government regulations and the fisheries crisis is exemplified in the case of Nova Scotia by individual transferable quotas (ITQs).

Communities need to organize themselves for collective empowerment. In so doing they define for themselves where they want to go and explore ways in which to get there. The diversity of the cases presented clearly shows that communities have to define for themselves, in their own circumstances, what is 'optimally' best for them. We would say that these case studies indicate that a 'guarded' but open attitude to the opportunities presented by globalization and modernization seems to be appropriate. A second strategy seems to rest in the investment and accumulation of social capital – networking, organizational formation, participatory democracy, and cooperative education – that is closely tied into outside opportunities. Third, reviving and converting cultural values that build interpersonal trust and group solidarity, networking, and organizational formation that strengthen relationships both within and beyond community boundaries are necessary ingredients. Fourth, strategies are needed for conflict resolution and negotiation that prevent conflicting interests from undermining trust.

While it is common to think of disembedding as something that is leading to the erosion of interpersonal trust and solidarity, sometimes the effects are the reverse. That is, lack of trust and solidarity may in themselves be the causes of disembeddedness. Thus, when there are conflicts of interest among community members, entrepreneurs, and newcomers, embeddedness must be negotiated to provide a solid base for trust and cohesiveness. This will also reduce the potential for factionalization and atomization that may come from the efforts by the community to diversify its economic base to become more globally competent.

The state must be involved in such a way as to support and protect community autonomy and control without engendering dependent relationships. This is not an easy prescription given certain inherent tendencies towards bureaucratic control that characterize any centralized state (Foresta, 1986). Communities have to be constantly vigilant. This is best accomplished through ongoing strategies that institutionalize public participation and accountability in as broad a manner as possible. Successful co-management systems in fishery jurisdictions such as Japan clearly indicate the vital importance of these factors in preserving community autonomy.

The experiment in community quotas in the community of Sambro is another case in point. To be realized and to become generally applicable, this management model needs government initiative and support. Before it can happen such institutional innovations need a solid base of legitimacy.

The concept of embeddedness can help government agencies, social researchers interested in coastal fishing communities, and community members alike, in more fully realizing what impacts orthodox management regimes have on local communities and what needs to be done to rectify the problems to the advantage of those communities whose existence now and in the future is closely linked to what happens with the fisheries.

Communities are not at all doomed to surrender to global forces that impinge upon them, and encapsulation is not the only strategy open to them. Global forces have a potential for generating change at the local level that connects the community to the external world in ways that in the long run may well be beneficial. But for this to happen, the new challenges must be confronted. The initiatives and strategies must come from the bottom up, rather than from the top down.

Conclusion

Community, Market, and State: Dilemmas in Fisheries Policies

In this final chapter we summarize the main argument in terms of relationships between community, market, and state. We then sketch two sets of outcomes that are probable in both our regions, considering the ongoing changes in the relationships among communities, markets, and states within a system that is becoming increasingly global. We discuss the politics of trade-offs: the complexity and multiplicity of goals in fisheries policies that governments are pursuing in both Norway and Atlantic Canada and the intricate policy-making structures that allow various stakeholders to participate in the decision-making process. We conclude with a review of the goals and basic values that should inform policy-making for the next century.

Modernization in the Fisheries

As argued in the Introduction, changes in the fisheries of Norway and Atlantic Canada can be analysed as shifts in the balance between three different institutional principles or logics, that is, community, state, and market (Figure 10). The long-term development trend in the fisheries on both sides of the North Atlantic is one of modernization, that is, a gradual shift from the community pole towards the state–market axis. This process, which was not gradual and smooth, but happened precipitously, sometimes stimulated by crises, has taken different paths in Norway and Canada. In Norway the reforms during the 1930s created a sector system in the fisheries that propelled petty capitalist fishers into a dominant position.

Created in an explicit attempt to shield fishermen and coastal communities from the international depression by way of mandated sales organizations, the MSO system can be located on the community–state axis. In the postwar period

Figure 10. Community, state, and market principles of organizations in the fisheries.

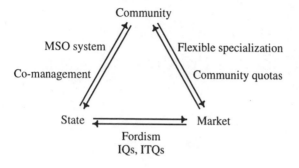

this sector system was challenged by the Fordist model, which belongs on the state–market axis in Figure 10.

As argued in Section 1 the MSO system withstood the industrialization campaign, and although the Fordist model gained considerable ground, the state–community axis remained predominant in Norwegian fisheries until well into the 1980s. By contrast, the scattered attempts at populist revolt in the Maritimes and Newfoundland during the 1920s and 1930s did not produce lasting structures, and the fisheries were left open to Fordist reorganization. But just as the MSO system did not become completely dominant in Norway, the Fordist model did not wipe out all other forms of industrial organization in Atlantic Canada. Here the traditional domestic commodity producers in the outports lingered on, protected by the Newfoundland government and the extension of unemployment insurance to the fisheries, both of which can be located on the state–community axis.

In the postwar period until the 1980s, then, the fisheries were largely dominated by the state pole. The major struggles in the fisheries sector have centred on whether state intervention should put priority on community or market concerns. Since then, however, the logic of state retreat and globalization has pushed the fisheries towards the community–market axis. As demonstrated in Section 3, it remains uncertain at which point the fisheries will come to rest between community and market principles of organization. On the one hand, there is a clearly identifiable trend towards neo-Fordist solutions, of large and often vertically integrated firms that have stepped into the global arena, leaving coastal people and communities to fend for themselves. On the other hand, we also see traces of post-Fordist solutions, of small, flexible firms, as well as networks of firms, closely embedded in coastal fishing communities.

One of the major causes for the changes in the balance among community, state, and market principles of organization in the fisheries in recent decades is the introduction of resource management. Hence, the retreat of the state is not so much a retreat from the fisheries, as such, but a retreat from regulation of trade and community support. Management of the fisheries resources was made possible by the extension of state jurisdiction; resource management is predominantly state management. But even though resource management until now has remained close to the state pole of our triangle, the ongoing debates about management reform, as demonstrated in Section 2, largely revolve around how community and market principles of organization can be utilized to make them more legitimate and efficient. On the one hand, the option of decentralization and delegation, of moving from consultation towards co-management, represents a movement outwards on the state–community axis. On the other hand, the institutionalization of private property rights in quotas, in the form of individually owned quotas that are freely transferred in markets, represents a movement outwards on the state–market axis. This trend, however, is more pronounced in Atlantic Canada than in Norway.

Dilemmas in Fisheries Organization

The community–state–market model can also be used to discuss a series of basic dilemmas that emerge when one juxtaposes the basic patterns identified in our three sections above. These dilemmas are, in the first instance, more impressive for the number of negative outcomes that are imaginable. We will try to link the findings from our three sections in ways that illuminate the very real possibility of diverse negative consequences before turning to some of the more positive alternatives.

If one begins with the community–market axis, one is immediately confronted by the manifold ways in which the global economy acts to uproot or disembed coastal communities in ways that are directly destructive or that leave the coastal communities vulnerable to negative incursions in the future. The global system does not only engage in 'cherry picking,' selecting those places and communities it finds most desirable, it also encourages the creation of neo-Fordist solutions that will in themselves possess only temporary stability. Alternatively, one has to appreciate that coastal communities, particularly traditional ones with little educational, social, or economic resources, are ill-prepared to participate in an advanced economic order, national or global, capitalist or not. The social network provided by the welfare state, which may prove unnecessary under a more rational organization of the fisheries, has provided buffers in Atlantic Canada and Norway that have left communities unequipped to deal with new economic circumstances.

The connections along the state–market axis are also rife with problems. Deregulation is a policy catchword. Deregulation is also an indicator of the trend towards state withdrawal and a growing inability to discipline the activities of larger fisheries organizations, be they national or international in origin. The withdrawal of the state also suggests a more limited public domain, a diminished capacity to provide public judgment on scientific issues, and a lower demand for individuals to act as citizens.

Privatization is another policy catchword. The introduction of individual quotas (IQs) or individual transferable quotas (ITQs) to accommodate the demands of the private sector for more definitive control of marine resources solves some short-term problems of overcapacity, while raising enduring questions about access rights, conservation, and enforcement. ITQs are also associated with increased economic stratification and social inequality. Further, the transition from state property in marine resources to private property may mean the surrender of public control and potential revenues.

The community–state axis also raises a series of difficulties. The general retreat of the state in the past few decades has been marked by a withdrawal of resources, both at the local level and in terms of the structures of regional fisheries management. The state is under particular pressure to eliminate subsidies to the fisheries because these subsidies underpin patterns of employment that are neither understood nor appreciated in urban areas. The idea that work cycles would be organized around the irregular availablity of a wild resource is unfamiliar to many in urban areas and even to some in the fishing industry itself. The widespread abandonment of fishing communities is a growing possibility. Further, different levels of government in Canada engage in a 'shell game' that continually transfers responsibility for cutbacks from the federal to the provincial to the municipal domain. This adds considerably to cynicism about the intentions and motivations of politicians throughout Atlantic Canada. On the other hand, the patterns of dependence which have been fostered by paternalistic state policies limit the alternatives coastal communities have for making realistic adjustments to the future. The idiocies of rural life – low levels of education, poor health facilities, and restricted economic resources – limit the options of governments, if one can presume some remaining goodwill in the circles of government.

The breakdown in the relationships tying state, market, and coastal communities together point towards social and ecological disintegration. However, breakdown in the existing order of things is also a prerequisite for the establishment of new ones. The current processes of transformation in the fisheries – globalization of markets, state retreat, coastal communities increasingly left to fend for themselves – open up new, and possibly better adaptations. The retreat of the state may not uniformly leave coastal communities more vulnerable to external forces. In the form of decentralization and delegation of management responsibility, for instance, state retreat may reinvigorate and empower coastal communities. Hence, proposals for co-management and community quotas presuppose that coastal communities take on larger responsibilities. Further, when retreating state agencies attempt to maintain authority relations in the absence of the funds to make the relations effective, regional protest movements will fill the gap being created by weakened governmental structures.

Substituting the market mechanism for bureaucratic procedures in the allocation of quotas, for instance, by way of ITQs, may not necessarily lead to social disruption either. Fordist mass production is not well suited to the instabilities of the fisheries and was, throughout most of the postwar period, kept alive only with heavy support from the state. If this is correct, ITQs, which presumably would benefit the most efficient actors, would favour flexible and embedded enterprises, not large-scale mass production. To a limited degree, we may also find that the market, for hard economic reasons, will find itself embracing principles of conservation, both for public relations in the short run ('green labelling') and stability in the long run through access to renewable resources. The move to restructuring state bureaucracies opens up the prospect that new forms of science may replace the 'big science' we associate with existing fisheries bureaucracies. One may be talking about science from new sources ('local knowledge'), or the reorganization of scientific paradigms to put more emphasis on complex ecological systems through the new science of 'chaos,' or more cautious approaches to existing predictive systems.

Further, globalization may imply greater openness of coastal communities to the international marketplace, but the decline of Fordist production may facilitate opportunities for the creation of production systems based on flexible specialization and re-embedding in local communities. One difficulty here is the prospect that neo-Fordist solutions may be the preferred solution in the new international markets. Given the less desirable social consequences of a more hierarchical, anonymous system, and the greater economic instability of neo-Fordist systems, advocates of flexible specialization and embeddedness may have some tough choices to make.

Coastal communities may benefit from the creation of defensive associations among themselves to cope with diminished state input and uncaring markets. These defensive associations may even become regional or subregional trading blocs intended to create more bargaining power for their constituents. Modern telecommunications make it possible for members of far-flung communities to create virtual communities that provide valuable information and support over distances that would have been insurmountable even a decade ago.

The Politics of Trade-offs

Policy-making for coastal communities and fisheries is ridden with irreducible dilemmas and cruel choices. Few solutions are easy or uncontroversial. More often than not, policy-makers must balance conflicting objectives. Perhaps more so than in other industries, economic efficiency in the fisheries must be reconciled with concerns such as social justice, regional development, resource conservation, and environmental protection. Policy-makers also have to balance what needs urgent attention in the short run, such as community support in crises, and what is required in the long run, such as strategies to enhance the viability of stock. These concerns are not always compatible. Remedies that are adequate and imperative for immediate problems may complicate or even rule out solutions that would be preferable in a longer perspective.

A striking feature of fisheries issues is that they involve ever more numerous and diverse groups of stakeholders. They range from individual fishermen, households, fish-plant workers, local communities, recreationists, and consumers, to well-organized lobbies, big corporations, animal-rights and environmental groups, and the public at large. Even government officials and scientists could be regarded as stakeholders.

Another complicating factor for policy-making is that fisheries policy is no longer the exclusive domain of national governments. Policy-makers often find themselves in a squeeze between domestic demands and international obligations. This is largely a consequence of free trade agreements and negotiated settlements on harvesting rights, in the context of a global economy. In some instances, the lack of international cooperation undermines the conditions for sound management at the national level, as national governments lack the jurisdiction required to manage the fishery in the interest of coastal communities. This is well demonstrated in the case of fish stocks that straddle lines of extended jurisdiction.

These features are found in both Norway and Atlantic Canada. In the years to come they are likely to intensify. Therefore, we need to ask how these dilemmas should be handled, given widespread agreement on the need for sustainable and responsible fisheries.

On both sides of the Atlantic there has been a tendency to neglect dilemmas and hard choices, such as between efficiency and coastal employment, or between short-term gains and long-term prospects, or to think of them as issues that have simple technical solutions. This tendency has become more evident in recent years. Notably, in the fishery, 'Scientific Management,' as advocated by Frederick Taylor almost a hundred years ago, is not only applied to industrial organization. In the fisheries, the Fordist approach has served as a model for the modernization of the industry from harvesting through processing to marketing. The same way of thinking has also characterized the management of natural resources. An underlying assumption of fisheries management is that the control of nature is within the range of human capacity, that negative social impacts of policies can be remedied through compensation schemes, and that a more 'rational,' science-based, and/or market-based approach would eliminate the need for government intervention.

Reflecting the logic of the 'Scientific Management' model, it is often argued that politics should be ruled out of the fisheries. What is wrong with fisheries management, according to this view, is the fact that politics have come to corrupt the management process and that social concerns have loomed larger than the bioeconomic imperatives. This is also the basic reason, according to this perspective, why fisheries in both areas are ridden with crises. Since politics are defined as the problem, two alternative solutions stand out: either to leave the decisions to the market or to vest the power to manage in the hands of scientists. Both would imply a depoliticization of the management process and lead to more rational procedures and efficient outcomes.

A major message of this book is that there are problems with this modernist approach to fisheries problems. Central is the fact that the viability of coastal communities and fishing as a way of life is threatened, because these concerns would be marginalized either by market forces or by fisheries management decisions. Social justice would be a non-issue, or one left to institutions outside the management domain. Delegating power to the scientists, management would become a question of research and expert knowledge and of designing a process where user groups and other stakeholders will have no legitimate place in decision-making. However, this is unlikely to serve either social or conservation objectives. We do not live in an ideal world where all the requirements of perfectly rational decisions can be met. Knowledge is uncertain and incomplete, objectives are ambiguous, conflicting, and unstable, power is fragmented, and decisions can be effectively challenged.

As we have shown, the fishery is a heterogeneous industry where policy making is basically about balancing conflicting interests and demands. Therefore, it should come as no surprise that user groups may object to well-intended and sensible decisions.

People act according to what they believe to be in their own interest, and these calculations do not always coincide with those of others or with those of the industry as a whole. For any policy-makers in the fishery, whether they are working in a government bureaucracy or a scientific establishment, these are facts that cannot be ignored.

Furthermore, technical solutions are not value free. Scientists work from assumptions that reflect a certain view on nature, people, and society. Neither is the market a politically neutral institution. It is a socially constructed, politically maintained institution functioning according to rules stemming from collective decision-making. The values it reflects are culturally specific, not naturally given. Also, as Lester Thurow (1980) maintained, relying on the market does not turn social justice into a non-issue. Rather it implies the a priori acceptance of the distributional effects that result, whatever they may be.

Social Values and Negative Goals

Planning in the fisheries has proven overly optimistic, crises have come abruptly, and outcomes have been contrary to what was intended. Thus, in coastal fishing communities, optimism has been replaced by pessimism, which in itself leads to disillusionment about the role of government in the fisheries. Our analysis of the fisheries, which frequently returns to the biological and economic uncertainties endemic to the industry, suggests that it may be more fruitful to specify what has not worked, or is not likely to work. Therefore, following Ottar Brox (1995), we wish to talk about 'negative goals,' that is, about paths we think should not be pursued.

That requires talking about positive values. Any planning process should start by clarifying the basic values that inform its policies. What is it that we care about and think is most important? What social qualities should not be sacrificed in the modernization process? Is the abandonment of communities and fishing as a way or life a price we are willing to pay?

Given the existence of coastal fishing communities with long histories of dependence on development of the fisheries in both regions, policies and decisions that do not respect the individual and collective choices of people in these communities must be viewed with scepticism. Further, given the common property and public interest dimensions of marine resources, governments have a particular incentive to ensure the integrity and survival of coastal communities. Moreover, the diversity of the human condition, and the practical commitment to participatory democracy and communicative action prevalent in small communities are patterns that must be supported if the basic values of the industrial democracies themselves are to be sustained. What should not be overlooked, however, is the ambiguous and controversial nature of policies and decisions in the fishery and that policy-making is essentially about making continuous trade-offs between mutually exclusive values and objectives.

The first, and most obvious, 'negative goal' is to recognize that many of the Fordist structures, along with their parallel developments in the administration of the fisheries and conventional government science, are liabilities that will need to be radically restructured if economic, administrative, and scientific tasks are to be accomplished in a new era. It is wise to assume that change, frequently large scale and abrupt, is a defining characteristic of the industry. The process of globalization ensures that the pace of change is likely to increase in range, and perhaps magnitude, for the immediate future. The resource fluctuations and economic cycles that characterize the industry demand institutional systems that are 'built low to the ground,' and that will be reasonably robust in the face of fragile bases of natural resource and uncontrollable international market forces.

Part of this restructuring requires a reconsideration of the subsidy systems that have traditionally adhered to the Fordist system of production. While public attention, and criticism, has largely focused on the individual forms of assistance that help fishers and their families smooth out the flow of income over a calendar year, grants and loans to encourage the development of large-scale harvesting and processing capacity to handle the presumed abundance in offshore stocks have been significant both in Atlantic Canada and Norway. Further, when Fordist systems get themselves into trouble in the fishery, as inevitably they do, their political clout has been effective in persuading national politicians to provide generous bail-out packages. Such enterprises are not exposed to the full rigours of the marketplace. The usual litany runs – too much has already been invested, too many jobs are at risk, and the next election is coming.

A corollary of this discussion of subsidies is that we must stop 'blaming the victim' exercises during periods of crisis. The individual subsidy packages associated with the fisheries may themselves be regarded as part of the cost the welfare state paid for allowing Fordist expansion in a better time. People will be surprised at how little long-term attachment there is to state subsidies, once the immediate crisis has been dealt with. In other words, the 'social fishery' had always been a Fordist fishery. The prospects for a more rational fishery in the future are considerably better than most think.

A third negative aim is to avoid the conception of nature as simple, controllable, and stable that is inherent in current models of resource management and in Fordist systems of production. Much of the wisdom accumulating as a result of the recent fisheries crises indicates that marine resources are complex, fluctuating, and perhaps even chaotic. We are going to have to be open to the prospect that new production systems will demand new forms of science.

A fourth negative aim must be to strip away some of the false appearance of scientific neutrality from much of the biological and economic modelling that dominates thinking in fisheries policy. Bioeconomic models are replete with normativity. The real romantics of the industry may be biologists and economists who assume that relatively simple, linear, steady-state models will produce ecologically and socially responsible resource management. We must recognize that neo-classical economists are politicians par excellence, and that politics, given the dilemmas alluded to earlier, are an integral part of the fishery.

Avoiding undesirable outcomes will require changes in established policy-making procedures. To begin with, consensus will have to be reached as to what these outcomes are to be. This is a question of improving dialogue within the policy process, for instance, by moving decisions away from the market and towards the community pole in our diagram. In conventional processes of planning, decisions are made within the framework of corporatist arrangements – leaving communities and political institutions at the margins of the policy process. A stronger emphasis on negative goals should lead to a more significant role for community representatives in decision-making, be they local politicians or 'street-level' bureaucrats. Only then will the values that are inherent to fishing as a way of life have a chance to be regarded as something not to be sacrificed in the process of modernization, and only then would the importance of a fishery that is truly embedded in local communities be fully understood.

Outcomes are largely a matter of political choice tempered by institutional frameworks and shaped by socially constructed values and objectives. Consequently, we need to ask which institutional designs and planning practices may help avoid undesirable outcomes and realize important objectives. We have argued that this, in turn, will force us to be more specific and explicit as to the social values underpinning both scholarly analysis and policy recommendations. There is no easy way out of the dilemmas facing the coastal communities and fishing industries of Norway and Canada. They confront policy-makers with difficult sometimes mutually exclusive political choices. If, as we move into a new century, we acknowledge this, then the larger community must more fully engage itself in discussing the core values and objectives of the fisheries and the development of coastal communities and in deciding what the outcomes should be.

References

Aegisson, Gunnar. (1993). 'Industrial Change and Development from Macrolevel Organizational Perspectives,' PhD thesis, Department of Industrial Management and Work Science, Norwegian Institute of Technology. University of Trondheim.

Agarwal, B. (1994). 'Gender and Command over Property: A Critical Gap in Economic Analysis and Policy in South Asia.' *World Development* 22: 1455–78.

Alexander, David. (1977). *The Decay of Trade: An Economic History of the Newfoundland Saltfish Trade, 1935–1965*. St John's: Institute for Social and Economic Research, Memorial University of Newfoundland.

Allen, J.C., and D.A. Dillman. (1994). *Against All Odds: Rural Community in the Information Age*. Boulder: Westview Press.

Andersen, Raoul (1974). 'North Atlantic Fishing Adaptations: Origins and Directions.' In G. Pontecorvo (ed.), *Fisheries Conflicts in the North Atlantic*. Cambridge: Ballinger Press, pp. 15–33.

Andersen, Raoul (ed.). (1979). *North Atlantic Maritime Cultures*. The Hague: Mouton.

Andersen, Raoul, and Cato Wadel (eds.). (1972). *North Atlantic Fishermen: Anthropological Essays on Modern Fishing*. St John's: Institute of Social and Economic Research, Memorial University of Newfoundland.

Apostle, Richard, and Gene Barrett. (1972). *Emptying Their Nets: Small Capital and Rural Industrialization in the Nova Scotia Fishing Industry*. Toronto: University of Toronto Press.

Apostle, Richard, and Knut H. Mikalsen. (1993). 'The Politics of Fishery Management: Preliminary Comparisons of Regulatory Regimes in North Norway and Nova Scotia.' *Scandanavian–Canadian Studies* 6: 47–74.

Apostle, Richard, and Svein Jentoft. (1991). 'Nova Scotia and North Norway Fisheries: The Future of Small-scale Processors.' *Marine Policy* 15: 100–10.

Appadurai, A. (1990). 'Disjuncture and Difference in the Global Cultural Economy.' *Theory, Culture and Society* 7: 295–310.

Arbo, Peter, and Bjørn Hersoug. (1994). *Fiskerinæring og 1994 regional Utvikling i Finnmark: Russetorsken – en ny Redning østfra?* Notat Nr. 26. Tromsø: NORUT-Samfunn, A/S.

Arnstein, Sherry. (1969). 'A Ladder of Citizen Participation.' *Journal of the American Institute of Planners* 35: 216–24.

Aronoff, Marilyn. (1993). 'Collective Celebration as a Vehicle for Local Economic Development: A Michigan case.' *Human Organization* 52: 368–79.

Bailey, Jennifer L. (1995). 'Troubled International Waters: Bringing the State (System) Back in,' paper presented before the XVI Congress of the European Society for Rural Sociology. Prague, Czech Republic. 31 July – 4 August.

Bannister, Ralph. (1989). 'Orthodoxy and the Theory of Fisheries Management: The Policy and Practice of Fishery Theory Past and Present.' MA thesis, Atlantic Canada Studies, Saint Mary's University, Halifax.

Barnes, T.J., and R. Hayter. (1992). '"The Little Town that did": Flexible Accumulation and Community Response in Chemainus, British Columbia.' *Regional Studies* 26: 647–64.

Barnett, Jane. (1990). 'A Review of the U.S. Farm-raised Catfish Industry and Its Implications for Canadian Groundfish Exporters.' Economic and Commercial Analysis Report no. 52. Ottawa: Department of Fisheries and Oceans.

Barrett, Gene. (1993). 'Flexible Specialization and Rural Community Development: The Case of Nova Scotia.' *North Atlantic Studies* 3: 50–60.

– (1992). 'Mercantile and Industrial Development to 1945.' In Richard Apostle and Gene Barrett (eds.), *Emptying Their Nets*. Toronto: University of Toronto Press, pp. 39–60.

– (1984). 'Capital and the State in Atlantic Canada: The Structural Context of Fisheries Policy between 1939 and 1977.' In Cynthia Lamson and Arthur J. Hanson (eds.), *Atlantic Fisheries and Coastal Communities: Fisheries Decision-Making Case Studies*. Halifax: Dalhousie Oceans Studies Programme, pp. 77–104.

– (1979). 'Underdevelopment and Social Movements in 1979 the Nova Scotia Fishing Industry to 1938.' In Robert J. Brym and R. James Sacouman (eds.), *Underdevelopment and Social Movements in Atlantic Canada*. Toronto: New Hogtown Press, pp. 127–60.

Barrett, Gene, and Anthony Davis. (1984). 'Floundering in Troubled Waters: The Political Economy of the Atlantic Fishery and the Task Force on Atlantic Fisheries.' *Journal of Canadian Studies* 19: 125–37.

Barth, Fredrik. (1981). *Process and Form in Social Life*. London: Routledge and Kegan Paul.

Barth, Fredrik (ed.). (1978). *Scale and Social Organization*. Oslo: Universitetsforlaget.

Bates, Stewart. (1944). *Report on the Canadian Atlantic Sea-Fishery*. Royal Commission on Provincial Development and Rehabilitation. Halifax: Province of Nova Scotia.

Beccatini, Giacomo. (1990). 'The Marshallian Industrial District as a Socio-economic Notion.' In F. Pyke, G. Becattini, and W. Sengenberger (eds.), *Industrial Districts and Inter-firm Co-operation.* Geneva: International Institute of Labour Studies, pp. 37–51.

Berger, Peter L., Birgitte Berger, and Hansfried Kellner. (1974). *The Homeless Mind.* Harmondsworth: Penguin.

Beverton, R.J.H. (1952). 'Some Observations on the Principles and Methods of Fishery Regulations,' unpublished paper, available from author.

Beverton, R.J.H., and S.J. Holt. (1957). *On the Dynamics of Exploited Fish Populations.* London: Chapman and Hall.

Binkley, Marian, and Victor Thiessen. (1988). 'Ten Days a "Grass Widow" – Forty-eight Hours a Wife: Sexual Division of Labour in Trawlermen's Households.' *Culture* 8: 39–50.

Bollman, Ray D. (ed.). (1992). *Rural and Small Town Canada.* Toronto: Thompson.

Brochmann, Bjøern. (1981). 'Virkinger på lang sikt aw statsstoette til Fiskeriene.' *Sosialekonomen.* 2: 23–9.

– (1980). 'Disktrikts-Norge boer ikke opprettholdes gjennom Sysselsetting som er verre å gjøren ingen Ting!' *Fiskets Gang* 66 (49):767–91.

Brox, Ottar. (1995). *Dit Ir Ikke Vil.* Gjøvik: Exil.

– (1989). *Kan Bygdenæringene Bli Lønnsomme?* Oslo: Gyldendal.

– (1984). *Nord-Norge: Fra Almenning til Koloni.* Tromsø-Bergen-Oslo: Universitetsforlaget.

– (1972). *Newfoundland Fishermen in the Age of Industry: A Sociology of Economic Dualism.* St John's: Institute of Social and Economic Research, Memorial University of Newfoundland.

– (1966). *Hva Skjer i Nord-Norge?.* Oslo: Pax.

Brusco, Sebastino. (1982). 'The Emilian Model: Productive Decentralisation and Social Integration.' *Cambridge Journal of Economics* 6: 167–84.

Bulmer, M.I.A. (1975). 'Sociological Models of the Mining Community.' *Sociological Review* 23: 61– 92.

Burkenroad, M.D. (1953). 'Theory and Practice of Marine Fishery Management.' *Journal du Conseil Permanent International pour l'Exploration de la Mer* 18: 300–10.

Carter, Roger. (1988). 'Co-operatives in Rural Newfoundland and Labrador: An Alternative?' In P. Sinclair (ed.), *A Question of Survival: The Fisheries and Newfoundland Society.* St John's: Institute of Social and Economic Research, Memorial University of Newfoundland, pp. 203–28.

Cashin, R. (1993). *Charting a New Course: Towards the Fishery of the Future.* Report of the Task Force on Incomes and Adjustments in the Atlantic Fishery. Ottawa: Department of Fisheries and Oceans.

Chesnais, F. (1993). 'Globalisation, World Oligopoly and Some of Their Implications.' In M. Humbert (ed.), *The Impact of Globalisation on Europe's Firms and Industries*. London: Pinter, pp. 12–21.

Christensen, Pål. (1995). *Ishavsfolk, Arbeidsfolk og Fintfolk. 1995 Tromsø Gjennom 10000 År*, Vol. 3. Tromsø: Tromsø Kommune.

– (1991). '"En havenes forpester – et kjempestinkdyr:' Om Trålspørsmålet i Norge før 2. Verdenskrig." *Historisk Tidsskrift* 4: 622–35.

– (1986). *De Sprikende Interesser. Stat, Klippfiskeksportører, Tilvirkere og Fiskere 1926–37*. Working Paper. Tromsø: Universitetet i Tromsø.

Christensen, Pål, and Abraham Hallenstvedt. (1990). *På Første Hånd. Norges Råfisklag Gjennom Femti År*. Tromsø: Norges Råfisklag.

Christensen, T., and M. Egeberg. (1994). 'Noen trekk ved Forholdet mellom Organisasjonene og den offentlige Forvaltningen.' In T. Christensen and M. Egeberg (eds.), *Forvaltningskunnskap*. Oslo: Tano, pp. 175–204.

Christie, Nils. (1982). *Hvor Tett et Samfunn?* Oslo: Universitetsforlaget.

Ciriacy-Wantrup, S.V., and R.C. Bishop. (1975). '"Common Property" as a Concept in Natural Resources Policy.' *Natural Resources Journal* 15: 713–27.

Clement, Wallace. (1986). *The Struggle to Organize: Resistance in Canada's Fishery*. Toronto: McClelland and Stewart.

Coady, L.W. (1993). 'The Groundfish Resource Crisis: Ecological and Other Perspectives on the Newfoundland Fishery.' In Keith Storey (ed.), *The Newfoundland Groundfish Fisheries: Defining the Reality*. Conference Proceedings, March. St John's: Institute of Social and Economic Research, Memorial University of Newfoundland, pp. 56–77.

Cohen, Anthony P. (1985). *The Symbolic Construction of Community*. London: Tavistock.

– (1975). 'The Definition of Public Identity: Managing Marginality in Outport Newfoundland Following Confederation.' *Sociological Review* 23: 93–119.

Cohen, Erik. (1979). 'A Phenomenology of Tourist Experiences.' *Sociology* 13: 179–202.

Coldevin, Axel. (1938). *Næringsliv og Priser i Nordland 1700–1880*. Det Hanseatiske Museums Skrifter Nr. 11. Bergen: Johan Griegs Boktrykkeri.

Coleman, James S. (1988). 'Social Capital in the Creation of Human Capital.' *American Journal of Sociology* 94 (Supplement): 95–121.

– (1974). *Power and the Structure of Society*. New York: Norton.

Connelly, Patricia, and Martha MacDonald. (1990a). 'A Leaner, Meaner Industry: A Case Study of "Restructuring" in the Nova Scotia Fishery.' In B. Fairley, C. Leys, and

J. Sacouman (eds.), *Restructuring and Resistance From Atlantic Canada.* Toronto: Garamond, pp. 131–50.
- (1990b). 'Class and Gender in Nova Scotia Fishing Communities.' In B. Fairley, C. Leys, and J. Sacouman (eds.), *Restructuring and Resistance From Atlantic Canada.* Toronto: Garamond, pp. 151–70.

Cooke, Philip, Frank Moulaert, Eric Swyngedouw, Oliver Weinstein, and Peter Wells. (1992). *Towards Global Localisation.* London: Routledge.

Copes, Parcival. (1986). 'A Critical Review of the Individual Quota as a Device in Fisheries Management.' *Land Economics* 62: 278–91.
- (1972a). 'The Fishermen's Vote in Newfoundland.' *Canadian Journal of Political Science* 3: 579–604.
- (1972b). *The Resettlement of Fishing Communities in Newfoundland.* Ottawa: Canadian Council on Rural Development.
- (1964). *Government Assistance, Productivity and Income in the Fishing Industry of Newfoundland.* St John's: Memorial University.

Crowley, R.W., B. MacEachern, and R. Jaspensi. (1993). 'A Review of Federal Assistance in the Canadian Fishing Industry.' In L.S. Parsons and W.H. Lear, *Perspectives on Canadian Marine Fisheries Management.* Ottawa: National Research Council of Canada, pp. 339–67.

Crowley, R.W., and H. Palsson. (1992). 'Rights Based Fisheries Management in Canada.' *Marine Resource Economics* 7: 1–21.

Dahl, R.A. (1986). *Democracy, Liberty and Equality.* Oslo: Norwegian University Press.

Dahmani, M. (1987). *The Fisheries Regime of the Exclusive 1987 Economic Zone.* Dordrecht: Martinus Nijhoff.

Davis, Anthony. (1984). 'Property Rights and Access Management in the Small Boat Fishery: A Case Study from Southwest Nova Scotia.' In Cynthia Lamson and Arthur Hanson (eds.), *Atlantic Fisheries and Coastal Communities: Fisheries Decision-Making Case Studies.* Halifax: Dalhousie University, Ocean Studies Programme, pp. 133–64.

Davis, Anthony, and Leonard Kasdan. (1984). 'Bankrupt Government Policies and Belligerent Fishermen Responses: Dependency and Conflict in the Southwest Nova Scotia Small Boat Fisheries.' *Journal of Canadian Studies* 19: 108–24.

Davis, Anthony, and Victor Thiessen. (1988). 'Public Policy and Social Control in Atlantic Fisheries.' *Canadian Public Policy* 14: 666–77.

DeWitt, Robert L. (1969). *Public Policy and Community Protest: The Fogo Case.* St John's: Institute of Social and Economic Research, Memorial University of Newfoundland.

Drivenes, Einar-Arne. (1982). 'Fiskarbonden og de Nordnorske 1982 Bygdesamfunn.' In Ragnar Nilsen, Jan Einar Reiersen, and Nils Aarsæther (eds.), *Folkemakt og Regional Utvikling.* Oslo: Pax, pp. 101–16.

Durkheim, Emile. (1964). *The Division of Labour in Society*. New York: Free Press.
Dyer, Christopher, and James McGoodwin. (1994). *Folk Management in the World's Fisheries: Lessons for Modern Fisheries Management*. Niwot: University of Colorado.
Eikeland, Sveinung. Robuste Fiskevær? Selinvasjonenes 1993 Virkninger for Finnmarkskysten. Alta. NIBR-rapport no. 9.
Elkin, S.L. (1986). 'Regulation and Regime: A Comparative Analysis.' *Journal of Public Policy* 6: 51–71.
Eriksen, Erik O., and Knut H. Mikalsen. (1990). *Staten, Nord-Norge og Fiskeripolitikken*. Tromsø: Institute of Social Science, University of Tromsø.
Etzioni, Amitai. (1988). *The Moral Dimension: Towards a New Economics*. New York: Free Press.
Eythorsson, Einar. (1991). 'Ressurser. Livsform og lokal Kunnskap: Studie av et Fjordsamfunn i Finnmark.' MA thesis, Institute of Social Science, University of Tromsø.
Finlayson, A.C. (1994). *Fishing for Truth: A Sociological Analysis of Northern Cod Stock Assessment from 1977 to 1990*. St John's: Institute of Social and Economic Research, Memorial University of Newfoundland.
Fisheries Council of Canada. (1994). *Building a Fishery that Works: A Vision for the Atlantic Fisheries*. Ottawa: Fisheries Council of Canada.
Flora, Cornelia, Jan Flora, Jacqueline Spears, and Louis Swanson. (1992). *Rural Communities: Legacy and Change*. Boulder: Westview Press.
Food and Agricultural Association. (1994). *Yearbook of Fisheries Statistics*. Rome: FAO.
Foresta, Ronald A. (1986). 'The Administrative Face of Natural and Historic Preservation.' *Applied Geography* 6: 309–24.
Francis, J. (1993). *The Politics of Regulation: A Comparative Perspective*. Oxford: Blackwell.
Freudenburg, W.R. (1992). 'Addictive Economies: Extractive Industries and Vulnerable Localities in a Changing World Economy.' *Rural Sociology* 57: 305–32.
Friedman, Jonathan. (1994). *Cultural Identity and Global Process*. London: Sage.
– (1990). 'Being in the World: Globalization and Localization.' *Theory, Culture and Society* 7: 311–28.
Frionor. (1971). *Frionor 25 År*. Oslo: Frionor.
Gaventa, J., B.E. Smith, A. Willingham (eds.). (1990). *Communities in Economic Crises: Appalachia and the South Philadelphia*. Philadelphia: Temple University Press.
Gerhardsen, Gerhard Meidell. (1949). *Salted Cod and Related Species*. FAO Fisheries Study no. 1. Washington, DC: Food and Agricultural Organization.
Gerrard, Siri (ed.). (1992). *Skårunge, Skoleelev eller Arbeidssoker?* Tromsø: Norges Fiskerihøgskole.

- (1975). 'Arbeidsliv og Lokalsamfunn: Samarbeid og Skille mellom Yrkesgrupper i et Nord-norsk Fiskevær,' MA thesis, Institute of Social Science, University of Tromsø.
Gherardi, S., and A. Masiero. (1990). 'Solidarity as a Networking Skill and a Trust Relation: Its Implications for Cooperative Development.' *Economic and Industrial Democracy* 11: 553-74.
Giddens, Anthony. (1990). *The Consequences of Modernity*. Oxford: Polity Press.
Giske, Anders. (1978). *Klippfiskomsetninga i Støypeskjeia: Hovedoppgave i Historie*. Bergen: University of Bergen.
Glaser, Barney G., and Anselm L. Strauss. (1980). *The Discovery 1980 of Grounded Theory: Strategies for Qualitative Research*. New York: Aldine.
Goffee, R., and R. Scase. (1982). "'Fraternalism' and 'Paternalism' as Employer Strategies in Small Firms.'" In Graham Day, with Lesley Caldwell [et al.] (eds.), *Diversity and Decomposition in the Labour Market*. London: Gower, pp. 107-24.
Gordon, H.S. (1954). 'The Economic Theory of a Common-Property Resource: The Fishery.' *Journal of Political Economy* 62: 124-42.
- (1953). 'An Economic Approach to the Optimum Utilization of Fishery Resources.' *Journal of the Fisheries Research Board of Canada* 10: 442-57.
Gough, J. (1993). 'A Historical Sketch of Fisheries Management In Canada.' In L.S. Parsons and W.H. Lear (eds.), *Perspectives on Canadian Marine Fisheries Management*. Canadian Bulletin of Fisheries and Aquatic Sciences 225. Ottawa: Ontario. National Research Council of Canada, pp. 5-53.
Gouldner, Alvin. (1964). *Patterns of Industrial Bureaucracy: A Case Study of Modern Factory Administration*. New York: Free Press.
Government of Canada. (1993). *Fisheries Management: A Proposal for Reforming Licensing, Allocation and Sanctions Systems*. Communications Directorate. Ottawa: Department of Fisheries and Oceans.
- (1991). *Review of the Atlantic Fisheries Consultative and Advisory Processes: Reflections on Improving the Advisory and Consultative Processes of the Atlantic Fisheries Sector*. Interim Report. Ottawa: Department of Fisheries and Oceans.
- (1990). *A Review of the Scotia Fundy Region Consultative Process*. Resource Management Branch, Fisheries and Habitat Management, Scotia-Fundy Region. Halifax: Department of Fisheries and Oceans.
- (1976). *Policy for Canada's Commercial Fisheries*. Ottawa: Environment Canada.
- (1964). *Commission of Enquiry into the Atlantic Salt Fish Industry*. Finn Commission. Ottawa: Government of Canada.
- (1953). *Newfoundland Fisheries Development Committee, Report*. Walsh Commission. St John's: Newfoundland Fisheries Development Committee.
- (1928). *Report of the Royal Commission Investigating the Fisheries of the Maritime*

Provinces and the Magdalen Islands MacLean Commission. Ottawa: F.A. Ackland, Printer to the King's Most Excellent Majesty.

Government of Norway. (1994). *Høringsnotat om Lov om Retten til å Delta i Fisket.* Oslo: Ministry of Fisheries.

- (1992–3). *Innstilling S. Nr. 50. Innstilling frå Sjøfarts og Fiskerikomiteen om Struktur - og Reguleringspolitikken overfor Fiskeflåten.* Oslo: Stortinget.
- (1992). *Forhandlinger i Stortinget*, Nr. 130. Oslo: Stortinget.
- (1991–2). *St. Meld. Nr. 58. Om Struktur og Reguleringspolitikken overfor Fiskeflåten.* Oslo: Ministry of Fisheries.
- (1976). *St. Meld. Nr. 81. (1975–76) Oppsynet med Fiskeri- og Petroleumsvirksomheten: Etablering av en Kystvakt.* Oslo: Ministry of Defence.
- (1963). *St. Prp. Nr 143. (1963–64) Forhøyelse av Bevilkning på Statsbudsjettet for 1964 under Kap. 1531, Pristilskott, Post 72, Til Støtte av Torske – og Sildefisket og Bevilkning på Statsbudsjettet for 1964 under Kap. 1076, Pristilskudd m.m. ny Post 72, Til Støtte for Effektiviseringstiltak i Fiskerinæringen.* Oslo: Ministry of Fisheries.
- (1959). *St. Meld. Nr 71. (1959) Innstilling fra Torskefiskeutvalget 1957.* Oslo: Ministry of Fisheries.
- (1948). *St. Meld. Nr 3. (1948) Råfisklovens Gjennomføring.* Oslo: Ministry of Fisheries.
- (1947). *St. Meld. Nr 10. (1947) Om Nasjonalbudsjettet 1947.* Oslo: Ministry of Finance.
- (1937). *Innstilling om Fiskerienes Lønnsomhet. Instilling VIII (Hovedinnstilling) fra Komiteen til Behandling av Forskjellige Spørsmål Vedrørende Fiskeribedriften. (Income Task Force.)* Oslo: Ministry of Commerce.
- (1932). *Ot. Prp. Nr. 58. (1932) Om Utferdigelse av en Lov om Adgang for Kongen til å Treffe Foranstaltninger til Ordning av Utførsel av Klippfisk.* Oslo: Ministry of Commerce.

Grant, Ruth Fulton. (1934). *The Canadian Atlantic Fishery.* Toronto: Ryerson Press.

Granovetter, Mark. (1990). 'The Old and the New Economic Sociology: A History and an Agenda.' In R. Friendland and A.F. Robertson (eds.), *Beyond the Marketplace.* New York: Aldine de Gruyter, pp. 89–112.

- (1985). 'Economic Action and Social Structure: The Problem of Embeddedness.' *American Journal of Sociology* 91: 481–510.

Granovetter, Mark, and Richard Swedberg (eds.). (1992). *The Sociology of Economic Life.* Boulder: Westview Press.

Great Britain. (1993). *Report of the Newfoundland Royal Commission 1933.* Amulree Commission. London: Government of Great Britain.

Grønås, O., J. Halvorsen, and L. Torgersen. (1948). *Problemet Nord-Norge: Statistisk Undersøkelse av Nord-Norges Andel and Nasjonalinntekten i 1939.* Bodø: Studieselskapet for Nordnorsk Næringsliv.

Haché, J.-E. (1989). *Report of the Scotia-Fundy Groundfish Task Force.* Ottawa: Department of Fisheries and Oceans.

Hall, Peter. (1992). 'The Movement from Keynesianism to Monetarism: Institutional Analysis and British Economic Policy in the 1970s.' In Sven Steinmo, Kathleen Thelen, and Frank Longstreth (eds.), *Structuring Politics: Historical Institutionalism in Comparative Analysis.* New York: Cambridge University Press, pp. 114–54.

Hallenstvedt, Abraham. (1993). 'Ressurskontroll i norsk Fiskeriforvaltning.' In J.H. Williams (ed.), *Ressursforvaltning og Kontroll – Kontrollpolitikk og Virkemidler i Fiskeriene i de Nordiske Land og EF. Nordiske Seminar- og Arbeidsrapporter.* Copenhagen: Nordisk Ministerråd, pp. 119–52.

– (1990). 'Fiskeindustriens Organisering.' In Einar Moxnes (ed.), *Fiskeindustriens Organisering og Rammevilkår.* Oslo: Ministry of Fisheries, pp. 101–27.

– (1982). *Med Lov og Organisasjon.* Oslo, Bergen and Tromsø: Norwegian University Press.

Hallenstvedt, Abraham, and Bjørn Dynna. (1976). *Fra Skårunge til Hvedsmann.* Trondheim: Norges Fiskarlag.

Handegård, Odd. (1982). 'Statens Fiskarbank – Administrasjon og Klientideologi.' In Knut H. Mikalsen and Bjørn Sagdahl (ed.), *Fiskeripolitikk og Forvaltningsorganisasjon.* Tromsø: Universitetsforlaget, pp. 61–100.

Hardin, Garrett. (1968). 'The Tragedy of the Commons.' *Science* 162: 1243–8.

Harrop, Martin (ed.). (1992). *Power and Policy in Liberal Democracies.* Cambridge: Cambridge University Press.

Harvey, David. (1989). *The Condition of Postmodernity.* Oxford: Blackwell.

Head, C. Grant. (1976). *Eighteenth Century Newfoundland.* Carleton Library No. 99. Ottawa, McClelland and Stewart in association with the Institute of Canadian Studies, Carleton University.

Henderson, Paul, and David Francis. (1993). *Rural Action: A Collection of Community Work Case Studies.* London: Pluto Press.

Hersoug, B. (1983). 'Fiskeriplanlegging – Offentlig Styring eller Politisk Pliktøvelse?' In B. Hersoug (ed.), *Kan Fiskerinæringa Styres?* Oslo: Novus Forlag, pp. 19–78.

Hersoug, B., and D. Leonardsen. (1979). *Bygger de Landet?* Oslo: Pax.

– (1976). *Distriktspolitikk og Ideologi: En Historisk/Sosiologisk Analyse av Arbeiderpartiets Distriktspolitikk i Tiden 1935–1972. Magisteravhandling i Sosiologi.* Tromsø: University of Tromsø.

Hoel, Alf Håkon, Svein Jentoft, and Knut H. Mikalsen. (1991). *Problems of User-Group Participation in Norwegian Fisheries Management.* Occasional Papers A56. Tromsø: Institute of Social Sciences, University of Tromsø.

Holling, C.S. (1986). 'Resilience of Ecosystem: Local Surprise and Global Change.' In

E.C. Clark and R.E. Munn (eds.), *Sustainable Development of the Biosphere.* Cambridge: Cambridge University Press, pp. 292–317.

Holm, Petter. (1995). 'The Dynamics of Institutionalization: Transformation Processes in Norwegian Fisheries.' *Administrative Science Quarterly* 40: 398–422.

- (1991). 'Særinteresser vs Almeninteresser i Forhandlingsøkonomien.' *Tidsskrift for Samfunnsforskning* 32: 99–119.

- and Leigh Mazany. (1995). 'Changes in the Organization of the Norwegian Fishing Industry.' *Marine Resource Economics* 10: 299–312.

Holt, Robert T., and John E. Turner. (1970). *The Methodology of Comparative Research.* New York: Free Press.

Holton, Robert. (1992). *Economy and Society.* London: Routledge.

Hønneland, Geir B. (1993). *Makt og Kommunikasjon i Kontrollen med Fisket i Barentshavet: Hovedoppgave i Statvitenskap.* Tromsø: University of Tromsø.

House, J.D. (ed.) (1986). *Fish vs Oil: Resources and Rural Development in North Atlantic Societies.* St. John's: Institute of Social and Economic Research, Memorial University of Newfoundland.

Hurley, Geoff, and David Gray. (1988). *Fisheries Management in Atlantic Canada.* Scotia-Fundy Region. Halifax: Department of Fisheries and Oceans.

Hutchings, Jeffrey A. and Ransom A. Myers. (1994). 'What Can Be Learned from the Collapse of a Renewable Resource? Atlantic Cod, *Gadus morhua*, of Newfoundland and Labrador.' *Canadian Journal of Fisheries and Aquatic Sciences* 51: 2126–46.

Immergut, Ellen. (1992). 'The Rules of the Game: The Logic of Health Policy-making in France, Switzerland and Sweden.' In Sven Steinmo, Kathleen Thelen and Frank Longstreth (eds.), *Structuring Politics: Historical Institutionalism in Comparative Analysis.* New York: Cambridge University Press, pp. 57–89.

Inglis, Gordon. (1985). *More than Just a Union: The Story of the NFFAWU.* St John's: Jesperson.

Innis, H.A. (1954). *The Cod Fisheries: The History of an International Economy.* Toronto: University of Toronto Press.

Japan, Ministry of Fisheries. (1993). *Annual Report.* Tokyo.

Jentoft, Svein. (in press). *The Invisible Cod: Fishermen's and Scientists' Knowledge – Commons in a Cold Climate: Reindeer Pastoralism and Coastal Fisheries in North Norway.* Casterton Hall: Parthenon.

- (1993). *Dangling Lines: The Fisheries Crisis and the Future of Local Communities – the Norwegian Experience.* St John's: ISER Books, Memorial University of Newfoundland.

- (1984). 'Hvor Sårbare er Fiskerimiljøene?.' In Svein Jentoft and Cato Wadel (eds.), *I Samme Båt: Sysselsettingssystemer i Fiskerinæringen.* Oslo: Universitetsforlaget, pp. 149–77.

Jentoft, Svein, and Anthony Davis. (1993). 'Self and Sacrifice: An Investigation of

Small Boat Fisher Individualism and Its Implication for Producer Cooperatives.' *Human Organization* 52: 356–67.

Jentoft, Svein, and Knut K. Mikalsen. (1994). 'Regulating Fjord Fisheries: Folk Management or Interest Group Politics?'. In Christopher Dyer and James McGoodwin (eds.), *Folk Management in the World's Fisheries*. Niwot: University of Colorado Press, pp. 287–316.

– (1987). 'Government Subsidies in Norwegian Fisheries: Regional Development or Political Favouritism?' *Marine Policy* 11: 217–28.

Jentoft, Svein, and Trond Kristoffersen. (1989). 'Fishermen's Co-management: The Case of the Lofoten Fishery.' *Human Organization* 48: 355–65.

Jentoft, Svein, and Cato Wadel. (1984). *I Samme Båt: Sysselset-tingssystemer i Fiskerinæringen*. Oslo: Universitetsforlaget.

Johansen, Johan. (1972). *Trålfiske og Trålerdebatt i 1930-årene. Hovedoppgave i Historie*. Trondheim: Universitetet i Trondheim.

Johansen, Steinar. (1994). '"Ikke bare fanges – men også selges": En Organisasjonsteoretisk Studie av Markedsorientering i Fire Fiskeindustribedrifter,' MA thesis, Institute of Social Science, University of Tromsø.

Jónsson, Ô.D. (n.d.). *Geopolitics of Fish: The North-Atlantic Situation and Development Potentials for Fisheries Dependent Communities*. Reykjavik: Fisheries Research Institute, University of Iceland, Mimeo.

Kearney, John. (1989). 'Co-management or Co-optation? The Ambiguities of Lobster Fishery Management in Southwest Nova Scotia.' In Evelyn Pinkerton (ed.), *Cooperative Management of Local Fisheries*. Vancouver: UBC, pp. 85–102.

Kernaghan, Kenneth, and David Siegel. (1991). *Public Administration in Canada*. Scarborough: Nelson.

Kimber, Stephen. (1989). *Net Profits: The Story of National Sea*. Halifax. Nimbus Publishing.

Kirby, Michael. (1984). 'Restructuring the Atlantic Fishery: A Case Study in Business-Government Relations.' Paper presented at the School of Business, Dalhousie University, 1 March.

– (1982). *Navigating Troubled Waters: A New Policy for the Atlantic Fisheries*. Report of the Task Force on Atlantic Fisheries. Ottawa: Supply and Services Canada.

Knutsen, Nils Magne. (1988). *Nessekongene: De Store Handelsdynastiene i Nord-Norge*. Oslo: Gyldendal.

Kvavik, R. (1976). *Interest Groups in Norwegian Politics*. Oslo: Universitetsforlaget.

Krasner, S.D. (1983). 'Structural Causes and Regime Consequences: Regimes as Intervening Variables.' In S.D. Krasner (ed.), *International Regimes*. Ithaca: Cornell University Press, pp. 1–21.

Lamson, Cynthia, and Arthur Hanson (eds.). (1984). *Atlantic Fisheries and Coastal*

Communities: Fisheries Decision-Making Case Studies. Halifax: Dalhousie Ocean Studies Programme, Dalhousie University.

Larkin, P.A. (1977). 'An Epitaph for the Concept of Maximum Sustainable Yield.' *Transactions of the American Fisheries Society* 106: 1–11.

Laxer, Gordon. (1991). 'Class, Nationality and the Roots of the Branch Plant Economy.' In Gordon Laxer (ed.), *Perspectives in Canadian Economic Development.* Oxford: Oxford University Press, pp. 228–66.

Lelè, S.M. (1991). 'Sustainable Development: A Critical Review.' *World Development* 19: 607–23.

Lien, Børre. (1975). *Findus og Norsk Fiskeripolitikk 1943–1956: Hovedoppgave i Historie.* Tromsø: University of Tromsø.

Lind, Christopher. (1995). *Something's Wrong Somewhere: Globalization, Community and the Moral Economy of the Farm Crisis.* Halifax: Fernwood Press.

Lipset, Seymour Martin. (1971). *Agrarian Socialism: The Cooperative Commonwealth Federation in Saskatchewan.* Berkeley: University of California Press.

Loucks, L. (1995). 'Coastal Community-based Decision-making: Values for Sustainable Coastal Zone Management.' MA thesis, Atlantic Canada Studies, Saint Mary's University, Halifax.

– (1993). 'Life Patterns: Communities, Ecology and Connections,' unpublished paper, Atlantic Canada Studies, Saint Mary's University, Halifax.

MacCannell, D. (1973). 'Staged Authenticity: Arrangements of Social Space in Tourist Settings.' *American Journal of Sociology* 79: 589–603.

MacInnes, Daniel, and Anthony Davis. (1990). *Representational Management or Management of Representation? The Place of Fishers in Atlantic Canadian Fisheries Management.* Antigonish: Sociology and Anthropology, St Francis Xavier University.

MacInnes, Daniel, Svein Jentoft, and Anthony Davis (eds.). (1996). *Social Research and Public Policy Formation in the Fisheries: Norwegian and Canadian Experiences.* Halifax: Institute of Ocean Studies, Dalhousie University.

MacLeod, Greg. (1986). *New Age Business: Community Corporations That Work.* Ottawa: Canadian Council on Social Development.

MacNaghten, P., and Urry, J. (1995). 'Towards a Sociology of Nature.' *Sociology* 29: 203–20.

Manuel, P.M. (1992). 'A Landscape Approach to the Interpretation, Evaluation and Management of Wetlands.' PhD thesis, Dalhousie University, Halifax.

March, James G., and Johan P. Olsen. (1989). *Rediscovering Institutions: The Organizational Basis of Politics.* New York: Free Press.

Marchak, M. Patricia. (1988–9). 'What Happens When Common Property Becomes Uncommon?' *B.C. Studies* 80: 3–23.

Marchak, Patricia, Neil Guppy, and John McMullan (eds.). (1987). *Uncommon*

Property: The Fishing and Fish-Processing Industries in British Columbia. Toronto: Methuen.

Matthews, David Ralph. (1993). *Controlling Common Property: Regulating Canada's East Coast Fishery.* Toronto: University of Toronto Press.

– (1976). *'There's no Better Place Than Here': Social Change in Three Newfoundland Communities.* Toronto: Book Society of Canada.

Maurstad, Anita, and Jan H. Sundet. (in press). 'The Invisible Cod: Fishermen's and Scientists' Knowledge.' In Svein Jentoft (ed.), *Commons in a Cold Climate: Reindeer Pastoralism and Coastal Fisheries in North Norway.* Casterton Hall: Parthenon.

McCay, Bonnie J. (1988). 'Fish Guts, Hair Nets, and Unemployment Stamps: Women and Work in Co-operative Fish Plants.' In Peter Sinclair (ed.), *A Question of Survival: The Fisheries and Newfoundland Society.* St John's: Institute of Social and Economic Research, Memorial University of Newfoundland, pp. 105–36.

– (1979). '"Fish is Scarce": Fisheries Modernization on Fogo Island, Newfoundland.' In R. Andersen, (ed.), *North Atlantic Maritime Cultures.* The Hague: Mouton, pp. 155–89.

– (1978). 'Systems Ecology, People Ecology, and the Anthropology of Fishing Communities.' *Human Ecology* 6: 397–422.

– (1976). 'Appropriate Technology and Coastal Fishermen of Newfoundland,' PhD thesis, Columbia University, Ann Arbor, University Microfilms.

McCay, Bonnie, and James Acheson (eds.). (1987). *The Question of the Commons: The Culture and Ecology of Communal Resources.* Tucson: The University of Arizona Press.

McCay, Bonnie J., Richard Apostle, Carolyn Creed, Alan Finlayson, and Knut Mikalsen. (1995). 'Individual Transferable Quotas (ITQs) in Canadian and US Fisheries.' *Ocean and Coastal Management* 28(103): 85–116.

McCay, Bonnie J., and Svein Jentoft. (1996). 'From the Bottom Up: Participatory Issues in Fisheries Management.' *Society and Natural Resources* 9(3): 327–50.

McLuhan, Marshall. (1960). 'Media Log.' *Explorations in Communication, An Anthology.* E.S. Carpenter and Marshall McLuhan (eds.), Boston: Beacon Press, pp. 180–3.

Midré, Georges, and Anne Solberg. (1980). 'Sosial Integrasjon og Sosial Kontroll.' In Iens E. Høst and C. Wadel (eds.), *Fiske og Lokalsamfunn.* Oslo, Universitetsforlaget, pp. 179–85.

Mikalsen, Knut. (1987). *Reguleringspolitikk og Styringsproblemer.* Tromsø, Institute of Social Sciences, University of Tromsø.

– (1982). 'Lovgivning, Eiendomsrett og Næringspolitikk.' In Knut H. Mikalsen and Bjørn Sagdahl (eds.), *Fiskeripolitikk og Forvaltningsorganisasjon.* Tromsø: Universitetsforlaget, pp. 174–201.

– (1977). 'Lovgivning og Interessekamp.' MA thesis, Oslo, Institute of Political Science, University of Oslo.
Mikalsen, Knut, and B. Sagdahl (eds.). (1982). *Fiskeripolitikk og Forvaltningsorganisasjon*. Oslo, Bergen, and Tromsø: Norwegian University Press.
Moldenæs, T. (1993). *Legitimeringsproblemer i Fiskerinæringen: En Studie av Utformingen av en Stortingsmelding*. Tromsø: NORUT-Samfunnsforskining, University of Tromsø.
Moran, Emilio F. (1982). *Adaptability: An Introduction to Ecological Anthropology*. Boulder: Westview Press.
Murray, Fergus. (1983). 'The Decentralization of Production – the Decline of the Mass-Collective Worker?' *Capital and Class* 19: 74–99.
Myers, R.A., Hutchings, J.A., and N.J. Barrowman. (1996). 'Hypotheses for the Decline of Cod in the North Atlantic.' *Marine Ecology Progress Series* 138: 293–308.
Nåtti, Janko. (1993). 'Atypical Employment in the Nordic Countries: Towards Marginalization or Normalization?' In Thomas B. Boje and Sven E. Olsson Hart (eds.), *Scandanavian is a New Europe*. Oslo: Universitetsforlaget, pp. 171–215.
Neis, Barbara. (1991). 'Flexible Specialization and the North Atlantic Fisheries.' *Studies in Political Economy* 36: 145–75.
Newfoundland. (1937). *Report of the Commission of Enquiry Investigating the Seafisheries of Newfoundland and Labrador*. St John's: Government of Newfoundland.
North, D.C. (1990). *Institutions, Institutional Change and Economic Performance*. Cambridge: Cambridge University Press.
NOS. (Norges Offisielle Statistikk) *Historisk Statistikk 1994*. Oslo: Central Bureau of Statistics (Statistisk Sentralbyrå).
NOS. (1978). *Fishery Statistics 1976*. Oslo: Central Bureau of Statistics.
NOS. (1973). *Fishery Statistics 1971*. Oslo: Central Bureau of Statistics.
Norway, Director of Fisheries. (1928). *Torskefiskeriene og Handelen med Klippfisk og Tørrfisk: Årsberetning vedkommende Norges Fiskerier*, no. 2. Bergen: John Griegs Boktrykkeri.
Nozick, Marcia. (1992). *No Place Like Home: Building Sustainable Communities*. Ottawa: Canadian Council on Social Development.
Ommer, Rosemary E. (1989). 'The Truck System in Gaspé, 1822–77.' *Acadiensis* 19: 91–114.
– (1981). '"All the Fish of the Post": Resource Property Rights and Development in a Nineteenth-Century Inshore Fishery.' *Acadiensis* 10: 107–23.
O'Neill, Mollie. (1981). *The Fishery in Norway: Efforts to Reconcile the Fishermen's Needs, Community Development and National Management*. Dartmouth, Nova Scotia: Fisheries and Oceans Canada.
Ostrom, Elinor. (1990). *Governing the Commons: The Evolution of Institutions for Collective Action*. New York: Cambridge University Press.

References 359

Ostrom, Vincent, and Elinor Ostrom. (1977). 'A Theory for Institutional Analysis of Common Pool Problems.' In G. Hardin and J. Baden (eds.), *Managing the Commons*. San Francisco: Freeman, pp. 157–72.

Otnes, Per. (1980). *Fiskarsamvirket i Finnmark*. Arbeidsnotat 143. Oslo: Institute of Sociology, University of Oslo.

Paine, Robert. (1972). 'Entrepreneurial Activity without Its Profits.' In Fredrik Barth (ed.), *The Role of the Entrepreneur in Social Change in Northern Norway*. Oslo: Universitetsforlaget, pp. 33–55.

Parsons, L.S. (1993a). 'Shaping Fisheries Policy: The Kirby and Pearse Inquiries – Process, Prescription and Impact.' In L.S. Parson and W.H. Lear (eds.), *Perspectives on Canadian Marine Fisheries Management*. Ottawa. National Research Council of Canada, pp. 385–409.

– (1993b). *Management of Marine Fisheries in Canada*. Canadian Bulletin of Fisheries and Aquatic Sciences 225. Ottawa: National Research Council of Canada.

Paulgård, Gry. (1993). 'Nye Tider – nye Tonar? Eit Fiskerisamfunn i Endring.' In Kåre Heggen, Jon Olav Myklebust, and Tormod Øia (eds.), *Ungdom i Lokalmiljø*. Oslo: Samlaget, pp. 39–56.

Pearse, Peter. (1982). *Turning the Tide: A New Policy for Canada's Pacific Fisheries*. Commission on Pacific Fisheries Policy. Vancouver: Ministry of Supply and Services.

Pentland, H. Clare. (1991). 'The Transformation of Canada's Economic Structure.' In Gordon Laxer (ed.), *Perspectives on Canadian Economic Development*. Toronto: Oxford University Press, pp. 193–226.

Perlin, A.B. (1959). *Story of Newfoundland*. St John's: Dicks.

Perrucci, C.C., R. Perrucci, D.B. Targ, and H. Targ. (1988). *Plant Closings: International Context and Social Costs*. Hawthorne, NY: Aldine de Gruyter.

Perry, S.E. (1987). *Communities on the Way: Rebuilding Local Economies in the United States and Canada*. Albany: State University of New York Press.

Perulli, P. (1990). 'Industrial Flexibility and Small Firm Districts: The Italian Case.' *Economic and Industrial Democracy* 11: 337–53.

Phyne, John. (1990). 'Dispute Settlement in the Newfoundland Inshore Fishery: A Study of Fishery Officers' Responses to Gear Conflicts in Inshore Fishing Communities.' *Maritime Anthropological Studies* 3: 88–102.

Pinkerton, Evelyn (ed.). (1989). *Co-operative Management of Local Fisheries: New Directions for Improved Management and Community Development*. Vancouver: UBC Press.

Piore, Michael J., and Charles Sabel. (1984). *The Second Industrial Divide: Possibilities for Prosperity*. New York: Basic Books.

Pocius, Gerald. (1991). *A Place to Belong: Community Order and Everyday Space in Calvert, Newfoundland*. Athens: University of Georgia Press.

Polanyi, Karl. (1944). *The Great Transformation*. New York: Rinehart.
Portes, A., and J. Sensenbrenner. (1993). 'Embeddedness and Immigration: Notes on the Social Determinants of Economic Action.' *American Journal of Sociology* 98: 1320–50.
Powell, Walter, and Paul Dimaggio. (1991). *The New Institutionalism in Organizational Analysis*. Chicago: University of Chicago Press.
Pross, Paul. (1992). *Group Politics and Public Policy*. Toronto: Oxford University Press.
Pross, Paul, and Susan McCorquodale. (1987). *Economic Resurgence and the Constitutional Agenda: The Case of the East Coast Fisheries*. Kingston, Ontario, Institute of Intergovernmental Relations, Queen's University.
Pyke, Frank, and Werner Sengenberger. (1991). *Industrial Districts and Local Economic Regeneration*. Geneva: International Institute of Labour Studies.
Ragin, Charles C. (1987). *The Comparative Method: Moving Beyond Qualitative and Quantitative Research*. Berkeley: University of California Press.
Rahman, M.D. Anisur. (1993). *People's Self-Development: Perspectives on Participatory Action Research*. London: Zed Press.
Revold, Jens. (1980). 'Fiskarsamvirke: Modernisering og Fellesorganisering,' MA thesis, Institute of Social Science, University of Tromsø.
Robertson, Roland. (1993). *Globalization: Social Theory and Global Culture*. London: Sage.
Robinson, Guy. (1990). *Conflict and Change in the Countryside: Rural Society, Economy and Planning in the Developed World*. London: Belhaven Press.
Rokkan, S. (1966). 'Numerical Democracy and Corporate Pluralism.' In R.A. Dahl (ed.), *Political Oppositions in Western Democracies*. New Haven: Yale University Press, pp. 89–105.
Rossvær, Viggo. (1994). *Avstigning på Sørvær*, parts 1 and 2. Tromsø: Institute of Social Science, University of Tromsø.
Rostow, W.W. (1960). *The Stages of Economic Growth: A Non-Communist Manifesto*. New York: Cambridge University Press.
Rudeng, Erik. (1989). *Sjokoladekongen. Johan Throne Holst – en Biografi*. Oslo: Universitetsforlaget.
Sabatier, P.A. (1977). 'Regulatory Policy-making: Toward a Framework of Analysis.' *Natural Resources Journal* 17: 415–60.
Sacouman, R. James. (1979). 'Underdevelopment and the Structural Origins of Antigonish Movement Co-Operatives in Eastern Nova Scotia.' In Robert J. Brym and R. James Sacouman (eds.), *Underdevelopment and Social Movements in Atlantic Canada*. Toronto: New Hogtown Press, pp. 107–26.
Sagdahl, Bjørn. (1992). *Ressursforvaltning og Legitimitetsproblemer*. Report No. 15/92–20. Bodø: Nordland Resarch Institute.

References

- (1982). 'Struktur, Organisasjon og Innflytelsesforhold i Norsk Fiskeripolitikk.' In Knut H. Mikalsen and Bjørn Sagdahl (eds.), *Fiskeripolitikk og Forvaltningsorganisasjon*. Tromsø: Universitetsforlaget, pp. 15–46.
- (1973). 'Trålpolitikk og Interessekamp,' MA thesis, Institute of Political Science, University of Oslo.

Sanger, C. (1987). *Ordering the Oceans: The Making of the Law of the Sea*. Toronto. University of Toronto Press.

Sayer, Andrew, and Richard Walker. (1992). *The New Social Economy: Reworking the Division of Labour*. Cambridge, Mass: Blackwell.

Schrank, W.E. (1994). 'Extended Fisheries Jurisdiction and the Current Crisis in Atlantic Canada's Fisheries,' St John's, Memorial Univerisity of Newfoundland, Mimeo.

Schrank, W.E., N. Roy, R. Ommer, and B. Skoda. (1992). 'An Inshore Fishery: A Commercially Viable Industry or an Employer of Last Resort?' *Ocean Development and International Law* 23: 335–67.

Scott, A.J. (1988). 'Flexible Production Systems and Regional Development: The Rise of New Industrial Spaces in North America and Western Europe.' *International Journal of Urban and Regional Research* 12: 171–85.

Sinclair, Peter. (1990). 'Poor Regions, Poor Theory: Towards Improved Understanding of Regional Inequality.' Saskatoon, University of Saskatchewan, Sorokin Lecture, 25 January.

- (1984). 'The State Goes Fishing: The Emergence of Public Ownership in the Newfoundland Fishing Industry,' paper presented at the Annual Meetings of the Atlantic Association of Sociologists and Anthropologists, Fredericton, University of New Brunswick.

Smith, M. Estellie. (1990). 'Chaos in Fisheries Management.' *Maritime Anthropological Studies* 3: 1–13.

Smith, Eivind. (1979). *Organisasjoner i Fiskeriforvaltningen*. Oslo: Tanum-Norli.

Smith, Tim D. (1994). *Scaling Fisheries: The Science of Measuring the Effects of Fishing, 1855–1955*. Cambridge: Cambridge University Press.

Solhaug, Trygve. (1976). *De Norske Fiskeriers Historie 1815–1880*. Oslo: Norwegian University Press.

Steele, D.H., R. Andersen, and J.M. Green. (1992). 'The Managed Commercial Annihilation of Northern Cod.' *Newfoundland Studies* 8: 34–68.

Steen, Sverre. (1957). *Det Gamle Samfunn*, vol. 4 in *Sverre Steen, Det Frie Norge*. Oslo: Cappelen.

- (1930). *Det Norske Folks Liv og Historie Gjennem Tidene*, vol. 5. Olso: Aschehoug.

Stinchcombe, Arthur L. (1990). *Information and Organizations*. Berkeley: University of California Press.

- (1968). *Constructing Social Theories*. Chicago: University of Chicago Press.

Storey, Keith (ed.). (1993). *The Newfoundland Groundfish Fisheries: Defining the Reality*. Conference Proceedings, St John's, Institute of Social and Economic Research, Memorial University of Newfoundland.

Strand, T. (1978). 'Staten og Kommunane: Standardisering, Hjelp og Sjølvhjelp.' In J.P. Olsen (ed.), *Politisk Organisering*. Oslo: Universitetsforlaget, pp. 143–84.

Streeck, Wolfgang, and Philippe C. Schmitter. (1988). 'Community, Market, State – and Associations? The Prospective Contribution of Interest Governance to Social Order.' In W. Streeck and P.C. Schmitter (eds.), *Private Interest Government: Beyond Market and State*. London: Sage, pp. 1–29.

Swedberg, Richard. (1991). 'Major Traditions of Economic Sociology.' *Annual Review of Sociology* 17: 251–76.

Taylor, Frederick W. (1967). *The Principles of Scientific Management*. New York: Norton. 1st edition 1911.

Thelen, Kathleen, and Sven Steinmo. (1992). 'Historical Institutionalism in Comparative Politics.' In Sven Steinmo, Kathleen Thelen, and Frank Longstreth (eds.), *Structuring Politics: Historical Institutionalism in Comparative Analysis*. New York: Cambridge University Press, pp. 1–32.

Thiessen, Victor, Anthony Davis, and Svein Jentoft. (1992). 'The Veiled Crew: An Exploratory Study of Wives' Reported and Desired Contributions to Coastal Fisheries Enterprises in Northern Norway and Nova Scotia.' *Human Organization* 51: 342–53.

Thompson, J.D. (1967). *Organizations in Action*. New York: McGraw-Hill.

Thompson, W.F. (1950). *The Effect of Fishing on Stocks of Halibut in the Pacific*. Washington: University of Washington Press.

Thormar, Sigmar. (1984). 'Export Organizations of the Newfoundland and Icelandic Sea Fisheries,' MA thesis, Carleton University, Ottawa.

Throne-Holst, Leif. (1966). *Fisk og Fiskeindustri i Nord-Norge*. Oslo: Dreyer.

Thurow, Lester C. (1980). *The Zero-Sum Society: Distribution and the Possibilities for Economic Change*. New York: Basic Books.

Tönnies, Ferdinand. (1963 [1887]). *Community and Society*. New York: Harper Torchbook.

Tracy, Michael. (1989). *Government and Agriculture in Western Europe 1880–1988*. New York: Harvester Wheatsheaf.

Tuohy, C.J. (1992). *Policy and Politics in Canada. Institutionalized Ambivalence*. Philadelphia: Temple University Press.

Urry, J. (1995). *Consuming Spaces*. London: Routledge.

Vik, Harald. (1982). *Finotro A/S 1952–1982*. Hånningsvåg: Finotro.

Vogel, D. (1986). *National Styles of Regulation: Environmental Policy in Great Britain and the United States*. Ithaca: Cornell University Press.

Wadel, Cato. (1980). 'Lokalsamfunn, Hushold og Bedrift: Noen aktuelle Næringspoli-

tiske Koordineringsproblemer i Kyst-Norge.' In Iens L. Høst and C. Wadel (eds.), *Fiske og Lokalsamfunn*. Oslo: Norwegian University Press, pp. 311–40.
- (1973). *'Now, Whose Fault is That?' The Struggle for Self-Esteem in the Face of Chronic Unemployment*. St John's: Institute of Social and Economic Research, Memorial University Printing Services.
- (1969). *Marginal Adaptations and Modernization in Newfoundland: A Study of Strategies and Implications of Resettlement and Redevelopment of Outport Fishing Communities*. St John's: Institute of Social and Economic Research, Memorial University of Newfoundland.

Walters, Carl, and Jean-Jacques Maguire. (1996). 'Lessons for Stock Assessment from the Northern Cod Collapse.' *Reviews in Fish Biology and Fisheries* 6(2): 125–37.

Watt, John W. (1963). *A Brief Review of the Fisheries of Nova Scotia*. Halifax. Province of Nova Scotia.

Weber, Max. (1978). *Economy and Society*. Berkeley: University of California Press.

Williamson, H. Anthony. (1975). 'The Fogo Process in Communication.' In *Training for Agriculture and Rural Development*. Rome: Food and Agricultural Organization, pp. 93–8.

Williamson, O.E. (1985). *The Economic Institutions of Capitalism*. New York. Free Press.
- (1975). *Markets and Hierarchies: Analysis and Antitrust Implications*. New York. Free Press.

Wille-Strand, Erling. (1941). *Motorstriden i Lofoten: 1941 Hovedoppgave i Historie*. Oslo: Universitetet i Oslo.

Wilson, James, and Peter Kleban. (1992). 'Practical Implications of Chaos in Fisheries: Ecologically Adapted Management.' *Maritime Anthropological Studies* 5: 67–75.

Wilson, James A., Ralph Townsend, Peter Kelban, Susan McKay, and John French. (1990). 'Managing Unpredictable Resources: Traditional Policies Applied to Chaotic Populations.' *Ocean and Shoreline Management* 13: 179–97.

Winson, Anthony. (1988). 'Researching the Food Processing–Farming Chain: The Case of Nova Scotia.' *Canadian Review of Sociology and Anthropology* 25: 520–59.

Young, F.W., and T.A. Lyson. (1993). 'Branch Plants and Poverty in the American South.' *Sociological Forum* 8: 433–50.

Young, O. (1994). *International Governance: Protecting the Environment in a Stateless Society*. Ithaca and London: Cornell University Press.

Ørebech, P. (1986). *Reguleringer i Fisket*. Tromsø: Marinjuss.
- (1984). *Norsk Fiskerirett*. Tromsø: Universitetsforlaget.

Øyen, Else, (ed.). (1990). *Comparative Methodology: Theory and Practice in International Social Research*. London. Sage.